COMMON SPIDERS OF NORTH AMERICA

The purpose of the American Arachnological Society is
to further the study of Arachnids worldwide, foster closer
cooperation and understanding between amateur and
professional arachnologists, and to publish the Journal of
Arachnology. http://www.americanarachnology.org/

COMMON SPIDERS
of NORTH AMERICA

Richard A. Bradley

Illustrations by Steve Buchanan

Sponsored by the American Arachnological Society

UNIVERSITY OF CALIFORNIA PRESS
Berkeley   Los Angeles   London

University of California Press, one of the most distinguished university presses in the United States, enriches lives around the world by advancing scholarship in the humanities, social sciences, and natural sciences. Its activities are supported by the UC Press Foundation and by philanthropic contributions from individuals and institutions. For more information, visit www.ucpress.edu.

University of California Press
Oakland, California

First paperback printing 2019

Library of Congress Cataloging-in-Publication Data

Bradley, Richard Alan.
    Common spiders of North America / Richard A. Bradley ; illustrations by Steve Buchanan ; sponsored by the American Arachnological Society.
        p.   cm.
    Includes bibliographical references and index.
    ISBN 978-0-520-27488-4 (cloth : alk. paper)
    ISBN 978-0-520-31531-0 (pbk. : alk. paper)
    ISBN 978-0-520-95450-2 (ebook)
    1. Spiders—North America—Identification. I. Buchanan, Steve.
II. American Arachnological Society. III. Title.
QL458.41.A1B73   2013
595.4'8097—dc23   2012018390

Manufactured in China

23 22 21 20 19
10 9 8 7 6 5 4 3 2 1

The paper used in this publication meets the minimum requirements of ANSI/NISO Z39.48-1992 (R 2002) (*Permanence of Paper*). ♾

Cover illustration: Steve Buchanan.

TO MY PARENTS, MARJORIE AND WESLEY BRADLEY,
FOR FOSTERING A DEEP RESPECT AND CURIOSITY ABOUT
THE NATURAL WORLD,

AND MY BROTHER, DAVE,
FOR SHARING MANY JOYFUL HOURS
EXPLORING AND CELEBRATING WILD PLACES.

CONTENTS

My introduction to spider biology began over 30 years ago with a course taught by Donald Lowrie while I was in graduate school at the University of New Mexico. He ignited in me a fascination for these little animals, above and beyond mere identification. Herbert and Lorna Levi produced a wonderful guide to terrestrial invertebrates in 1968; it was focused on spiders and is still widely used today. The genesis of the current book was the demand for a more complete guide to the common spiders of North America. The purpose of this volume is to foster an appreciation of spiders. Familiarity with these small predators will enrich your life and open a door to a miniature world of wonder.

This book could never have been completed without assistance from many people. I thank Bruce Leech and Susan Ward for many years of their help navigating the spider literature. I thank Lynda Behan and Robin Taylor for their patient help with editing and revision. Their advice and suggestions substantially improved the work. Photographs were provided by Todd Blackledge (Figs. 21, lower left; 22, left), Jonathan Coddington (Fig. 25, left), John Maxwell (Fig. 28, right), Brent Opell (Fig. 27, lower right), Nathan Quesinberry (Fig. 6), Lenny Vincent (Fig. 27, upper right), Joseph Warfel (Fig. 3, left), and Annie Whitney (Fig. 28, left). All other photographs are my own.

I thank illustrator Steve Buchanan for his willingness to collaborate on this project. He translated my suggestions, through many cycles of revision, into accurate and beautiful illustrations. It was a great joy to work with him over the past three years. I thank Herb Levi for reading early drafts of the manuscript and providing helpful suggestions. Gail Stratton read the entire wolf spider text and provided suggestions, corrections, and enthusiastic support. Allen Brady generously shared his knowledge of wolf spiders. Michael Draney read the material on sheet weavers, reviewed the sheet weaver illustrations, and made helpful suggestions. I thank Rod Crawford for sharing his extensive knowledge of the Linyphiidae as well as his help with information about other spiders of the Western states. Lenny Vincent provided access to his extensive library of reference photographs and his knowledge of spider fauna in the Southwest.

Bruce Cutler read the jumping spider accounts and provided a number of corrections and helpful suggestions. G. B. Edwards shared his extensive knowledge, access to his database, arranged loans of material, reviewed the jumping spider illustrations, and provided many helpful corrections. Mark Stowe shared his expertise on orb-weaving spiders—in particular, the bolas spiders. He also edited and revised the section on these fascinating animals. Rick Vetter shared his knowledge of the interactions between spiders and people. He also edited the sections on spider bites, *Latrodectus*, and *Loxosceles*. Allen Dean generously provided unpublished information about the biology and distribution of many species. Joel Ledford cheerfully shared his knowledge of many spiders, particularly the leptonetids and telemids. Mark Harvey, Victor Fet, Lorenzo Prendini, and Bill Shear shared their knowledge of arachnid biodiversity for the estimates in the table.

I gratefully acknowledge the many suggestions (unsigned) regarding the identification chapter from arachnology volunteers working with Paula Cushing at the Denver Museum of Nature and Science. Many of my arachnological colleagues and other naturalists reviewed sections of text and draft illustrations and gave generously of their time and expertise. These included Lina Almeida, Joe Beatty, Robb Bennett, David Bixler, Todd Blackledge, Jason Bond, Joanne Bovee, Sandra Brantley, Don Buckle, Karen Cangialosi, Don Cameron, Jim Carrel, Jon Coddington, Fred Coyle, Sarah Crews, Charles Dondale, Nadine Dupérré, Rosemary Gillespie, Hank Guarisco, Marshal Hedin, Maggie Hodge, Gustavo Hormiga, Bernhard Huber, Beth Jakob, Dan Jennings, Thomas Jones, Kenn Kaufman, Kim Kaufman, Sara Klips, Robin Leech, Steve Lew, Wayne Maddison, Sam Marshall, Yuri Marusik, Pat Miller, Frances Murphy, Brent Opell, Pierre Paquin, Matt Persons, Kevin Pfeiffer, Norm Platnick, Tom Prentice, Jon Reiskind, David Richman, Michael Roberts, Nina Sandlin, Jeff Shultz, Derek Sikes, Joey Slowik, Ryan Stork, Tamas Szuts, Darrell Ubick, George Uetz, and Dustin Wilgers. They enriched this book and provided many suggestions and corrections. Any remaining errors are mine alone.

I thank Joe Warfel for congenial companionship in the field and his expert advice on photography. I thank Bill Hickman for his hard work and enthusiasm on a variety of spider projects over the years, but particularly for his constant sunny disposition. I thank the following curators for sharing their expertise as well as providing access to specimens under their care: Jon Coddington (United States National Museum), Rod Crawford (University of Washington Burke Museum), Paula Cushing (Denver Museum of Nature and Science), Allen Dean (Texas A&M University Insect Collection), G. B. Edwards (Florida Collection of Arthropods), Charles Griswold (California Academy of Sciences Entomology), Janet Kempf (National History Museum of Los Angeles County), Hans Klompen (Ohio State Acarology Laboratory), Laura Leibensperger and Herbert Levi (both at Museum of Comparative Zoology), and Pat Miller (Mississippi Entomological Museum). I am indebted to the American Arachnological Society for their encouragement and financial support—especially to Paula Cushing, who was instrumental in promoting this project during her tenure as president. Special thanks are due to an anonymous arachnophile whose generous donation supported the production of the color plates.

Most of all, I am grateful to Amy Tovar, without whose loving support this book would never have been completed.

Finding, Studying, and Identifying Spiders

Spiders

Naturalists know that the world is full of interesting organisms. Yet for most people, naturalists included, a large portion of our lives is lived in human-altered, domesticated environments. We don't notice the enormous variety of life that is all around us. The lawn becomes an undifferentiated green carpet; the forest edge, a green wall. Even more invisible to us is the microworld. Down among the thatch of dead grass and leaves at the soil surface is a world teeming with bacteria and fungi that break down and recycle dead tissues, creating the basis for the living soil. The microscopic organisms feed a plethora of others. By kneeling down and parting the grass with our hands or a stick, we get a glimpse into this microcosmos. Peering closely, we will likely observe one of its most active inhabitants: spiders. Spiders are ubiquitous in soil-surface communities. This abundance lives literally under our feet, yet we seldom notice.

As a biology teacher, and when presenting talks in my community, I am impressed that most people are unaware of the great variety of living organisms that share our world. They seem genuinely surprised to learn that there are so many different kinds of wild animals, even in cities and towns. When leading field trips to local parks, I can usually find dozens of species of small creatures during a relatively short walk. Children particularly embrace these experiences and relish the act of discovery. Where I can, I pass out sweep nets and set the children loose into abandoned fields or meadows. Their enthusiasm and energy yield nets full of colorful and surprising tiny beasts. We dump the contents onto white sheets on the ground and gather around to inspect the catch. Of course, many insects fly out of the nets as soon as they are opened, and some of the spiders also make a mad dash to escape. The remaining animals often feign death, remaining motionless for a while. After a few moments, they too begin to move away toward the edge of the sheet. Having spread themselves out, the organisms' true complexity and colorful diversity become evident. It is not unusual to find a few dozen spiders, including many different species.

Armed with some simple skills, you can find spiders of many kinds, even in the most mundane places. Some spiders have become regular occupants of our cities and parks. Some spiders inhabit buildings, but many more live outside. Venture outdoors and an abundance of spiders can be found in even a small residential yard. I recently compiled a list of all the types of spiders that I have found in our rural yard in central Ohio over the past 20 years. This list includes more than 160 different species of spiders representing 20 families—all this in one modest backyard acre!

NORTH AMERICAN SPIDER FAUNA

Worldwide, there are more than 42,000 spider species (Platnick 2012). The tropics host the greatest variety of spiders; nearly a quarter of the world's 110 families of spiders are found exclusively or primarily in the tropical realm. The focus of this book is the spiders of North America north of Mexico. No one knows exactly how many different types of spiders exist here, because relatively few scientists study spiders. Our knowledge of them is incomplete. At last count, a total of 3,807 species of spiders had been found in this region. Nearly the same number has been found in Europe, a substantially smaller geographic area. Judging from the continuing rate of discovery of new species in North America, it is reasonable to believe that the current list represents only a fraction of the true diversity. Why? Many of the species of spiders are small and inconspicuous. Other species are so similar that only a specialist using a microscope can distinguish them.

Within the region covered in this book, North America north of Mexico, the areas with the greatest diversity of spiders are along the southern boundary. Florida has many tropical and subtropical species. California and Texas are both large states with considerable topographic and climatic variety, thus it is not surprising that they host many spider species. Although many spiders have adapted to existence in the arid regions of the Southwest, humid habitats and forests typically harbor the

highest numbers of spiders. This guide provides information on most of the larger, more conspicuous, and common species of spiders found in North America north of Mexico. For some genera of spiders, many species look the same. In these cases, one species from the group is illustrated. Representatives of all 68 families of spiders found in the region studied here are included. Nearly half of the species in North America north of Mexico never grow to more than 3 mm (about an eighth of an inch) long. Many of these tiny spiders live near the soil surface, often in the loose layers of fallen leaves or under rocks, and are rarely observed. Others inhabit caves or crevices that are rarely visited. This guide is biased to include those larger spiders that a curious naturalist is likely to encounter. In total, 469 species representing 311 genera are included.

THE ROLE OF SPIDERS IN NATURE

As small or intermediate-sized predators in the food web, spiders are pivotal to the normal function of both natural and human-modified habitats. One study concluded that spiders accounted for more than half of the predator biomass in the forest floor community. Most spiders are considered generalist predators because they will capture and consume a variety of different kinds of prey. Some spiders are specialists that feed on only one type of prey. Because of their small size, the remarkable abundance of spiders is the key to their importance. Density estimates of spiders range from hundreds to millions of spiders per acre. Of course, these are mostly tiny spiders. Despite their bad reputation with the general public, spiders are mostly beneficial to humans. Scientific studies have established that spiders play an important role by eating insects in natural environments as well as in agricultural croplands. Although spiders cannot completely control explosive pest outbreaks, their presence probably dampens fluctuations in pest insect populations in farm fields and orchards.

Each natural environment has its own community of small animals, including many kinds of spiders. Consider the tiny spaces under the leaf litter of a deciduous forest: here you will find an abundance of tiny sheet-weaving and hunting spiders that prey on the resident mites and springtails. These spiders themselves will become food for larger spiders, ants, beetles, and perhaps wasps that also forage among the fallen leaves. On first glance, deserts seem to be inhospitable places. A closer look reveals that life abounds. In the dry brush of a desert wash in Arizona, sweeping with a net captures large numbers of small spiders searching for their insect prey. Up close, these are sometimes marvelously colored and beautiful animals, such as jumping spiders, but we overlook them in favor of the larger lizards and birds that also forage in this habitat. At night this same desert comes alive with both spider and scorpion predators.

Under the canopy of a soybean field in the Midwest live both sedentary cobweb weavers and active hunting lynx and jumping spiders. Near the ground are the tiny webs of numerous sheet-weaving species. Running on the soil surface are a variety of wolf spiders—small, medium, and large. Each of these is both predator and prey. They feed on small insects but may also eat spiders smaller than they are. In turn, they fall prey to large predatory spiders and insects, such as wasps and praying mantises.

Among the reedy growth at the edge of a freshwater pond can be found numerous horizontal orb webs of longjawed orbweavers waiting for emerging mosquitoes. Perhaps a fishing spider clings to a twig of emergent vegetation, patiently waiting for an unwary minnow, tadpole, or aquatic insect nymph to stray within reach. Along the muddy shore, small pirate wolf spiders stalk their insect prey. For each of these situations, the most numerous and important small predators in the ecosystem may be the spiders. They are inconspicuous but crucial components in every terrestrial environment. Tiny species, often living near the soil surface, are usually the most diverse residents. Pick any spot and the first small predator that you encounter will probably be a spider.

SPIDERS IN BUILDINGS

A few spiders are closely associated with human-created habitats. These spiders can tolerate dry conditions indoors and may become locally abundant. Most of the spiders in buildings reside in relatively undisturbed or dark areas under furniture, in corners, closets, basements, cellars, garages, and warehouses. Outbuildings (such as barns and sheds) sometimes support large numbers of spiders. Spiders in buildings provide a service by capturing and devouring insect pests. Most will live out their lives without ever encountering their human landlords.

SPIDER LIFE HISTORIES

The lives of most spiders are relatively short. The majority of species live less than a year. The seasonal timing of reproduction in spiders usually conforms to one of a few basic patterns. One pattern, characteristic of large orbweavers, begins with emergence from the egg case in the spring. The young spiders grow throughout the summer and mature in the autumn. Egg cases are laid at this time, before the first hard frosts in the North, and the adults usually die before winter (see egg cases in color plates). The population continues through the winter as eggs or tiny juveniles in egg cases destined to emerge in spring. In a second common pattern, maturity is achieved in the spring, eggs are produced in late spring or early summer, and growth extends through the autumn. Individuals spend the winter as adults or subadults. Subadult spiders are one molt away from maturity. In the first appropriate weather of early spring, spiders become active, the subadults molt into adults, and courtship and mating occurs. Other spiders actually mature and reproduce during the winter months. This pattern is common among dwarf sheet weavers that live close to the ground. Even in the North, where winter is cold, courtship may happen under the insulating cover of snow near the soil surface. Some multigenerational populations of spiders, particularly in warmer climates, can mature at any season and adults are found year-round.

Courtship and Mating

Some of the most fascinating aspects of spider behavior are their courtship and mating. Soon after a male spider molts into his mature form, he prepares for mating in a behavior called "charging the palps." The palps (short for pedipalps) are small leglike appendages at the front of the spider's head (see Fig. 7). In adult male spiders the last segment (the tarsus) of the palps is modified into a receptacle for the transfer of his sperm. Charging the palps entails spinning a small web, sometimes a single silk line, and depositing the seminal fluid containing sperm as a droplet onto this sperm web. The male then repositions himself and draws the droplet up into one of his palps and then the other.

Most male spiders live a relatively brief period as adults. It has been estimated that the average lifespan of an adult male orbweaver is only about one week, and they spend this time searching for females. In some species adult males may not eat during this period. For many spiders the males use special odors called pheromones to locate the females. Chemical cues may be present in the silk web, or draglines, of females. For example, some wolf spider males find the females by following her silk draglines. When a male encounters a female, he usually begins to court her immediately. This courting may involve a combination of plucking the web, vibrating or waving his palps and legs to provide a visual signal, or producing soft sounds by stridulation (rubbing two body parts together). Some spiders will tap the surface of the substrate (leaf, ground) with their legs or their body to produce the courtship sounds. This vibratory signal is usually transmitted to the female through the substrate. Some species of diurnal (day-active) spiders—in particular,

FIGURE 1. A pair of filmy dome spiders (*Neriene radiata*) mating. The male is above with the embolus of his palp inserted and the hematodocha is inflated, as sperm is pumped into the female's gonopore (photo by author).

the members of the jumping spider family— use brightly colored body parts in visual courtship displays. It is important to remember that male spiders run the risk of being confused with prey, attacked, and eaten by the (usually larger) female. Elaborate courtship behavior may make this less likely.

If the female accepts the male, the two will mate. In some species the male has specific structures that permit him to grasp the female, or her jaws, during mating. In some crab spiders the male spins a few thin silk lines over the female as a mating veil. The male then reaches out to the female and inserts a syringe-like structure called the embolus (part of the palp) into her reproductive opening and pumps in the seminal fluid. In some spiders thin-walled sacs in the palps, called the hematodochae, inflate and assist in insertion and insemination (Fig. 1). Female spiders have bilaterally symmetrical reproductive structures, so after the first mating, the male may change sides and inseminate the other side. In some spiders the male will pause to recharge his palps between matings. The entire process may last a few moments, or sometimes it may extend for hours. The female stores the seminal fluid, containing the sperm, in special internal structures called spermathecae. When the female

FIGURE 2. Female spiders with their egg cases: (upper) *Misumessus oblongus;* (lower) *Pirata minuta* (photos by author).

lays eggs, she adds sperm from the spermathecae to fertilize them.

After mating, the pair usually separates relatively quickly. Typically the males provide little parental care. Females continue to feed, adding yolk to the eggs. The abdomen swells noticeably in such gravid females. At the appropriate time the female spins a special platform of silk upon which she deposits the eggs. The eggs are fertilized as they pass out through the oviduct. She then covers the eggs with a series of layers of soft silk padding and a tough outer covering, and finally she attaches this egg case in a suitable location. The shape and color of the egg case is often distinctive (Fig. 2). A few are so characteristic that the species that made them can be identified from the egg case alone. Some spiders leave the eggs to hatch, and the young emerge on their own. The female may remain with the egg case, guarding it until she dies. A few spider females carry the egg case with them until the young emerge. For exam-

ple, wolf spiders carry the egg case attached to the spinnerets (Fig. 2, lower); the nursery web spiders and cellar spiders carry the egg sac with their jaws. For spiders that lay eggs in the late autumn, the eggs are capable of surviving the freezing weather and emerge in the spring.

Growth and Development

When the young emerge from their eggs, they remain within the silken egg sac and continue to develop. Spiders grow by shedding their "skins" (called an exoskeleton). Each shedding of the exoskeleton is called a molt. Depending on species and sex, spiders can molt about five to ten times before reaching adulthood. Each developmental stage between successive molts is referred to as an instar. The young spiders, called spiderlings, undergo their first molt within the silken egg sac and then emerge as second-instar spiderlings. The spiderlings often remain together in a tight cluster for some hours before dispersing. In many spiders the young disperse by an aerial flight called ballooning. Ballooning is not powered flight; rather, it is passive, relying on a breeze more like the flying of a kite. The tiny spiderlings produce strands or loops of silk by walking back and forth near the tip of a grass blade, branch, or other exposed structure. Sometimes the spider pulls out a small amount of silk and points its abdomen skyward, the gentle breeze pulling out more and more silk until the spider is lifted into the air.

Ballooning usually occurs when there are gentle rising air currents, often early on a sunny morning. Catching their silk threads, the currents loft the tiny spiders skyward. Often the "flight" is short, but on occasion small spiders, even adults, can be carried high into the atmosphere during ballooning. Such long-distance dispersal is responsible for colonization of vacant territory with a fresh population of spiders. It represents a gentle "rain" of spiders that can occur almost anywhere. A number of scientific studies have shown that the first arrivals in an area after a catastrophe such as a flood, volcanic eruption, ash flow, landslide, or fire are usually spiders (Crawford, Sugg, and Edwards 1995).

Spiderlings search for suitable environments particular to their species. For web builders, the physical characteristics of the vegetation or other substrate are often critical to the choice of an appropriate place to build a web. For burrowing spiders, it may be the soil texture or moisture. After establishment, young spiders forage for food to grow. Growth in all arthropods, including spiders, involves a series of molts. Each molt is the process of replacing the exoskeleton with a fresh, slightly larger one. In spiders this size, increase is initially accomplished by movement of body fluids between body compartments. The process begins internally with the dissolution of the inner layers of the old exoskeleton, freeing it from the body. The spider then forces body fluids into the front part of the body (cephalothorax) until it splits at the front edge, then around the sides. This pressure forces the soft vulnerable spider through this opening, and as its body is lifted out of the old exoskeleton, it pulls out its legs and palps last. All of this is usually accomplished with the spider hanging suspended on a silk line.

In some large spiders, such as tarantulas, the molt begins with the spider resting on its back rather than hanging from a thread. As the new exoskeleton hardens, the spider maintains blood pressure in the cephalothorax and legs to stretch them into a slightly larger size than previously. The abdomen has a flexible covering, and it is this area that appears shrunken during the transfer of fluid to the expanded cephalothorax and legs. Just after molting, the body proportions are a bit unusual, with the abdomen being small. A few meals will solve this discrepancy, and the spider's proportions are quickly restored.

SPIDER SILK

A key feature of spiders is their ability to spin silk. The silk of spiders is produced by glands in the abdomen and secreted from special fingerlike appendages at the end of the abdomen called spinnerets. Each spinneret has many microscopic spigots through which the silk is secreted. The silk itself is composed almost entirely of protein. This protein is in a liquid form within the body of the spider, but as it is pulled out, the molecular structure of the silk is rearranged and the silk becomes a solid yet elastic fiber. The spider does not eject the silk; rather, it must be pulled out either by attaching a silk line then moving the spinneret itself, tugging the silk out with a leg, or moving the body away from where the silk is attached. In addition to these fibrous proteins, some spiders make liquid adhesive glue that may be added to the silk as it is produced.

Spider silk has remarkable properties. It is exceptional both in strength and elasticity, and there is more than one type of silk, each of which has its own properties and specific uses (Hayashi, Shipley, and Lewis 1999). One of the most basic uses of silk by spiders is the production of a single line of silk dragline that is played out wherever the spider walks. The spider may catch itself in mid-fall or it may use the line to climb back up to where it fell. For many web-building spiders, which drop from their web when disturbed, the dragline often serves this purpose. Before moving on to a new location, it is not uncommon for the spider to produce a sort of anchor of silk. This is a tiny patch of adhesive silk that serves to fix the silk line. Another common use of silk is to spin a cocoon or silk-lined retreat. All spiders employ silk in constructing their egg case. Perhaps the most important use of silk for many spiders, however, is to produce the capture snare or web. Spiders' webs are distinctive of the species, and the characteristics of webs are useful for identification.

Some spiders have an unusual silk-spinning organ called a cribellum. The spiders that possess a cribellum produce a distinctive type of silk. The cribellum has hundreds of tiny spigots, each producing an impossibly thin strand of silk. Combined, these fibers become a fluffy tangle that is produced around a pair of relatively thick, straight silk lines. This is accomplished with the aid of a special comb of spines on the hind legs that are used to manipulate the cribellar silk as it is produced. This special comb is called the calamistrum. The silk from the cribellum, referred to as cribellate silk, has unique properties. It is adhesive without glue. The large number of strands are drawn out and teased into irregular yarnlike fibers. The abundance of microscopic lines adhere to any irregular surface, even one that appears quite smooth to us, such as the shiny wing covers of a beetle. Because each fiber is composed of a teased fluff of lines, the webs of cribellate spiders look different. Sometimes "hackled" or "hackle-banded" is used to describe this sort of line. The lines look thicker or more reflective, sometimes in webs with simple thin lines combined with thicker cribellate sections. Under certain lighting conditions, the cribellate silk may even appear bluish.

SPIDER BEHAVIOR

Spiders are good at waiting. Many spiders will stop moving when they notice a potential predator, such as an observant human. A stationary spider is often difficult to see, unless the coloration contrasts with the background. Most spiders choose to rest in places where their colors match the background, rendering them camouflaged. When we search for spiders, we most often notice them if they are exposed in a web or because they are moving. If you stay still and watch a motionless spider long enough, it may resume its normal behavior. The common behaviors of spiders that we are likely to observe are related to finding food or searching for and courting a mate. To understand what we are seeing, it is important to think in terms of the spider's world.

SPIDER SENSES

To understand spider behavior, we must try to perceive the world as spiders do, through their very different senses. This is indeed challenging because humans primarily use sight and sound, while spiders are first and foremost creatures of touch. Spiders sense subtle vibration, typically substrate-borne vibration. What does this mean? Spiders are covered with exquisitely sensitive sensory organs that are adapted to reception of vibratory clues about their environment. The place where the spider is standing, or hanging, is the substrate. The spider notices tiny vibrations felt through sensors in and on its eight legs and sometimes the palps. If the substrate is solid—such as a leaf, branch, rock, or the ground—the vibrations are very subtle. Spiders have been shown to sense the faint signals of walking insects many inches away and can even estimate the distance and direction of potential prey (Barth 2002). Webs are a sort of extension of the spider's body. Vibrations are transmitted efficiently to the legs through the thin taut strands of the web lending web-building spiders remarkable sensitivity to disturbance of the web. These spiders can even distinguish what sort of prey has entered the web, merely by the pattern of vibrations. Thus spiders use their legs to "hear" the world they are standing on. By extension, the webs of spiders expand their sensory capabilities out into the environment.

Spiders also sense air movement with their hairs. The most sensitive hairs are the trichobothria, fine microscopic hairs borne in pits and attached to a flexible membrane of the exoskeleton. These trichobothria can feel air movement so well that they are sometimes referred to as sensors of touch-at-a-distance. Spiders can detect the approach of potential prey, or a dangerous predator, well before it comes in contact with them. This sense is limited, however, and sound waves are perceived only to the extent that they jiggle the trichobothria. There is modest evidence that they can also sense airborne chemical compounds—that is, essentially smell—with the hair receptors or tiny pit organs of the legs and palps. The clearest demonstration of odor reception is the perception of sexual perfumes called pheromones. Some spiders have been shown to respond to the sexual pheromones of others of their species from distances as great as 1 meter. This distance, for a spider, would be roughly equivalent to a city block in human terms. Spiders' chemoreceptors are also sensitive to direct contact, a sense similar to taste.

Although useful, sight is only an important sense for some spiders. They do have eyes, typically eight, but the eyes of most spiders are small, and their vision is adapted to sense pattern and movement. For many types of spiders, vision is used to detect the threatening approach of a potential predator. The presence of many eyes, each with a different visual field, ensures that spiders can detect movement from almost any direction. They often do have eyes in the back of their heads! It truly is difficult to sneak up on a spider without it noticing our approach. The activity patterns of spiders are influenced by the light cycle of day and night. Most spiders are active primarily at night. Like many nocturnal animals, nocturnally active spiders possess a reflective layer, the tapetum, at the back of their eyes. This has the function of reflecting the low light of the moon or stars at night back through the sensory retina, thus amplifying its signal and increasing its sensitivity in the dark. Wolf spiders are famous for having large eyes that enable night vision.

The structures of the spiders' eyes reveal much. The front lens of the eyes is a part of the exoskeleton and fused in place. Among the eight eyes, the most interesting differences are between the anterior median eyes (AME), or principal eyes, and the other eyes. Spiders can shift their gaze by moving the retina within the head well behind the lens. Thus the AME are the only ones that can search the visual field without the spider moving its body, making them a little like human eyes. This fact is dramatically illustrated by the jumping spiders, which have unusually large and acute AME. For them, the long tubular internal structure acts much like a miniature telephoto lens. Closely watching the extraordinary eyes of jumping spiders scanning their environment, one can actually see a pattern of shadow shift within the eye tube. Of course, the jumping spiders are also famous for moving their bodies to face any object of interest, be it a potential prey, mate, or even a human observer. This habit of returning the human gaze with their own binocular visage is one of the features of jumping spiders that provide their appeal to many human watchers.

So how good is spider vision? The answer depends on the type of spider. Some spiders can barely form a simple image, using their vision mostly to avoid predators. A few cave spiders are blind. Others, most famously the jumping spiders, are primarily visual creatures. Their eyes are both large and acute. Jumping spiders, and perhaps others, are also capable of sensing color. Color vision ability is employed by female jumping spiders to watch the bright and colorful courtship displays of males.

SPIDER INTELLIGENCE

For most invertebrates, including spiders, much of their behavior is based on preprogrammed behavioral patterns that we call instinct. Nevertheless, it is clear that many spiders learn about their environment and this information is incorporated into decisions they make. They can be trained to respond to particular stimuli for a positive reward and will ignore stimuli associated with a negative or unpleasant experience. Spiders also learn spatial position information: they can orient themselves and return to familiar locations such as their web, retreat, or burrow.

Some spiders exhibit flexible behavior, which can be modified to suit subtly different situations. Typically these spiders are considered more intelligent. Examples include jumping spiders that employ spatial memory and planning in their stalking of prey at a distance (Tarsitano and Jackson 1997). They use an approach path that is out of view of the prey until the last possible moment before they pounce. They even alter their tactics for different prey. By contrast, the web-building spiders'

FIGURE 3. (Left) *Clubiona riparia*, folded-leaf retreat; the spider folds the leaf twice, using silk to fasten the edges (photo by Joseph Warfel). (Right) *Phidippus audax* and cocoon (photo by author).

behavioral program seems less flexible, yet their ability to adapt the detailed web-building protocol to local conditions is remarkable. Their reaction to disturbance at the web varies depending on the signal. Sometimes it is the struggling of a potentially suitable prey organism that will elicit an attack. Other times it is the approach of a potential threat, such as a hunting wasp. In this instance the spider will quickly drop from the web or vibrate it violently. For an adult female spider, it might be the arrival of a potential suitor. Each of these situations provides a complex suite of stimuli that must be received, interpreted, and acted upon, often quickly.

SPIDER MOVEMENT

Spiders have excellent climbing ability. They can move through almost any habitat. Equipped with specialized tufts of hair or claws on the tips of their legs, spiders can scale the smoothest surfaces with ease. The spiders' silk dragline is left behind as a continuous record of their journey. On occasion I've been able to follow the silk trail and find the spider, hunting for its next meal. If a wandering spider finds itself at the end of the trail—for example, the tip of a branch—it can just lower itself down on a silk line until it finds a new place to continue

searching. A number of spiders are fast runners. They exploit this ability to dash away to the nearest cover and then remain motionless until the threat moves on. Many spiders habitually walk, or even run, on the water's surface. These spiders are usually not immersed and floating; rather, they exploit a waxy water-repellant covering their lightweight bodies, allowing them to stay dry and to remain on top of the water surface. Watching a large spider on the water, you may notice that the legs are resting in shallow dimples, caused by the spider's weight pushing down and the water's surface tension pushing back up.

SPIDER RETREATS

Second only to their patience, spiders possess remarkable hiding ability. Some hide in plain sight, possessing bodies that are camouflaged against the background where they choose to rest, combined with their habit of remaining still. Other spiders squeeze into such tiny spaces as cracks in the soil or the bark on trunks of trees, gaps in the gravel, or debris at ground level. Many spiders spin silken cocoons or retreats where they spend their inactive period, usually the day (Fig. 3, right). The retreat may be a simple thin cocoon under a leaf or branch or under the eaves of a building. Others build elaborate camouflaged spaces that are covered with bits of local environmental debris such as dead leaves, soil, flakes of lichen, even body parts of prey they have recently eaten. Some sac spiders are famous for folding over large blades of grass and sewing them into strong, resilient, waterproof chambers (Fig. 3, left). Web-building spiders often build small silk retreats near their hunting webs.

HOW SPIDERS HUNT

Spiders capture their prey either by grasping it with their front legs or biting their prey. Some spiders have such strong jaws that they can hold the prey as they inject venom. Others make a quick bite then back off and wait for the venom to take effect. After the prey is paralyzed, the spider moves in to feed. A third approach is to wrap the prey in silk to immobilize it, and then bite. Such spiders approach a potential prey item, turn so that their spinnerets are pointing at the target, then use their legs to rapidly draw out silk from the spinnerets and toss it onto the prey, a wrapping attack. For some spiders that use a wrapping attack on a surface—such as a rock, the trunk of a tree, or the ground—the spider moves around the prey and its action produces a sort of wall or tent of silk over the victim.

Spiders feed exclusively on liquid material. Once the prey has been immobilized, the spider regurgitates digestive fluids into the prey. The enzymes in this fluid liquefy the internal contents of the prey. The spider then sucks the liquid back out, digestive secretions and digested prey materials together. They may repeat this cycle of spitting-and-sucking for a few minutes to an hour or more. Some spiders use the teeth on their jaws, along with the moveable fangs to crush and manipulate their prey as the feeding progresses. Spiders that eat this way leave only a small pellet of indigestible material. Others do not crush the prey and feed only from small holes produced by the fangs; for these spiders the prey remains are a hollow, but still recognizable, empty husk after the spider has finished its meal.

SPIDERS AS PREDATORS

Web-builders

The most familiar predatory strategy of spiders is to build a silk snare. Just over half (54 percent) of North American spider species use a web to capture their prey. The webs of spiders come in a great variety of types. Some are complex, with many functional components. Others can be as simple as a series of trip lines emanating from a retreat where the spider hides.

Ambushers

Many spiders dispense with building a web and employ a sit-and-wait ambush strategy. Remaining absolutely still, the spider is not noticed until it is too late. Of course this works best in a spot where prey may likely be found. Spiders can afford to use this time-consuming method because they expend little energy while waiting. One group of crab spiders, the flower spiders, are sit-and-wait hunters that take advantage of the attractiveness of flowers. They wait at flowers for the arrival of pollinators and often capture them. Flower spiders rest with their front four legs outstretched. These legs are well armored with heavy spines that assist in grasping the prey until it can be bitten. The venom acts quickly and as soon as the prey has been subdued, the legs release and the meal is held by the fangs alone.

Burrowers

Some spiders dig burrows as a retreat. Burrows often provide protection from predators and a favorable cool and moist microclimate. For some, the burrow is merely a retreat. These spiders leave the burrow and wander in search of food, usually under the cover of darkness. Other burrowing spiders remain in the burrow, or near its entrance. If danger threatens, they can quickly disappear into the burrow. If potential prey happen by, a quick lunge is sufficient for capture. Several families of spiders build trapdoors to cover the burrow entrance, which can be closed when the spider is not hunting. The door serves to hide the burrow and protect the spider from predators.

Stalkers

Some spiders stalk their prey, a bit like a cat. These species move through the vegetation, or over the ground, periodically stopping to sense potential prey. They may feel the vibrations made by the prey as they walk by, or the spider may actually see the prey moving. The best-known visually hunting spiders are the jumping spiders. These diurnal spiders approach stealthily to within range, then pounce. Another group of stalking diurnal hunters are the active and agile lynx spiders. They run and jump over the vegetation with long thin legs. When they notice prey, they leap on it using their spiny legs as a capture basket.

Among the stalkers, the nursery web spiders are known for their relatively large size and their preference for humid or aquatic habitats. Several species of this group feed at the water's edge and from the water surface. They submerge below the surface to capture prey or to elude potential predators with some difficulty, because spiders are buoyant and will float to the surface if they release their grasp. They typically crawl below the water surface by holding onto aquatic vegetation. A number of other spiders are commonly found hunting at the water surface, including the long-legged water spider of Arizona and several species of pirate wolf spiders.

Wandering Hunters

The wandering hunters include spiders that typically hunt in vegetation, usually at night. These spiders move quickly, sometimes running, along the branches and leaves. They capture prey by running up onto them. The pale ghost spiders, sac spiders, and prowling spiders feed in this way. These families share an adaptive behavior that supports their relatively active running lifestyle. They actually feed on plant nectar and other sugary secretions, both from the flowers and special nectar glands on the other parts of plants. This sugary solution provides them with the caloric energy to power their constant motion. In the Southern states, another group, the huntsman spiders, also forage in a similar way over the surface of plants, rocks, or the walls of buildings.

On the ground, too, there is a suite of wandering hunters. These include the familiar wolf spiders, ground spiders, wandering spiders, the antlike runners, as well as a number of smaller families. Most of the wolf spiders hunt on the ground or in low vegetation near the ground. They have excellent vision, and although many hunt during the day, more individuals can be found active at night. Their hunting strategy involves short movements alternating with periods of rest. It is while the spider is still that it detects the movement of potential prey. At night this detection is probably by sensing the vibrations caused by the prey moving over the substrate.

NAMES FOR SPIDERS

Scientists called taxonomists assign names to all organisms based on their biological relationship to one another. These names are recognized and applied worldwide, no matter what the language of the scientist. The names consist of seven parts or categories, which range from the broadest and most inclusive, to the narrowest and least inclusive; the categories are kingdom, phylum, class, order, family, genus, and species. Individual organisms that are members of the same species (and there are more than a million described species of organisms) are members of the same genus, the same family, the same order, and so on, and are the most closely related and can mate and reproduce. All members of the same genus are less closely related; all members of the same family even less closely related and so on, up to members of the same kingdom. In the case of spiders, they are members of the kingdom Animalia, which includes all animals, from elephants to mosquitoes. Spiders are arthropods (members of the phylum Arthropoda), which are animals that possess exoskeletons and whose limbs and feet are jointed. This phylum also includes the insects and thereby contains more member species than any other phylum. The phylum Arthropoda contains several classes, one of which is Arachnida. Spiders are in this class, along with mites, ticks, scorpions, and harvestmen (called daddylonglegs by some)—all of which are arachnids. Among the arachnids are the "true spiders," and these compose the order Araneae, the subject of this book.

Typically, in naming organisms, we could move from the category "order" to the final three categories—family, genus, and species—but to properly categorize spiders, we need two more groups within the order Araneae. One group consists of the suborders, of which there are two—the Mesothelae, an odd group of spiders with segmented abdomens, and all the rest of the spiders without segmentation, the Opisthothelae. All North American spiders fall into the Opisthothelae. Then the Opisthothelae are divided into two infraorders, Mygalomorphae and Araneomorphae. The mygalomorphs are mostly large burrowing spiders, such as the tarantulas. These spiders account for fewer than 4 percent of spiders in North American north of Mexico. The other 96 percent are members of the infraorder Araneomorphae.

The family name is what we use to refer to groups of spiders. For example, what are popularly called wolf spiders are in the family Lycosidae, also called lycosids. Likewise, people know jumping spiders, which are in the family Salticidae, or salticids. Crab spiders, in the family Thomisidae, are also known as thomisids. Some family groupings are controversial among scientists. There are differences of opinion based on which evidence for relationship is considered most important. For example, a current controversy concerns the placement of certain genera in the Amaurobiidae or Agelenidae. Such disagreements are a normal part of systematics but can be frustrating or confusing to nonspecialists. Finally, all organisms, spiders included, have what is called a formal scientific binomial name, its genus and species. An example among spiders of a genus is *Phidippus*, which is one genus in the jumping spider family Salticidae. North of Mexico, the genus *Phidippus* contains 47 species, of which *audax* is one. *Phidippus audax* is the formal binomial name of an impressive jumping spider popularly known as the bold jumper. To summarize, the full scientific name of the bold jumper is described below:

phylum Arthropoda
  class Arachnida
    order Araneae
      suborder Opisthothelae
        infraorder Araneomorphae
          family Salticidae
          (jumping spiders)
            genus *Phidippus*
              species *audax*
              (bold jumper)

There are a few points to make about the use names. One is that the formal binomial name (genus, species) is always italicized, such as *Phidippus audax*. Second is that the genus name is capitalized, the species is not. Third, in formal scientific practice, the name of the scientist who named it and in what year is listed after the italicized name. The naming scientist is called the "author." If the species has never been recategorized, and therefore its name has never been changed, the author's name is not in parentheses, and it and the year of description follow the name of the species. If, however, a species has been recategorized, then the original author's name and year of naming will appear in parentheses—for example, *Phidip-*

pus audax (Hentz, 1845), which was originally placed in the genus *Attus* by Nicolas Hentz back in 1845.

Finally, some common spiders have been given English names that can refer to just a single species or to an entire family. For example, a member of the single species *P. audax* is known as the bold jumper. However, the entire family of Araneidae is known as the orbweavers. In cases where many families exhibit a shared distinctive behavior, such as the seven families that build somewhat similar circular webs, I identify them by that behavior collectively, the orb-shaped web group.

Finding and Studying Spiders

Spiders can be found almost anywhere. The key to finding them is careful observation. Typically spiders sit motionless, particularly after being disturbed by an approaching human. To find spiders, stay still for a moment and look closely. If you are patient, a spider may reveal itself by resuming its normal behavior. If you have never searched for spiders before, your yard or a nearby park may be a good place to start. It's helpful to keep a small clear plastic container available in your pocket, just in case you want to capture a spider for close observation. The habitat, time of day, weather, and the season all influence what kind and how many spiders you will be able to find. If at first you don't find many, don't get discouraged.

FOLLOW THE WEBBING

One of the key features of "noticing" is paying attention to webbing. Silk strands or complete spider webs are often the first clue to a spider's presence. Check under or within the web before assuming that it is empty, because a motionless spider may be hiding there in plain sight. A few

moments after disturbance, an exposed spider will sometimes dash for cover, especially web-building species that often have a nearby retreat where they spend their inactive periods. When disturbed in a web, the spider quickly returns to this retreat. Careful examination of a flat orb can reveal a line of silk leading to the spider hiding in a nearby curled leaf or similar place. Gently touching a spider in its retreat will sometimes "flush" it out into the open. This may backfire because some spiders will quickly drop to the ground. These spiders either curl up feigning death or scamper away quickly, but in either case they are often very difficult to find after they have reached the ground. Many times the spider will have left an invisible silk line; after a few moments the spider will climb back up to the web.

Spiders that construct funnel-shaped retreats are also difficult to expose. One useful technique is to simulate the vibrations of a potential prey item. For this ruse to work, you need to wait a few moments after you have disturbed the spider because it will usually freeze for some time before resuming its normal foraging. Using a thin flexible blade of grass or something similar, tickle the webbing or drag the blade slowly across the surface. The spider may come out to investigate. For some spiders the buzz of a tuning fork can be a powerful attractant. Set the fork into vibration, then gently touch the end to the web. The periodic vibrations may resemble the buzzing wings of a captured fly. What spider could resist that?

COVER IS EVERYTHING

During the day, looking under bark, logs, or rocks will often reveal spiders. It is important to emphasize that these animals are found in these places because of the dark, moist microclimate created by these objects on the ground's surface. When we turn over a log or rock, we cause a disturbance that can be quite destructive to these microenvironments. Exposure to the dry air, sunshine, and potential predators can be deadly to the animals that live there. We must all make a habit of returning the log, rock, or other material as close as possible to its original position after a search, considerably reducing the harmful effects of our actions. Indoors, spiders can often be found in cellars, basements, and garages. Even an infrequently used closet is a good place to look. Glance into corners and up to where the wall meets the ceiling. Spiders tend to seek out quiet hidden spots. Under the furniture is another great place to look.

NIGHT SEARCHING

Many spiders are more active at night than during the day. Even a short excursion around the outside of a building will often reveal a few spiders. For some people, searching for spiders at night is a bit scary. It is true that at night we are in their world. We have poor night vision; the nocturnal creatures have the advantage. A good LED headlamp is a useful and inexpensive tool. With this light on your forehead, your hands are free for holding a magnifying lens, camera, or small containers for capturing spiders. If you have only a flashlight, try holding the light near the side of your head at eye level. You may see tiny bright reflections, called eyeshines, from spiders and insects. This effect is even more obvious with a headlamp. The eyes of nocturnal animals have a reflective layer to amplify the low moonlight or starlight. When we shine a light on such an eye, a point of light is reflected back. If you approach the source of the eyeshines quietly, you may discover the animal, often a spider. The best technique is actually to look in the distance. If you search too closely, the general illumination is so bright that it drowns out the delicate eyeshines. In the distant dim illumination, the little points of light are more obvious.

USING A MIST SPRAY

You may have noticed that spider webs are sometimes really obvious and beautiful on a dewy morning. The moisture condensing on the webs makes them stand out. You can simulate this effect in drier conditions by spraying a fine mist of water onto a likely patch of vegetation with a household spray bottle. Almost any mister will work well, even an empty container of household glass cleaner or some similar product. The finer the mist droplets the better; the focused squirting types are useless. A tiny atomizer may work, but these are best for searching in a small area. Be careful to rinse the container out several times with water so that you aren't spraying something noxious. Fill the mister with plain tap water and then walk to an area with dense vegetation, or even a patch of unkempt lawn. If there is air movement, spray into the air a distance away upwind of your target area and let the mist drift onto the vegetation. If you spray hard directly on the vegetation, you may dislodge the delicate webbing or even wash it away. Calm conditions are best, but a slight breeze usually isn't a problem. Just continue to produce a fog and wait as it drifts onto the plants. Often strands of webbing, and sometimes an entire web, will become visible as if by magic. Once you have the silken clue, it is easier to search for the spiders. Single lines of silk are often draglines left by a spider. Follow them to find a spider hiding nearby.

DUSTING WEBS

Some observers prefer using cornstarch (fine white powder) instead of a water mist. This is either used in a plastic bottle with a narrow opening, such as an empty eye drops bottle or an old sock. If the bottle is squeezed, it will produce a puff of powder. This coats the web and makes it much more obvious. The other method is to fill an old sock with cornstarch. Shaking the sock will produce a cloud of powder that will coat any nearby webbing.

OBSERVING SPIDERS

A good magnifying glass is a useful tool. The features of spiders are often small, and a magnifying lens will reveal many details. It is now possible to purchase binoculars that focus very closely. These are sometimes sold for observing butterflies, but they work wonderfully for spiders, too. One advantage to using binoculars is that you will be far enough away from the spider that it may not even be aware of you, and it will behave normally. As a dedicated amateur, you can make valuable additions to our knowledge of spiders. Often you have more time to devote to careful observation than professionals, who tend to focus on one particular experiment or project. Taking good notes about the habitat preferences or behavior of particular species is an important way that you, as a citizen-scientist, can contribute to the body of knowledge about spiders.

PHOTOGRAPHING SPIDERS

Modern digital cameras usually have a "flower mode" for close-up photography. Even a fairly crude photo can reveal considerable detail that escapes our casual looking. Better still, the photo can be saved and later compared directly to the illustrations. The specific techniques of close-up macrophotography are beyond the scope of this book. A good clear photograph can serve as a tangible record of a spider. Be sure to jot down a note about where and when you took the photo.

CATCHING SPIDERS

Be sure that you have permission to catch or collect spiders from the owner of the property where you are searching. Some parks and reserves do not permit disturbing plants or animals; in others a special collecting permit may be required. With care, you can place a small container in the path of a rushing spider and capture it. If you notice the spider before it moves, use a small soft paintbrush or pipe cleaner to nudge the spider into your container. If the spider is in a web or vegetation, it is often helpful to hold the container below the spider as it will often drop when disturbed. Pipe cleaners are useful to prod spiders out of cracks in the soil, bark, or among rocks. The soft bristles of the pipe cleaner will usually avoid damaging or crushing the spider. A variety of small transparent plastic vials or even small rectangular clear plastic packaging boxes are useful complements to a spider-hunting excursion. With the spider in a container, it is often possible to get an excellent view with a magnifying glass. Binoculars, held backward, make excellent magnifiers. For the enthusiast, small field microscopic viewers with excellent optics are available.

Another way to catch tiny spiders without damaging them is to use a device called an aspirator or "pooter." This device has a stiff tube connected to a chamber, usually of clear plastic. A second, flexible tube is also connected to the chamber. While holding the stiff tube near the spider, suck on the flexible tube and you can draw it into the chamber. A fine screen is used so that there is no danger of swallowing the spider. Many biological supply companies sell pooters of this kind. You can either observe the spider directly in the pooter's chamber or transfer it into another better observing container.

Sweeping

Using a heavy-duty insect net (or even a pillow-case stretched over a stiff wire clothes hanger), sweep through the top foot of vegetation or tall grass. Don't try this where there are cacti, brambles, or roses, because the net will snag and the spiders will escape. Remember to use quite a bit of vigor as you sweep or the spiders may fall off the vegetation in front of your net. After a dozen or so sweeps, dump the contents of the net onto a flat sheet and capture the spiders as they run away. Be prepared to act quickly! If you sort carefully through the remaining material, you may find a number of spiders curled up and "playing possum" among the debris. It is not uncommon for a sweep sample to capture a dozen species representing four to seven families after a few minutes of work. This is one of the best methods of capturing such active hunters as jumping, lynx, nursery web, sac, or ghost spiders. You may also capture small web-building species.

Beating

This method is much like sweeping. Spread a large cloth sheet under a bush or the low branches of a tree. Grab the branches and give them a vigorous shake, alternately striking them with a stick or stiff branch. Spiders (and other creatures) will be dislodged and fall onto the sheet. Be ready, because the spiders will move quickly. Some spiders may return to the branch up a silk line that they produced as they fell. Another approach is to hold an umbrella, upside down, under the branches. Spiders that fall into the umbrella often slide to the middle and are easy to locate. It is important to note that beating and sweeping do not work well in wet conditions. If there is heavy dew or if it has rained recently, the spiders will adhere to the wet cloth and are often damaged or killed during the sweeping. They are also much less likely to tumble into the net. Avoid sweeping in such conditions.

Pitfall Trapping

One of the most effective methods of capturing ground-living spiders is pitfall trapping. Any smooth-sided container buried flush with the ground surface will work. Some people place a funnel at the top of the container. Inside the pit you may want to put a second cup so that you can remove the contents without disturbing the edge of the pit, because the pit's edge is the key to success. If a spider detects a lip or a ridge, it is likely to walk around rather than tumble into the trap. Rain can also be a problem, so where rain is likely, you'll want to place a cover over the pit propped up about an inch or so above the edge. A smaller gap between trap and lid will reduce accidental captures of vertebrates, such as small toads or shrews. Sometimes pits are left "dry," but most collectors use automotive antifreeze in the pits. The antifreeze will kill and preserve the captured spiders with minimal evaporation. Both methods have potential problems. If you leave pits dry, you may find only one large (well-fed) spider or centipede; smaller ones will have been killed and/or eaten. If you use wet pits, the fluid may attract wildlife that could be poisoned. Antifreeze is available in two forms: ethylene glycol–based and propylene glycol–based. Propylene glycol is preferred because it is not as toxic to mammals. If you do use a toxic liquid in your pits, you must cover the pits with wire screen or other barriers to keep wildlife out.

If you put out pit traps, you should be prepared to check them frequently. Some nocturnal raiders may dig up and destroy your pits. Raccoons, skunks and opossums are infamous for this behavior. To prevent this, cut a piece of chicken wire screening about 2 feet square. Place the wire screen over your pit and pin down the four corners with tent stakes or heavy bent wires. The mesh is wide enough that small creatures (such as spiders) will pass through undisturbed, and the edge is usually far enough from the trap that it will discourage inquisitive raccoons from attempting to tunnel into the trap. Remember to remove the pit and refill your hole when you are finished collecting.

Litter Sampling

Using gloves, collect a large amount of leaf litter. You can either examine the sample immediately or store it in a black plastic trash bag or similar container for later. Be sure to include the leaves and duff at the surface of the soil because this will have many small animals. Keep the bag of litter cool until you are ready to extract the spiders. A litter sample left in the sun will quickly bake and the dead curled-up spiders will be impossible to find. When you are ready to search, the sample can be spread out on a light-colored sheet and examined for spiders. Just toss a handful at a time onto the sheet. Some spiders will run immediately, others will hesitate or play dead. Being patient is the key to success. A small soft-bristled paintbrush is a useful tool for nudging spiders into a container for later study.

Biologists often extract spiders from leaf-litter samples using a device called a Berlese funnel, available through biological supply companies. This consists of a container in which the litter is gently placed with a screen at the bottom, over a funnel. The screen is usually either a quarter-inch or half-inch mesh hardware cloth. Window screen is much too fine-meshed. A light bulb (incandescent, not fluorescent) is suspended above the leaf-litter sample. Be careful with this method that the bulb is far enough away from the leaves that they won't be singed or burned. There is a danger of fire if you are not careful. Small creatures will gradually flee from the bright, dry conditions and move down toward the screen. Eventually they will drop through the screen into the funnel and then the collecting container, usually a small jar filled with alcohol. Of course if you choose to use alcohol, the spiders will be killed. On the other hand, dry containers can also be used to collect the spiders alive but must

be monitored constantly or the spiders will escape. The complete extraction of a sample may take several days.

KEEPING SPIDERS IN CAPTIVITY

It is sometimes helpful to keep spiders in captivity for observation. During general collecting, you may catch subadult spiders, which are one molt away from sexual maturity. If you keep them for a week or so in a container with appropriate moisture and food, they may molt into an adult that you will be able to identify. Avoid adding rocks or pebbles, which may shift and crush the spider when you pick up the container. It is not always easy to find food for small spiders. Fruit flies often work well. Pinhead crickets can be fed to larger, robust spiders, but they may actually kill and eat small spiders. Try to provide a variety of prey insects that are about half the size of the spider or smaller. Tiny moths that come to porch lights at night are another good choice. Ants should be avoided unless you are confident that the spider is an ant specialist. Many spiders will not capture or eat aggressive ants; also the ant may kill the spider.

The key to successfully maintaining live spiders in captivity is to provide a range of moisture in different parts of the container, from wet to dry. This way the spider can select an appropriate humidity. Alternately, you can slip a fresh leaf into the container to provide moisture and a perch for the spider. Remember to change the leaf if it starts to dry up or turns brown. Air circulation is not as important for spiders as for many other animals. Spiders have very low metabolic rates and require relatively little oxygen. Spiders may do just fine in a closed plastic container as long as you open it a few times a week to add food and perhaps a spritz of water from a mister or atomizer. Be stingy, it is easy to overdo with water. Spiders kept in overly wet conditions are often susceptible to disease.

MAKING A SPIDER COLLECTION

One of the great ironies of biology is that biologists spend a fair amount of their time killing the organisms that they study. Of course, most biologists avoid destructive collection whenever the objectives of their studies can be met in other ways, or if the populations of the organisms in question are vulnerable. In the case of spider study, collecting is still relatively frequent. There are two reasons for this: First, with the exception of a few large or distinctive species, most spiders cannot be identified to the species level without microscopic examination. Serious study of an organism usually begins with identification. Second, because of their relatively small size and high reproductive capacity, most spiders have high populations in suitable habitat. There are frequently hundreds to thousands of individuals per hectare. For those species, removal of a few individuals is unlikely to adversely affect the health of the population.

If you decide to collect spiders, be prepared to make an organized, properly labeled collection. At a minimum, the spiders should be preserved in an appropriate preservative, either isopropyl or preferably ethyl alcohol. About 70 percent alcohol is usually used, but a higher concentration may preserve the spider well. The problem with higher concentrations of alcohol is that the specimen may become stiff, fragile, and difficult to work with. Use a container that will seal tightly. Pill bottles from the pharmacy are usually not good because the loose-fitting lids will leak and the alcohol may evaporate quickly.

One essential practice is to put a label directly inside the container with the spider. Labels attached to the outside often fall off or get separated from the specimen. Also, you cannot use ordinary ballpoint ink on the label because it will dissolve in the alcohol solution. Surprisingly, you can use pencil, which is cheap, available, and permanent. In some art and office-supply stores, you can often find ink pens that are alcohol resistant. Many computer laser printers use a photocopy-like process that creates labels that look great for a few weeks but unfortunately deteriorate after some years when the heat-fused toner begins to fall off of the paper. Sometimes the entire letters fall off and show up at the bottom of the container, looking sort of like black alphabet soup! Inkjet inks are variable: some are excellent and not alcohol soluble, others will dissolve. If in doubt, do a test. The label should be no larger than about 1 inch by ½ inch so that it will fit in a small vial with the spider. The label should include the state, county, and town or distance and direction to the nearest town. The date is important. It is best to spell out an abbreviation for the month, such as Mar 1, 2009, rather than using 3/1/2009 because some people, including most scientists, write the month second, and 3/1/2009 could be confused with January 3. Include your name as collector. On the back of the label you can provide habitat information, such as "pine forest," "bushes in city park," or "under a rock."

Voucher Specimens

Vouchers are specimens with complete data, including at a minimum the date, the exact location, and the habitat where obtained, which can become a tangible record of the observation. They vouch for the existence of that spider in the particular situation. A good photograph can serve as a voucher, but even an excellent photo might not contain enough detail to identify a spider to the species level. To document a species, the voucher specimen must be an adult. You should probably release immature spiders at the time of capture. It is not always easy to tell if your spider is an adult. The best clues are the presence of enlarged ends on the palps of males (see Fig. 8). This may seem clear, but older immature males may have swollen palps, yet still not be adult. If you look carefully with a magnifying glass, a mature male's palps will have a complicated structure with a variety of points or projecting parts. In an immature male, the palps will be smooth without such points. Adult females will have a contrasting, usually shiny, dark epigynum underneath at the front of their abdomen. Some spider enthusiasts maintain immature spiders in captivity until they molt into identifiable adults.

If you wish to donate your collected spiders to an institution such as a university or museum, be sure to contact them in advance. Some institutions have limited staff and may not be able to accept your collection. Doing technical identification and cataloging of specimens is time consuming and expensive. Other institutions may have no spider expert available. Don't be surprised if an institution declines your offer of specimens. Still, if you contact enough local museums and universities, you will find a good home for a well-documented collection.

Identifying Spiders

At first when we find a little creature, we want to know what type it is. The first step is to find out if it is a spider at all, or perhaps some similar type of arthropod, such as a harvestmen. This step is really the *recognition* of a spider. We may also want to know what kind of spider we are looking at—for example, is it a wolf spider? This is identification to the level of the family. That information may be sufficient; because knowing what particular family a spider belongs to is the key to predicting its role in nature and many basic features of its biology. If we are willing to delve a bit deeper, we may try to identify the spider to the genus or even species level.

RECOGNITION OF SPIDERS

One of the first steps in learning about an unfamiliar group of organisms is to learn how to recognize and then identify them. Spiders are usually easy to recognize as spiders, although a few mimic ants. Most people have already learned how to distinguish a spider from an insect. Spiders have eight legs, insects six. But

of course there is more to it than that. There are a variety of other animals that resemble spiders; the arachnid relatives such as scorpions, mites, ticks, and harvestmen also possess eight legs. Spiders are placed in the order Araneae within the class Arachnida, one of 11 orders within this class found in North America (Fig. 4; Table).

Other common arachnids are sometimes confused with spiders. Perhaps the most frequently encountered animals that are misidentified as spiders are the harvestmen, order Opiliones. These include the common daddylonglegs of gardens and fields. Harvestmen can be distinguished from spiders by the fact that their body is fused into what appears to be one unit (Fig. 5). At the back of the body, the abdomen is divided into segments; sometimes visible as rings. Harvestmen have one conspicuous pair of eyes, often on a bump in the center of the head area. Spiders have two body regions without visible segments. Most spiders have six or eight eyes. Spiders also have a set of fingerlike appendages on the abdomen (spinnerets) with tiny spigots where silk is released. Some other arachnids can produce silk, but none pos-

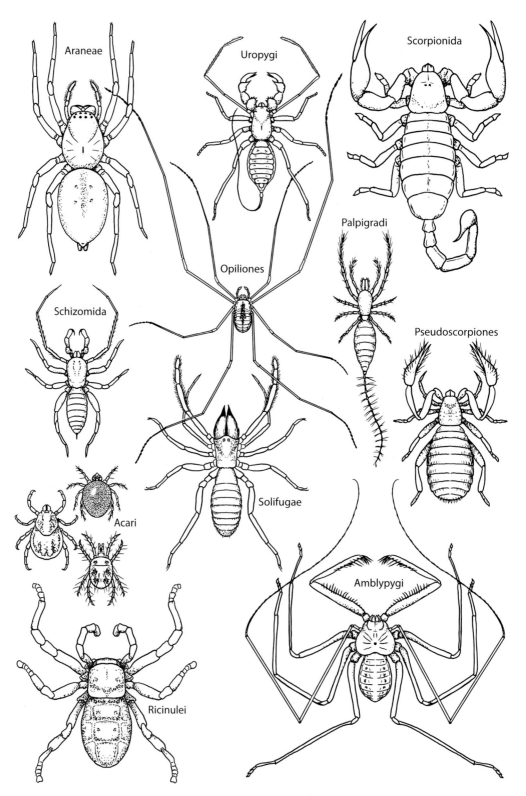

FIGURE 4. Examples of the 11 orders of Arachnida.

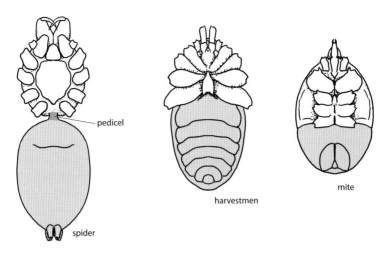

FIGURE 5. Ventral view of a spider, harvestmen, and mite.

The living orders of the class Arachnida.

Order	Common Names	Number of World Species	Number of North American Species[a]
Acari	mites, ticks	45,231	5,106
Amblypygi	tailless whipscorpions, whipspiders	203	5
Araneae	spiders	42,473	3,807
Opiliones	harvestmen, daddylonglegs	6,491	225
Palpigradi	micro whipscorpions	106	4
Pseudoscorpiones	pseudoscorpions	3,837	408
Ricinulei	hooded tick spiders	87	1
Schizomida	schizomids	287	11
Scorpionida	scorpions	2,000	90
Solifugae	sun spiders, camel spiders, windscorpions	1,187	175
Uropygi	whiptailed scorpions, vinegaroons, uropygids	128	1

[a]The North American list is for species occurring north of Mexico. The number of species is based on estimates of the number of described species. For each of these orders, the actual number of species is likely much higher.

sess spinnerets on the abdomen. Spiders are the only arachnids with a venomous bite.

Other arachnids with venom use different methods of delivering their poison. The venom in scorpions is delivered by a gland in their stinger. The pseudoscorpions produce venom in their pedipalps, but these tiny animals are much too small to envenomate humans or pets. The remaining groups of arachnids defend them-selves differently. Harvestmen have no venom and thus do not have a poisonous bite. They do have special stink glands on their bodies that emit a pungent substance thought to make them distasteful to predators. In the scorpions, pseudoscorpions, vinegaroons, and whipspiders the palps are modified into grasping pinchers or spiny raptorial structures that they use for both prey capture and defense (see Fig. 4).

FIGURE 6. A velvet mite (family Trombidiidae, order Acarina)—not a spider. In nature the mite is a very bright orange-red color (photo by Nathan Quesinberry).

Some species of larger mites are sometimes misidentified as spiders. The velvet mites are the group most frequently confused with spiders (Fig. 6). The distinctive bright red-orange color and the velvety body surface make these mites easy to recognize. Immature velvet mites feed as parasites on small invertebrates, and as adults they feed on insect eggs. Some tiny red mites called chiggers are troublesome to humans because of their itchy bites.

SPIDER ANATOMY FOR IDENTIFICATION

The basic anatomy that you will need to learn to identify spiders is relatively simple (Fig. 7). Some of the terms, however, are unfamiliar. Please know that putting in the effort necessary to learn these terms will greatly increase your ability to identify spiders.

The front part of a spiders' body is the cephalothorax, which functions like the head and thorax of insects. The back part of the body is the abdomen. There is a narrow stalk that connects them called the pedicel (see Fig. 5). The mouthparts, the appendages, and many sensory organs are on the cephalothorax. The most visible mouthparts are the chelicerae, which have a robust base and a moveable fang. The hollow fang contains a venom duct with an opening near the tip. To each side of the che-

licerae are the pedipalps, or just palps. These have one fewer segment than the legs and are used differently, but do look quite leglike. The palps have many sensory hairs and pits that are important for the spiders' senses of taste and smell. In adult male spiders the last segment of the palps is greatly modified to store the sperm and later to transfer the sperm to the female during mating. The palps of adult male spiders are swollen and appear something like little boxing gloves from a distance (Fig. 8).

The presence of these enlarged palps is usually the easiest way to tell the sexes apart. The structures of the male's palps are important for technical identification. The largest appendages are the four pairs of legs, usually referred to by Roman numerals (I, II, III, and IV). The top of the cephalothorax is covered with a plate of the exoskeleton called the carapace. At the front of the carapace there are usually four pairs of eyes, sometimes fewer. Each leg is composed of seven segments. Beginning near the body, these are the coxa (pl: coxae), the trochanter (pl: trochanters), the femur (pl: femora), the patella (pl: patellae), the tibia (pl: tibiae), the metatarsus (pl: metatarsi), and the tarsus (pl: tarsi). These leg segments have a hard tubelike exoskeleton and are separated from each other by a thin flexible cuticle that forms the joint. The surfaces of the segments are covered with many hairs and spines as well as microscopic sensory structures.

The pedipalps are similar in many ways to the legs but do not possess metatarsi, so they have only six segments. The coxae of the palps are modified on their medial (near the center of the body) region into mouthparts called endites (Fig. 9). The endites are often clothed with a dense brush of hairs. The brush of the endites along with similar ones on the labium (lower lip) and the rostrum (upper lip) surround the mouth opening. Each of these hairy structures forms part of a dense sieve. When spiders eat, they actually liquefy their prey by external digestion then suck the liquid meal into the mouth through this filterlike sieve.

The back part of the spider's body is the

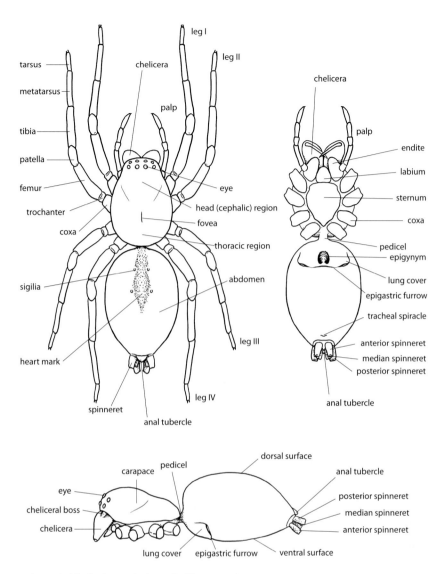

FIGURE 7. The basic terminology of spider external anatomy.

abdomen. The cuticle of the abdomen is usually more pliable than the cephalothorax. The size of the abdomen changes depending upon how recently the spider has eaten, or if the spider is a female with developing eggs. The top of the abdomen is often smooth, but it may have a series of tiny oval or circular indentations called sigilla. These sigilla are actually an external reflection of internal attachment points for muscles. In the center of the top of the abdomen there is an area directly over the heart that may have distinctive coloration: the heart mark. If there is a broader distinctive marking on this upper central part of the abdomen, it is often called the folium. On the underside there is a crease or fold running across the abdomen called the epigastric furrow (see Fig. 7).

In front of this fold on either side are slits that open into the book lungs. The book lung covers are sometimes distinctively colored or textured. In the midline of the fold are the reproductive openings, not obvious in males. In adult females there is usually a dark plate called the epigynum that covers the female

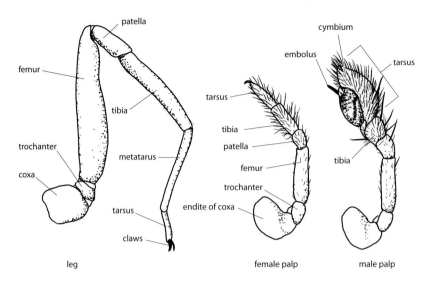

FIGURE 8. The segments of a spider's leg, the palp of an adult female, and the palp of an adult male (*Tibellus oblongus*). The detailed structure of the palp in male spiders is used in technical identification.

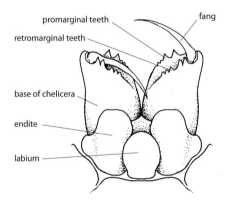

FIGURE 9. The terminology of the mouthparts of a spider.

reproductive ducts. Structures of the epigynum, including projections, pits, and other features, are important for mating. In one group of spiders the adult females lack an obvious epigynum. These features are also key for technical identification. There are tiny openings to the respiratory system on the underside of the abdomen. The position of these tracheal spiracles is important for technical identification and sometimes useful for live-spider identification. At the end of the abdomen are the spinnerets, fingerlike appendages used by the

spider to spin silk. There are two or three pairs of spinnerets.

THE PRINCIPAL FEATURES USEFUL FOR IDENTIFICATION OF LIVING SPIDERS

Identification of a spider can be simple; with experience, some are recognizable at a glance— for example, the golden silk orbweaver (*Nephila clavipes*, Plate 16) or the bold jumper (*Phidippus audax*, Plate 67). Unfortunately most spiders are not this easy to identify. Technical identification of a spider species usually involves close examination of the reproductive parts (genitalia) under a dissecting microscope or magnifying lens. Some spider genera include many species for which the genitalia are the only clear distinguishing features. For example, in the ground spider genus *Zelotes* (family Gnaphosidae) there are no fewer than 47 species in North America, most are nearly identical. For these spiders the field naturalist must be satisfied with confident placement in the correct genus. The availability of inexpensive high-quality digital cameras has made it possible to take excellent close-up photographs of tiny animals. With a clear macrophotograph one can often make a fairly confident identification. These are educated guesses; the only way to be certain of a species is for an expert to examine an adult under high enough magnification to reveal the details of the reproductive structures or to perform a genetic analysis.

One goal of this guide is to simplify the identification of spiders based on a few features of their body and behavior. The task of identification is made more complex by the fact that quite a few spiders have distinct color varieties or forms. Color is not as useful for identification as pattern, but even pattern can be deceptive. Sometimes forms look so dissimilar that people assume that they are different species. Also, the appearance of a spider changes throughout its life; immature spiders can look very different from adults. In some spiders the males look like small versions of females; but for most species the males and females are distinctive. Because males typically live only briefly, perhaps a few weeks as adults, most large mature spiders are probably females.

WEB OR BURROW ASSOCIATION More than half of the species of spiders in North America north of Mexico are typically found in their webs, and the webs come in a few basic types. Each species builds a web with consistent features. The presence of the spider at or within a burrow can also be a useful hint. Each type of spider has particular habitat preferences. Some spiders forage up in the vegetation, others exclusively on the ground. A few are found only in sandy areas, others only in forests or fields. Wandering spiders that are not associated with a web or burrow are often restricted to a particular hunting zone.

BODY SHAPE AND PROPORTIONS Size is variable and particularly difficult to estimate. Many spiders show extreme variation in size; some adults of the same species can be less than half as large as others. Shape is more useful for identification. For example, most of the longjawed orbweavers (family Tetragnathidae) have long thin abdomens, an easy way to distinguish them from most other spiders in the orb-shaped web group. Cobweb weavers (family Theridiidae) typically have a teardrop-shaped or spherical abdomen. When studying a spider, it is helpful to think in terms of *relative* shape or size: Is the cephalothorax as wide as the abdomen? Is it as long?

EYE ARRANGEMENT Eye number and arrangement are consistent and distinctive features (Fig. 10). Unfortunately the eyes are small on many spiders. Most spiders have four pairs of eyes, some fewer. The usual arrangement of the eyes is in two rows, four eyes in each row. The eyes nearest the front of the head are called the anterior eyes or anterior eye row (AER); those behind are the posterior eyes or the posterior eye row (PER). The eyes near the centerline of the body are referred to as median eyes; those that are further out toward the edge of the head are lateral eyes. Combining these terms we get the anterior lateral eyes (ALE), anterior median eyes (AME), posterior lateral eyes (PLE), or posterior median eyes (PME). The AME are called the principal eyes because of how they form during development. They may or may not be the most important eyes or those with the highest visual acuity. The other eyes are called secondary eyes. The principal eyes often appear darker than the others because they lack the reflective layer, or tapetum, found in the other eyes.

Eye arrangement on the cephalothorax is

FIGURE 10. The terminology of spider eyes.

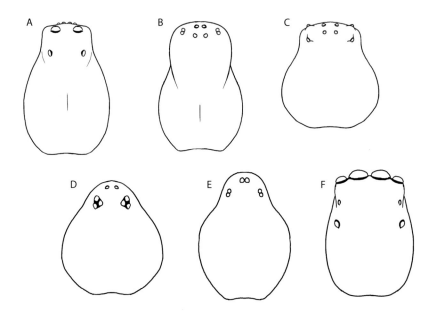

FIGURE 11. Dorsal views of carapaces with eye arrangements for several families of spiders. (A) Lycosidae. (B) Amaurobiidae. (C) Thomisidae. (D) Pholcidae. (E) Sicariidae. (F) Salticidae.

often distinctive (Fig. 11). The position of the rows of eyes are sometimes said to be procurved or recurved (see Fig. 10). Imagine a line running through the centers of each eye in the row. If this imaginary line is curved so that the outer eyes are shifted backward, the row is said to be recurved. Conversely, if the lateral eyes are shifted forward, the row is procurved. What constitutes an eye row is somewhat arbitrary. For example the eyes of wolf spiders are often described as being in three rows. The first row of four eyes (AER) is nearly straight, the second is a row of two eyes (PME), and the third is composed of the last two (PLE) well behind (Fig. 11A). These four posterior eyes could just as easily have been described as one greatly recurved row. The region delimited by the four median eyes is the median ocular quadrangle or median ocular area (MOA).

SPINNERET ARRANGEMENT The fingerlike structures on the abdomen that possess the silk-gland spigots are the spinnerets. Most spiders possess three pairs of spinnerets; however, some have only two pairs. For example, some mygalomorph spiders lack anterior spinnerets. The terminology of the spinnerets is based on their position (Fig. 12A). The pair in the front (looking at the underside of the abdomen) are the anterior spinnerets (AS). The next, usually smaller ones are the median spinnerets (MS), and the hindmost pair are the posterior spin-

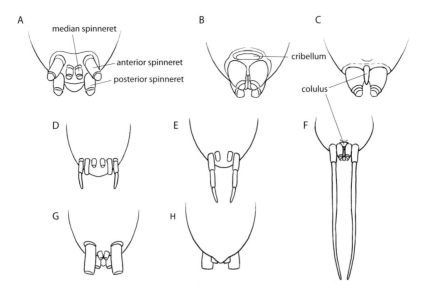

FIGURE 12. Examples and terminology of spider spinnerets. (A) Basic terms. (B) Spider with cribellum. (C) Spider with prominent colulus. (D) Hahniidae with all six spinnerets in a row. (E) Dipluridae with only four spinnerets. (F) Hersiliidae with very long spinnerets. (G) Gnaphosidae with cylindrical anterior spinnerets, ventral view. (H) Gnaphosidae, dorsal view.

nerets (PS). A few spiders have a spinneret-like plate called the cribellum (Fig. 12B). This plate may be divided into right and left halves, revealing its ancient roots as a paired structure. Many spiders that do not possess a cribellum have a small knoblike structure called the colulus (Fig. 12C). The colulus is often inconspicuous. Each spinneret has its own kinds and numbers of spigots and produces only certain types of silk. The spigots represent the outlet for silk from specific silk glands in the abdomen. Depending upon the kind of spider, there are several types of silk each produced by distinct abdominal silk glands.

LEG FEATURES All spiders normally possess eight legs. Sometimes legs lost due to injury are replaced at the next molt. Replacement legs may be somewhat smaller or abnormal in appearance. The key features of the legs that can be used for identification are related to either the relative lengths of the segments, the color pattern, spines, hairs, or claws on the legs. The leg formula is a listing of the legs arranged with the longest pair first; for example, a typical leg formula for crab spiders is I, II, IV, III. For other crab spiders and the somewhat similar huntsman spiders; it is usually the second pair of legs that it longest—for example, II, I, IV, III. Some spider families have legs with an unusual arrangement. The leg bases at the coxae are twisted so that the face of the leg that would normally be forward is turned upward. This has the consequence of converting the normal bends of the leg that form a sort of arch, going away from the body up then down to the tarsi at the tip, into a flat forward curve. These spiders' posture appears somewhat like a crab. This arrangement is called laterigrade legs (Fig. 13).

The spines and hairs on the body and legs of spiders are sometimes useful for identification. The smaller, thinner spines or microsetae are usually called hairs. The heavier ones, macrosetae, are called spines. Spines are often large enough to see, and their number and pattern of arrangement can be helpful identification clues. Some spiders possess many heavy spines, such as lynx spiders (family Oxyopidae), or a distinctive arrangement, such as pirate spiders (family Mimetidae). Others have uniform

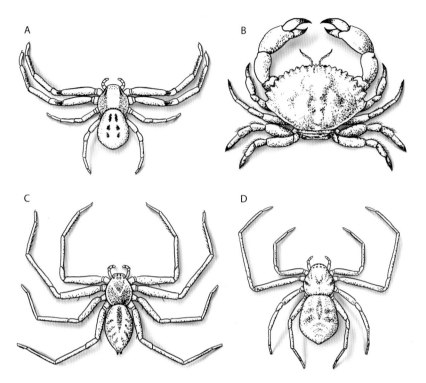

FIGURE 13. Examples of laterigrade legs. (A) Crab spider (*Misumenoides formosipes*). (B) Crab for comparison (*Cancer* sp.). (C) Huntsman spider (*Heteropoda venatoria*). (D) Running crab spider (*Ebo latithorax*).

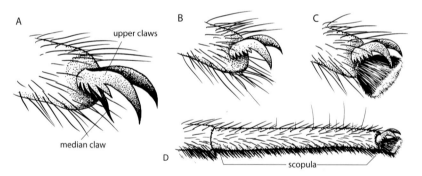

FIGURE 14. Terminology of claws and tufts. (A) Three-clawed spider. (B) Two-clawed spider. (C) Two claws with tuft. (D) Two claws with tuft and scopula.

short curved hairs without any heavy spines, such as recluse spiders (family Sicariidae).

It is a shame that the claws at the tips of spider legs are inconspicuous, because they are useful for distinguishing certain families. Spiders either have two or three claws at the tips of their tarsi. It is helpful that most of the two-clawed spiders also have claw tufts and and/or a dense brush of hairs (Fig. 14C). These dis-

tinctive brushes are called scopulae (Fig. 14D). Unlike the claws, tufts or scopulae can be seen with a good magnifying glass. The claw tufts often make the tips of the legs look a bit thicker and less delicate. Spiders with claw tufts and scopulae are expert climbers and can run rapidly over rough terrain or up vertical surfaces. If a spider demonstrates this sort of behavior, it is a good bet that it has tufts and/or scopulae.

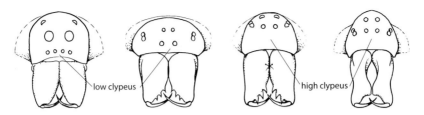

FIGURE 15. Spiders with a low clypeus: *Pirata* (left) and *Araneus* (right). Spiders with a high clypeus: *Pityohyphantes* (left) and *Emblyna* (right).

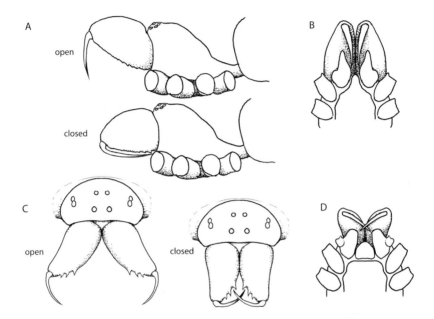

FIGURE 16. Comparison of cheliceral types. (A) Mygalomorph with large projecting chelicerae and downward striking fangs. (B) Mygalomorph, ventral view showing parallel fangs. (C) Araneomorph with pinching fangs. (D) Araneomorph, ventral view showing opposing fangs.

THE FACE The region between the anterior eye row and the edge of the carapace above the chelicerae is the clypeus (Fig. 15). Knowing the relative height of the median ocular area versus the height of the clypeus is often helpful. The usual way to define a "high" clypeus is that the height is greater than or equal to the height of the median ocular area. A high clypeus is typical of certain groups—for example, members of the families Linyphiidae (sheet weavers) and Theridiidae (cobweb weavers). This feature clearly distinguishes them from the orb-weaving families Tetragnathidae (longjawed orbweavers) and Araneidae (orbweavers) with their lower clypeus.

The structure of the chelicerae is sometimes useful in identification. Whether or not the chelicerae project forward, are spread at their ends (divergent), or are particularly heavy and swollen can also be helpful traits for identification. For example, the relatively large heavy and forward-facing chelicerae of the mygalomorph spiders provide a quick clue to this group (Fig. 16A). The bases of the chelicerae may be useful for identification. The chelicerae may be divergent, fused, have mastidia, or large spines. There are often teeth around the depressed groove where the folded fang rests (see Fig. 9). Tooth characteristics are usually too small to observe with the naked eye or a simple magnifying glass, but on a few spiders they are large enough to be obvious—for example, the conspicuous teeth of the longjawed orbweavers (Tetragnathidae) and some sheet weavers (Linyphiidae). The lateral edge of the fang bases may have a shiny structure called a boss. The cheliceral boss is a swollen structure that may support hinge-like movements of the fang bases; in some spiders it is distinctively colored and conspicuous. Most of these features are small, but they can be seen with a good magnifying lens.

CEPHALOTHORAX FEATURES The dorsal surface of the cephalothorax is covered by a large hard (sclerotized) plate, the carapace (Fig. 17). The head (cephalic) region is sometimes distinct, separated from the thoracic portion by cervical grooves. The thoracic region often bears external indications of internal anatomy—for example, the central fovea and the radial grooves, which are where internal muscles are attached.

PATTERN The markings of spiders can sometimes be useful for identification. The presence of regular banding on the legs, rings around leg segments, or lines along the legs can be distinctive. Spots, lines, and other patterns on the carapace of the cephalothorax or features of the folium on the abdomen may also be helpful. A band of color at the edge of the carapace is a marginal band; one just inside of this is referred to as a submarginal band.

COLOR Color can be helpful for spider identification, but it is not as useful as we might hope. The reason for this is that many spiders exhibit extreme variation in color. In addition to color varieties, some spiders can change color to match the background on which they are resting. Some species of crab spiders, which are ambush predators on flowers, can change their body color to match the flower they are hunting on. This color shift occurs over a few days. Other spiders can change the distribution of pigments under the cuticle quickly. Examples of this phenomenon are spiders that

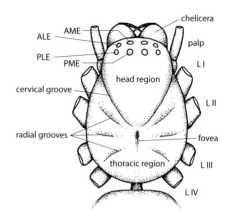

FIGURE 17. Terminology of the cephalothorax.

flush the abdomen with dark color as they drop from a web when they are alarmed. In some pale spiders the color of the abdomen can vary depending upon what the spider has been eating. Some of the most dramatic colors of spiders are the result of colorful scale-like hairs on the body. These scales are like those on the wings of butterflies, and they can rub off with age, thus changing the color or pattern visible on the spider.

USING THIS BOOK FOR IDENTIFICATION

There are two basic ways that you can find illustrations of a particular spider:

1. Go directly to the plates and search.
2. Limit your search to the most likely group by using the list below. After you have chosen one, use the key to families for that group. Once you are directed to a family, jump directly to searching the plates listed for that family.

BASIC GROUPS OF SPIDERS

One approach to identification of spiders is to group them by where they are found. This approach emphasizes features of the web or habitat where the spider is most likely to be encountered. It is important to remember that any spider may wander away from its web or

FIGURE 18. Various jumping spider (Salticidae) faces.

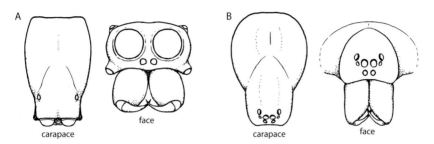

FIGURE 19. (A) Ogre face of *Deinopis*; dorsal view, face on. (B) Face of the wandering spider *Ctenus*; dorsal view, face on.

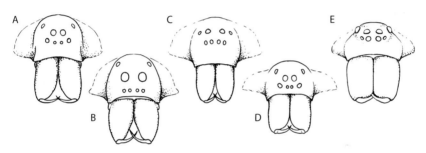

FIGURE 20. Faces of various hunting spiders. (A) *Dolomedes* (Pisauridae). (B) *Pardosa* (Lycosidae). (C) *Homalonychus* (Homalonychidae). (D) *Zoropsis* (Zoropsidae). (E) *Trechalea* (Trechaleidae).

burrow. These individuals are often adult males seeking mates. This section is not intended for identification of such vagrants. For them, consult the illustrations directly.

Before choosing a basic group you can save time by checking to determine if the spider belongs to one of two distinctive families:

Jumping spiders (family Salticidae) Jumping spiders can be recognized by the fact that they have huge anterior median eyes facing forward. Above and to the side of these are a pair of smaller eyes (see Figs. 11F and Fig. 18). The only other spider with huge eyes facing forward is the ogrefaced netcasting spider (*Deinopis*), but it actually has two small eyes between and in front of them (Fig. 19A). The jumping spiders are illustrated on Plates 58 to 69.

Wolf spiders (family Lycosidae) These ground hunters have an unusual eye arrangement. They have two large posterior median eyes above a row of four small eyes. Their posterior lateral eyes are large too but are much farther back on the head. The PME and PLE form a trapezoid on the top of the head (see Fig. 11A and Fig. 20B). The wolf spiders are illustrated on Plates 48 to 55.

Web Builders

Many spiders construct silken cocoons or retreats, even species that build no capture web. In most cases these cocoons are tubular with narrow openings at both ends (see Fig. 3B). The spider may use this as a temporary retreat when it is not active or as a protected place to construct the egg case. If the cocoon is not near a larger web, or if it is under a rock or other debris, you should skip to wandering spiders.

UNDER DEBRIS GROUP Thin sheets of webbing found under rocks, flat boards, logs, or other debris on the ground. If the web has a funnel-shaped entrance, you should also check the funnel-shaped web group. If you disturbed the web when you lifted the rock or debris a spider was hiding under, you should check both of these groups. If the spider has only six eyes it may be a recluse spider (family Sicariidae; Plate 76).

SPACE-FILLING WEB GROUP Three-dimensional tangles of lines filling an area of space, sometimes called cobwebs (Fig. 21, upper left). These webs are often under some sort of overhanging cover. The spider may rest in the middle of the tangle or have a retreat off to one side, typically above, the web.

SHEET WEAVERS GROUP The sheetweb weavers do not build a retreat or funnel-like structure as part of their web (Fig. 21, upper right). The sheet may be flat, dome-shaped, or bowl-shaped. The spider may either hang upside down directly under the web, or at the edge of the web, or move to a retreat near the web. Sheetweb weavers often include a loose series of knockdown threads above or below the sheet.

ORB-SHAPED WEB GROUP A flat frame of webbing with a circular orb in the middle (Fig. 21, lower left). The orb is a pattern of circular sticky lines that are actually produced as a spiral. The spiral strands are supported by a series of spoke-like radial strands. The center of the web may be a loose platform, with or without a hole that permits the spider to move from one side of the web to the other easily. A few species of orb-shaped web builders include an irregular tangle of silk lines near the orb.

FUNNEL-SHAPED WEB GROUP The web forms a sheet that extends into a retreat, a tube-like structure at one end of the web (Fig. 21, lower right). The spider hides within the tubular retreat and runs out onto the top surface of the web to capture prey that walk across, or fall on, the sheet. Spiders that build funnel-shaped webs often include a loose tangle of silk knockdown lines above the sheet. Other funnel-weaving spiders build a series of lines radiating away from their silk-lined retreat in a hole or other cover.

Burrowing Spiders

Spiders that dig burrows in the ground may use the burrow as a seasonal retreat or may center all of their activities at the burrow. The burrow entrance may be open, or it may have a lid. Other species apply a thin silk covering when they are not active, thus hiding the entrance that becomes covered by debris. Two principal types of spiders build burrows: wolf spiders and mygalomorph spiders. If the eyes are large and well separated, skip to the wolf spiders (family Lycosidae). If all eight of the eyes form a cluster at the front of the head and the spider has large protruding jaws, check the mygalomorph group.

Wandering Spiders

SIX-EYED SPIDERS GROUP Spiders with only six eyes in three groups.

JUMPING SPIDERS GROUP, FAMILY SALTICIDAE Spiders with huge anterior median eyes.

CRABLIKE SPIDERS GROUP Spiders that hold their legs curved like those of a crab (see Fig. 13).

SURFACE HUNTERS GROUP Spiders found wandering on rock walls, tree trunks, or the surface of ponds or streams.

FIGURE 21. Spider web types.
(Upper left) A space-filling web (*Theridion differens*).
(Upper right) A sheet web (*Florinda coccinea*).
(Lower left) An orb web (*Cyclosa conica*) (photo by
Todd Blackledge).
(Lower right) A funnel web (*Agelenopsis pennsylvanica*)
(photos by author, except where noted).

GROUND HUNTERS GROUP   Spiders found wandering on the ground, or found under a rock or other debris without a web.

FOLIAGE HUNTERS GROUP   Spiders found wandering in bushes, trees, or other vegetation.

MYGALOMORPH GROUP Spiders with huge protruding jaws and all eight eyes clustered in a small area. These spiders usually have four visible book lungs. Spinnerets extend beyond the end of the abdomen when viewed from above. Mygalomorphs are usually, but not always, large spiders. There are many burrowing spiders among this group.

TINY SPIDERS GROUP Spiders with a total body length of less than 3 mm (⅛ inch) that are found without webs, with loose tangles of silk, or small sheet webs.

Keys to Families within Groups

The idea behind a dichotomous key is that you are presented two choices, a couplet. Pick the one that most closely describes the features of the spider you are trying to identify. Go to the next choice indicated by a number. If you make a mistake, the number in the parenthesis after the first choice in the couplet is your guide to remembering how you arrived at this pair in the first place. If you reach a dead end, where neither choice seems to describe your spider, use this number to work your way back through the key and then take another direction. It is important to examine all of the features described in the couplet, but note that the spider may not match all of them. When you arrive at the name of a family, check the plates listed for that family.

UNDER DEBRIS GROUP • *Spiders That Build Thin Sheet Webs under Rocks, Logs, or Other Debris*

Plates 34–35

This group includes members of six families. The loose sheet of webbing associated with most of these spiders is attached to the debris. Such webs are usually disrupted when the rock or other material that the spider is hiding under is lifted. Sometimes the silk of the web is visible beyond the edge of the debris, providing a hint that there is a spider resident underneath. There are a number of common ground hunters (ground hunters group) that can often be found under rocks or debris in similar situations to the species included in this group. Some of these construct silken cocoons that they hide in during the day. The soft spiders of the family Cybaeidae (funnel-shaped web group) may also build flimsy webs under rocks, but these usually have a portion extending out from under the edge into the open. The recluse spiders (Sicariidae in the six-eyed spiders group; Plate 76) sometimes build thin sheet webs under debris, but they are also found wandering away from any web. Some spiders in other families that build space-filling or sheet webs are also found in rocky and debris-covered areas—for example, pholcids and linyphiids. Others are small, included in the tiny spiders group.

Key to Families of the Under Debris Group

1A   With six eyes. . . . . **meshweavers (Dictynidae in part *Blabomma, Yorima*; Plate 35)**

NOTE: Some cave forms of *Cicurina* in this family are eyeless or have only six eyes.

1B   With eight eyes. 2

2A (1)   Small spider with all six spinnerets forming a row; the carapace is dark reddish brown and very shiny. . . . . . **comb-tailed spiders (Hahniidae in part *Neoantistea*; Plate 35)**

2B   Spider not as above. 3

3A (2)   Spider relatively pale tan, light brown, or light amber without a cribellum
       . . . .**meshweavers (Dictynidae in part *Cicurina*; Plate 35)**

3B   Spider not as above, usually dark brown or dark reddish brown carapace with dark, sometimes nearly black abdomen. 4

4A (3)   Spider with a high clypeus, more than twice the height of the ocular area, without a cribellum . . . . **spurlipped spiders (Plectreuridae; Plate 34)**

4B   Spider with a low clypeus, about the same as the ocular area or even lower, with a divided cribellum. 5

5A (4)   Spider with very long, dark brown jaws. The chelicerae are more than twice as long as the width across the face where they are attached . . . . . **(Amphinectidae; Plate 34)**

5B   Spider not as above, jaws may be robust, but they are wide as well as long 6

6A (5)   Legs have few heavy spines but are clothed with hairs. Spider has a slight iridescence to the dark abdomen when viewed in strong light
       . . . . **rock weavers (Titanoecidae; Plate 34)**

6B   Legs have scattered heavy spines as well as four or five pairs of ventral spines on the anterior tibiae. . . . . **(Zorocratidae; Plate 35)**

SPACE-FILLING WEB GROUP • *Spiders That Build Three-dimensional Tangles or Cobwebs*

Plates 21–29

Cobwebs appear to be irregular tangles of lines that fill an area of space (see Fig. 21, upper left, and Fig. 22). Even though these webs may appear unorganized, they are carefully constructed and highly effective at catching prey. The matrix of silk lines, running in many directions, serve as a maze to flying insects. When these potential prey strike the web, they are often disoriented and may even be stopped. This fleeting moment is exploited by the resident spider that rushes out to wrap the prey with more silk and effect a capture. Some space-filling webs include taut sticky lines extending to the substrate. Such webs, referred to as gumfoot webs, are usually constructed by cobweb weavers (family Theridiidae; Fig. 22, right).

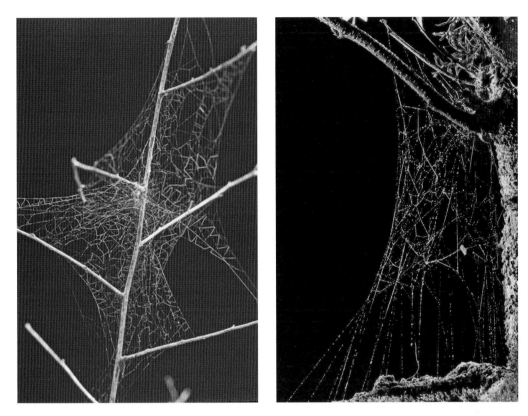

FIGURE 22. Examples of space-filling webs. (Left) Cribellate web of *Emblyna annulipes* (photo by Todd Blackledge). (Right) Gumfoot web of *Parasteatoda tepidariorum* (photo by author).

Key to the Families of the Space-filling Web Group

1A	Spider builds a tangled web among the branches of a shrub or cactus in the desert Southwest. The tangle contains a central "retreat" where the spider resides. The spider has only six eyes, is tan and brown, and is densely clothed with hairs. . . . .**desertshrub spiders (Diguetidae; Plate 22)**
1B	Spider not as above. 2

2A (1)	Spider is minute, less than 2 mm long, and builds a tiny three-dimensional web near the ground or in a cave or cavelike environment. . . . . **dwarf cobweb weavers (Mysmenidae; Plate 20)**
2B	Spider not as above. 3

3A (2)	Spider has relatively short legs; the longest are not more than two times the length of the body (the cephalothorax plus the abdomen). Spider has a cribellum. . . . . **meshweavers (Dictynidae; Plates 21, 22, 81)**
3B	Spider has much longer legs; the longest are often more than three times the length of the body. 4

| 4A (3) | The largest eyes are arranged in two groups of three. There may be two additional tiny anterior median eyes at the front of the carapace. The two triads are usually surrounded by dark pigment. The legs are all very long, at least four times as long as the body. The tarsi are flexible and often have a slight bend to them. Body shape is variable. . . . . **cellar spiders (Pholcidae; Plate 23)** |
| 4B | The eyes are usually in two rows of four. Anterior and posterior lateral eyes are usually close together. Spider has a globose teardrop-shaped abdomen. The tarsi are straight. Some species have different abdominal shapes, but these are species with reduced webs or that build no capture web. . . . . **cobweb weavers (Theridiidae; Plates 24–29)** or **cave cobweb spiders (Nesticidae; Plate 24)** |

SHEET WEAVERS GROUP • *Spiders That Spin Sheet Webs*

Plates 77–79

Spiders that build sheet webs are among the most diverse group of spiders. The two families (Linyphiidae and Pimoidae) include nearly 1,000 named species in North America, making it the group with the most representatives in North America north of Mexico. By far the largest family is the Linyphiidae. Most sheet weavers are small; some are among the smallest of all spiders. The webs of these spiders can take on many shapes. The sheet may be flat (Fig. 23, lower right), bowl-shaped (Fig. 23, left), or dome-shaped (Fig. 23, upper right). Many of these webs incorporate an irregular tangle of nonsticky lines above and/or below the sheet. The spider usually hangs below the sheet, near the center, but in a few species the spider may be found hanging near the edge of the sheet.

There are no clear features that are visible without a microscope to distinguish the sheetweb weavers (Linyphiidae) from the large hammockweb spiders (Pimoidae), thus no key is provided here.

ORB-SHAPED WEB GROUP • *Spiders That Build Orb-shaped Webs*

Plates 5–20, 47

In North America north of Mexico there are spiders representing seven families that build orb webs. The orb-shaped web in its most basic form is flat and essentially two dimensional (see Fig. 21, lower left, and Fig. 24). The web is suspended from a frame of strong, dry nonsticky threads attached to the surrounding vegetation or other supports. The orb itself consists of a series of dry radial threads, arranged much like the spokes of a bicycle wheel. Taking this wheel analogy a bit further, the central portion of the orb is referred to as the hub. The most important component of the web, from the standpoint of capture, is the spirally arranged circular part attached to the radial threads. This part is constructed of wet, sticky silk. This sticky silk is secreted from different glands (aggregate glands) than the other parts of the orb. This wet, sticky silk is effective glue, but it is stickiest when moist. After a few hours of exposure to dry air, the web dries. As a result, the spiral orb is often replaced with fresh sticky silk each day. Some species remove and replace the entire web. In other species only the sticky orb is replaced, while the majority of the frame and even many of the radial spokes are left intact. Spiders usually eat the old silk and have been shown to recycle the silk proteins quickly into new silk.

In one group, the members of the hackled orbweavers of the family Uloboridae, the adhesive is not wet, gluey silk but a series of very fine, dry threads produced by a special flat spinning plate called the cribellum (see Fig. 12). The dry threads are brushed into a delicate fluff of fibers by a

FIGURE 23 Examples of sheet webs. (Left) Web of bowl and doily spider, *Frontinella pyramitela*. (Upper right) Web of filmy dome spider, *Neriene radiata*. (Lower right) Web of hammock spider, *Pityohyphantes costatus* (photos by author).

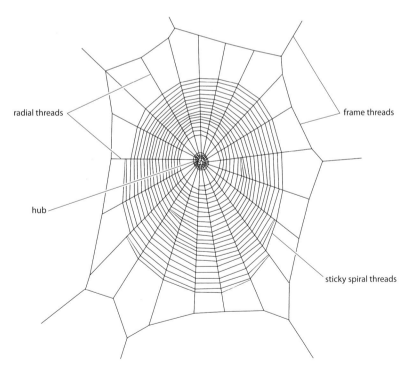

FIGURE 24. The features of an orb web.

FIGURE 25. Examples of unusual orb webs.
(Left) Cone-shaped web of the ray orbweaver, *Theridiosoma*.
(Right) Horizontal orb with out-of-plane radii of *Gertschanapis*
(photos by Jonathan A. Coddington).

special comb of bristles on the spider's fourth legs called the calamistrum. Not all orb-shaped webs are identical in structure. Spiders in the family Anapidae build a horizontal orb with many out-of-plain radii extending up from the orb (Fig. 25, right). Another unusual web is built by the ray orbweavers of the family Theridiosomatidae. Their web is pulled into a cone shape (Fig. 25, left).

The most distinctive variants are orbs with accessory structures. One example is a series of conspicuous white zigzag silk bands found in the webs of garden spiders (Fig. 26, left). These decorations are called stabilimenta. Another example is the accumulated material consisting of indigestible prey remains, bits of leaves, and silk that forms a messy-looking trashline down the center of some webs (Fig. 26, right).

More information about the orb-shaped web group can be found in a series of more than 25 monographs published by Herbert Levi and his students at Harvard, primarily in the *Bulletin of the Museum of Comparative Zoology*. These references provide details about the genera and species in the various families of orb-shaped web spiders. Much of the general information included in the species accounts in this book was obtained from these valuable sources. Another valuable resource is *The Orb-Weaving Spiders of Canada and Alaska* (Dondale et al. 2003), which is one of the most comprehensive summaries of these spiders in a single volume. For a list of these and other published references, consult *Spiders of North America: An Identification Guide* (Ubick et al. 2005).

Key to Families of the Orb-shaped Web Group

1A   Spiders are long and thin, almost sticklike, the posterior median eyes are huge (see Fig. 19A). Spider found in the Southeastern states.

. . . . **ogrefaced spiders (Deinopidae; Plate 47)**

1B   Spider not as above. 2

2A (1)   Spider builds a web with widely spaced spiral lines and the web is pulled into a cone shape (see Fig. 25, right). The small spider waits, holding onto a line connected to the center of the cone. . . . . **ray orbweavers (Theridiosomatidae; Plate 20)**

2B   Web not like this. 3

FIGURE 26. Orb web decorations. (Left) Stabilimentum of *Argiope aurantia*. (Right) Trash line of *Cyclosa turbinata* (photos by author).

3A (2) Spider builds a greatly reduced web, may not even resemble an orb. 4

3B   Spider builds a circular orb web. 7

4A (3) Spider builds an asterisk-shaped web, with a central hub suspended in a gap in the vegetation by a few radial lines but without a sticky spiral. The spider is hard to see because it is camouflaged and looks like a broken twig or bud.
. . . . **orbweavers (Araneidae in part *Ocrepeira*; Plate 6)**

4B   Spider and web not as above. 5

5A (4) Spider with a triangular-shaped web that looks like a pie slice–shaped part of an orb . . . .**hackled orbweavers (Uloboridae in part *Hyptiotes*; Plate 19)**

5B   Spider builds a single-line web or uses a single line with a glue droplet to hunt. 6

6A (5) Spider with a single-line web with a tuft, spider usually waits near tuft. Spider has only four eyes.
. . . .**hackled orbweavers (Uloboridae in part *Miagrammopes*; Plate 19)**

6B   Spider uses a single line with a glue droplet at the end to hunt. Spider has eight eyes. . . . . **orbweavers (Araneidae in part *Mastophora*; Plate 7)**

7A (3)   The web is oriented in a horizontal or tilted-horizontal plane. 9
7B   The web is oriented in a vertical or very nearly vertical plane. 8

8A (7)   The spider is very large and has black fuzzy tufts of setae at the ends of the
          femora and tibiae; the web has yellow silk.
          . . . . .longjawed orbweavers (Tetragnathidae in part *Nephila*; Plate 16)
8B   Spider not like this. . . . . orbweavers (Araneidae; Plates 5–17)

9A (7)   Spider is tiny, less than $^1/_{16}$ inch (2 mm) in length, and its web is less than 2
          inches (50 mm) in diameter, with the spiral rings closely spaced (usually less
          than 1 mm apart). The web is found near the ground among the leaf litter or in
          a cavelike environment. The web may have a few lines that are not in the same
          plane as the orb . . . . ground orbweavers (Anapidae; Plate 20)
          or dwarf orbweavers (Symphytognathidae; Plate 20)
9B   Spider is usually much larger, the spacing of the spiral rings is wider, usually at
          least 2 mm apart. 10

10A (9)   The web contains unusual hackled silk in the spiral portion and often has little
          tufts of silk or conspicuous white-line decorations.
          . . . . hackled orbweavers (Uloboridae; Plate 19)
10B   The web has ordinary sticky spiral lines and is without white decorations.
          . . . . .longjawed orbweavers (Tetragnathidae; Plates 16–18)

FUNNEL-SHAPED WEB GROUP • *Spiders That Build Funnel-shaped Webs*

Plates 3, 30–33, 57

A number of unrelated spider families spin webs with funnel-shaped retreats (see Fig 21, lower right; Fig. 27). The web has a wide opening or sheet that narrows into the area where the spider hides when not actively waiting for prey. Some spiders that spin tube-shaped webs, tent-shaped webs, and lampshade-shaped webs are included in this group. The most familiar are the funnel weavers or agelenids of the family Agelenidae (see Fig. 21, lower right, and Fig. 27, upper left). Hackledmesh weavers (Amaurobiidae) are the second most common members of the funnel-shaped web group and spin cribellate silk. Cribellate webs have a bedraggled, unkempt look (Fig. 27, upper right). The introduced *Badumna* (family Desidae) also builds a cribellate funnel web.

Two families of spiders build distinctive messy circular silk webs surrounding a central silk-lined tube (Fig. 28). The tube often extends into a tightly constricted crevice or hole. The outer web may include a number of radiating silk lines that function as trip lines. When a potential prey organism touches these, the vibratory signal triggers an attack from the spider waiting in the funnel. The larger of these two are the crevice weavers in the family Filistatidae (Fig. 28, left). The smaller tube-web spiders of the family Segestriidae may build their web in less conspicuous situations (Fig. 28, right).

The soft spiders of the family Cybaeidae are included in this group because some species build a funnel-like entrance to their flimsy webs that are usually found under rocks or debris. They could also have been placed in the under debris group. One genus in the comb-tailed spider family Hahniidae, *Calymmaria*, builds its characteristic webs in wet areas under logs, large flakes of bark, mats of moss, rock walls, rock overhangs, in caves, and under boards (see Fig. 27, lower left). The web takes the form of a sort of cone-shaped basket, wide at the top. Within or above the area enclosed by the silken cone, the spider spins a thin, flat silk sheet where the spider hangs.

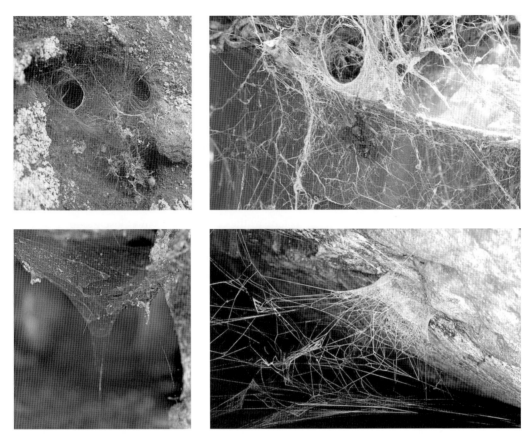

FIGURE 27. Examples of funnel-shaped webs. (Upper left) Web of *Coras* with two entrances (photo by author). (Upper right) *Badumna longinqua* near the entrance of her web (photo by Lenny Vincent). (Lower left) Photo of *Calymmaria persica* and her web (photo by author). (Lower right) Web of *Hypochilus pococki* (photo by Brent Opell).

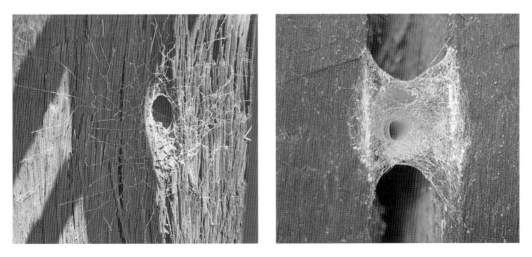

FIGURE 28. Tube webs. (Left) *Kukulcania* (photo by Annie Whitney). (Right) *Ariadna* (photo by John Maxwell).

The most unusual of all the funnel-shaped web builders are the lampshade weavers in the family Hypochilidae (Plate 57). The web is unique—a broad circular structure attached at its base and shaped something like an old-fashioned lampshade, which is narrowest in the middle (see Fig. 27, lower right). The spider rests against the rock face at the center of the web.

Key to Families in the Funnel-shaped Web Group

| 1A | Chelicerae with parallel fangs, with four book lungs on the underside of the abdomen, eyes are either in a close group on a prominence on the carapace, or anterior median eyes much smaller than other eyes. . . . . 2 |

1A   Chelicerae with parallel fangs, with four book lungs on the underside of the abdomen, eyes are either in a close group on a prominence on the carapace, or anterior median eyes much smaller than other eyes. 2

1B   Chelicerae with opposing fangs. 4

2A (1)   With very long spinnerets, nearly as long as the carapace, eyes in a close group on a prominence on the carapace. 3

2B   Spinnerets relatively short, legs long and thin, usually found in a peculiar lampshade-shaped web on or under a rock face, anterior median eyes much smaller than the other eyes. . . . . **lampshade weavers (Hypochilidae; Plate 57)**

3A (2)   A hard plate (scutum) at the front of the dorsal surface of the abdomen.
    . . . . **midget funnelweb tarantulas (Mecicobothriidae; Plate 3)**

3B   Without a hard plate at the front of the dorsal surface of the abdomen.
    . . . . **funnelweb spiders (Dipluridae; Plate 3)**

4A (1)   With only six eyes; builds a series of lines radiating away from a circular retreat, usually a tube-like structure (see Fig. 28, right) in an existing cavity such as a crack in a rock or hole in a log, even the end of a hollow stem. These spiders typically hold six legs forward at rest. . . . . **tubeweb spiders (Segestriidae; Plate 33)**

4B   With eight eyes. 5

5A (4)   Large posterior median eyes on the front of the face with a row of four smaller anterior eyes below them on the face (see Fig. 20B).
    . . . . **wolf spiders (Lycosidae in part *Sosippus*; Plate 32)**

5B   Spider not as above. 6

6A (5)   The eyes clustered on a prominence well back from the front of the carapace leaving a high clypeal space. Body covered with velvety hairs, in addition to spines in some; builds a circular retreat with an extensive tangle of radiating lines around the entrance (see Fig. 28, left) . . . . **crevice weavers (Filistatidae; Plate 4)**

6B   Spider and retreat not as above. 7

7A (6)   Without a cribellum. 8

7B   With a cribellum. 10

8A (7)   Posterior lateral eyes distinctly larger than the posterior median eyes; builds an inverted tent-shaped web, usually hanging under a rock ledge, log, or between layers of bark in humid areas.
    . . . . **comb-tailed spiders (Hahniidae in part *Calymmaria*; Plate 33)**

8B   Spider and web not as above. 9

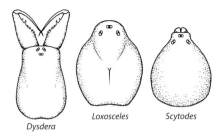

FIGURE 29. Eye arrangement of the common six-eyed spiders.

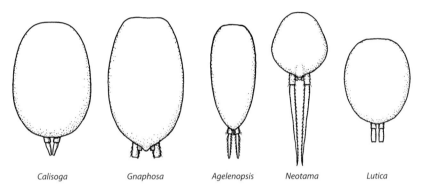

FIGURE 30. Appearance of spinnerets in spiders in which they are visible from above.

9A (8)   Spinnerets relatively long, posterior spinnerets longer than the anterior
         spinnerets (Fig. 30, Agelenopsis) . . . . **funnel weavers (Agelenidae; Plates 30–31)**
9B   Spinnerets relatively short, anterior spinnerets longer than the posterior
         spinnerets, usually with a loose funnel or insubstantial sheet leading to a thin
         silk retreat under a rock or in a moist area . . . . **soft spiders (Cybaeidae; Plate 33)**

10A (7) Dense tangled and messy-looking funnel-shaped web (Fig. 27, upper right),
         often in or around a human-altered environment, West Coast.
         . . . . **desids (Desidae in part *Badumna*; Plate 33)**
10B   Web is less dense, often a loose cribellate web with a retreat, widely distributed.
         . . . . **hackledmesh weavers (Amaurobiidae; Plate 32)**

SIX-EYED SPIDERS GROUP · *Spiders with Only Six Eyes*

Plate 76

These spiders are sometimes found around houses and buildings. They are different from each other in appearance, save the unusual characteristic of six eyes in three groups of two (Fig. 29). There are three families included in this group; the woodlouse spiders (Dysderidae), recluse spiders (Sicariidae), and the spitting spiders (Scytodidae). Each of these small families has distinctive features, and they are relatively easy to separate.

The tubeweb weavers (Segestriidae) also have six eyes in three groups of two, but they build distinctive tube retreats and hold six legs forward at rest (see funnel-shaped web group). A few other spiders have lost a pair of eyes, but these are web-building spiders, or small species found in caves. One cellar spider (*Spermophora senoculata*) has only six eyes. Similarly, the desertshrub spiders (Diguetidae; Plate 22) that are also found in the space-filling web group, have six eyes but live in cobwebs among desert cacti and other shrubs.

Key to Families in the Six-eyed Spiders Group

1A   Very large chelicerae, extending forward from the carapace (Fig. 29, Dysdera). Fangs are long, are folded back rather than opposing, and cross each other at rest. Body and legs are bright red or reddish brown; the abdomen is uniform pale. . . . .**woodlouse spiders (Dysderidae; Plate 76)**

1B   Spider not as above. 2

2A (1)   Spider has a peculiar cephalothorax that is much higher toward the back. The spider is not associated with silk and usually moves slowly. . . . . **spitting spiders (Scytodidae; Plate 76)**

2B   Spider has a relatively normal cephalothorax. There is a series of rows of setae and dark markings around the eyes, extending back covering the cephalic part of the carapace, narrowed in the middle and with a dark line extending toward the back of the cephalothorax, which somewhat resemble a violin. . . . .**recluse spiders (Sicariidae; Plate 76)**

JUMPING SPIDERS GROUP

Plates 58–69

The family Salticidae or jumping spiders has the most described species of any family of spiders in the world. There is considerable variation in the body size and shape of jumping spiders. Even so, all jumping spiders share one basic eye pattern. They have extremely large anterior median eyes (AME) and together with the anterior lateral eyes (ALE), they occupy a part of the carapace that faces forward (see Fig. 18).

CRABLIKE SPIDERS GROUP • *Spiders with Laterigrade Legs*

Plates 46, 70–75

This group of spiders holds their legs in a somewhat crablike posture referred to as laterigrade (see Fig. 13). The legs of these spiders often have femora with an oval cross-section. These femora lie flat with the broad front surface of each femur facing up. Because of this, the normal bends of the legs form a C-shaped curve. Most of these spiders also have flattened or somewhat flattened bodies, and many of them are relatively large. The most widespread and familiar examples are the crab spiders family (Thomisidae) and running crab spiders (Philodromidae).

Some crab spiders are quite colorful and live in flowers, where they ambush insect pollinators. The remaining members of the laterigrade group are colored in variegated somber shades of grays, tans, and browns that blend well into their environments. They are often found on flat surfaces such as large branches, trunks of trees, rocks, walls, or on the ground.

Key to Families in the Crablike Spiders Group

1A   Tarsi without dense claw tufts, legs without scopulae. 2

1B   Tarsi with dense claw tufts and usually dense scopulae. 3

2A (1)   The eyes, at least the lateral ones, on raised white or pale tubercles. There may be
heavy ventral spines on the tarsi and metatarsi of legs I, II. Legs I, II longer and
usually heavier than back legs. Some species that key this way are bright colored
(yellows, pinks, and reds) . . . . **crab spiders (Thomisidae; Plates 72–75)**

2B   The eyes not on raised tubercles, paired ventral spines on the tarsi, metatarsi,
tibiae of all legs. The tarsi are flexible and may appear bent. Leg IV longest, all
legs long and relatively thin. The one species in the region examined in this book
is usually found near streams limited to the desert Southwest of Arizona and
New Mexico. . . . . **longlegged water spiders (Trechaleidae; Plate 71)**

3A (1)   There appear to be six eyes in one row at the front of the face, the outer ones in
this row are quite small compared with the next innermost, which are large.
There are two additional large eyes facing sideways behind the ends of the front
row of eyes. Very flat spiders of the Southwest . . . . **flatties (Selenopidae; Plate 71)**

3B   Spider not as above. 4

4A (3)   The eight eyes are in two relatively straight rows. 5

4B   The eye rows, at least the posterior row, are slightly to noticeably recurved.
The legs are all about the same length, or the second pair is distinctly longer
than the rest. . . . . **running crab spiders (Philodromidae; Plate 70)**

5A (4)   The metatarsi have a soft three-lobed membrane at the distal end that permits
the tarsi to bend backward (hyperextend). This may be visible in the posture at
rest. Body and legs clothed with hairs, legs may have additional spines but not
usually paired ventral spines on anterior tibiae.
. . . . **huntsman spiders (Sparassidae; Plate 71)**

5B   The metatarsi normal. Three pairs of ventral spines on anterior tibiae.
. . . . **tengellids (Tengellidae in part *Lauricius*; Plate 46)**

SURFACE HUNTERS GROUP • *Spiders That Hunt on Flat Surfaces Such as Trunks or Rock Faces*

Plates 47, 56–57

This small group contains one large and common family, the nursery web spiders (Pisauridae), and one rarely encountered family, the longspinneret spiders (Hersiliidae). These two are placed together because they are most often found on the surfaces of logs, tree trunks, vegetation, rocks, or even the surface of ponds. The smaller nursery web spiders in the genus *Pisaurina* could just as easily have been included in the foliage hunter group. The larger fishing spiders in the genus *Dolomedes* often hunt at the edge of small streams and ponds but may also hunt for insect prey in vegetation when they are immature. The longspinneret spiders of the family Hersiliidae are easily distinguished from the nursery web spiders by their extraordinarily long posterior spinnerets (see Fig. 30 and Plate 47). These remarkable organs extend beyond the end of the abdomen by a length clearly greater than the length of the abdomen, often nearly as long as the entire body.

Other spiders, which can also be found on vertical surfaces, hold their legs in a distinctive crablike stance. In the South there are a few very large crablike spiders that are often found on the walls in buildings. These are huntsman spiders in the family Sparassidae. Spiders that adopt this laterigrade posture (see Fig. 13) have been placed together in the crablike spiders group.

GROUND HUNTERS GROUP • *Ground Hunters Not Associated with Webs*

Plates 39–55

It should be a surprise to no one that the ground is the most common place to find spiders that are not in webs. The great variety of ground hunters includes members representing 11 different families of spiders. Many of these spiders are nocturnal, but some large conspicuous ground hunters can also be found out during the day. Most of the nocturnal ground hunters hide under debris, such as rocks, logs, and leaves, during the day. A few of these spin a silk cocoon to rest in.

More information about spiders in the ground hunters group can be found in a series of more than 45 monographs published by Norman Platnick and his colleagues, primarily in the *Bulletin and Novitates of the American Museum of Natural History, New York*. These references provide details about the genera and species in the various families of ground active spiders. Two other excellent sources of information are *The Ground Spiders of Canada and Alaska (Araneae: Gnaphosidae)* (Platnick and Dondale 1992) and *The Wolf Spiders, Nurseryweb Spiders, and Lynx Spiders of Canada and Alaska (Araneae: Lycosidae, Pisauridae, and Oxyopidae)* (Dondale and Redner 1990). Much of the general information included in the species accounts of this book was obtained from these valuable sources.

In addition to the species described here, other spiders that live in webs under rocks or debris on the ground are included in the under debris group. Some pirate spiders (Mimetidae) may occasionally, or regularly in the case of *Ero*, be found under rocks or leaves on the ground. Nevertheless, these pirate spiders are treated with their relatives in the foliage hunters group.

Key to Families in the Ground Hunters Group

1A   The spider has large posterior median eyes (usually more than twice as large as smaller eyes), and four smaller eyes below them on the face. 2

1B   Eye pattern not as above. 3

2A (1)   With a high clypeus, the space between the anterior eye row and the top of the chelicerae, that is much taller than the space between the eye rows (see Figs. 15 and 20C). The posterior median eyes are the largest. The abdomen has a distinctive pentagonal shape. A ground spider of the desert Southwest, sometimes covers itself with sand grains.
. . . . **dusty desert spiders (Homalonychidae; Plate 46)**

2B   Not as above; also posterior lateral eyes quite large and so far back they appear to be a separate third row of eyes (see Figs. 11A and 20B).
. . . . **wolf spiders (Lycosidae; Plates 48–55)**

3A (1)   Spiders with an odd arrangement of eight eyes; two eyes in front, flanked by a group of three irregular-shaped eyes on each side (Fig. 31). Rare southern spider found on the ground at night, or under rocks. . . . . **prodidomids (Prodidomidae; Plate 45)**

3B   Spiders not as above. 4

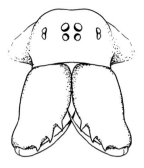

FIGURE 31. Eye pattern of *Prodidomus*.   FIGURE 32. Face of *Pachygnatha*.

4A (3)	Four eyes forming a trapezoid at the front of the face (see Fig. 19B). Eyes appear to be in three rows, the front row with two eyes, behind them a row of four, the middle two are largest, finally two widely separated larger eyes at the back. . . . . **wandering spiders (Ctenidae; Plate 39)**
4B	Spiders not as above. Eyes usually in two rows of four of approximately equal size, or with anterior median eyes that are somewhat larger than the rest. 5
5A (4)	With a cribellum, a relatively large spider with a variegated brownish coloration, found in central California . . . . **false wolf spiders (Zoropsidae; Plate 46)**
5B	Not like this. 6
6A (5)	With four eyes forming a trapezoid well separated from the lateral eyes. The lateral eyes are similar in size and grouped close together at each side; with a low clypeus. Spider has robust, somewhat divergent chelicerae with prominent teeth (Fig. 32). . . . . **thickjawed orbweaver (Tetragnathidae in part *Pachygnatha*; Plate 14)**
6B	Not like this. 7
7A (6)	Spinnerets are closely grouped and usually conical in shape, not spreading. A group of similar families that are difficult to separate based on characters visible with a simple magnifying glass. . . . . **sac spiders (families Corinnidae, Liocranidae, Miturgidae, Tengellidae; Plates 40, 45–46)**
7B	Anterior spinnerets are cylindrical in shape, usually longest 8
8A (7)	The spider has large jaws that protrude forward, two rows of four relatively equal-sized eyes, and a cribellum. Small spider of the intertidal zone in Florida. . . . . **desids (Desidae in part *Paratheuma*; Plate 47)**
8B	Spider not as above. 9
9A (8)	Spider without claw tufts. . . . . **zodariids (Zodariidae; Plate 47)**
9B	Spider with claw tufts. The posterior median eyes may be irregular in shape. The anterior spinnerets are relatively large, cylindrical in shape, and well separated from each other (see Figs. 12G,H, 30) . . . . . . **ground spiders (Gnaphosidae; Plates 41–44)**

Plates 36–38

Members of this group have been most frequently found running on foliage or branches at night. Some species may be encountered in the daytime, in particular the lynx spiders of the family Oxyopidae. The nocturnal members of the group usually hide during the day in silken cocoons. Most of these spiders are pale in color. The foliage hunters are searching for prey in a complex three-dimensional environment. Most of these spiders regularly leave a single thread of silk as a safety line wherever they wander. In the event that the spider is dislodged or falls, it can return via this silk line, often called a dragline. If you find single lines of silk strung through the foliage, you may be able to trace the path taken by one of these spiders.

Two families, the nursery web spiders (Pisauridae) and the longspinneret spiders (Hersiliidae), that are occasionally found on foliage are even more often found on the surface of tree trunks, logs, rocks, walls, or ponds. Members of these families are placed in their own group: the surface hunters.

Key to Families in the Foliage Hunters Group

1A   Tarsi lack tufts of hairs (claw tufts or scopulae), front two pair of legs longest, the patella and tibia as long as or longer than the femur, the legs very spiny with distinctive strong and slightly curved spines. Between each of these large spines on the tibia and metatarsi of the first two pairs of legs there are a series of smaller spines that increase in length from the base toward the tip of the leg segment (Fig. 33A). The first two pairs of legs are typically held with the femur vertical or folded back, and the remainder of the leg is extended forward (Fig. 33B). Relatively slow walking spiders. They can sometimes be found on the ground or under rocks.
. . . . **pirate spiders (Mimetidae; Plate 38)**

1B   Spider not as above. 2

2A (1)   Spider has six to eight pairs of conspicuous spines on the undersides of the front two pairs of tibiae. The tibiae are usually darker than the other leg segments. With two parallel dark lines on the carapace extending from the eyes back to the posterior end of the carapace . . . . **zorids (Zoridae; Plate 37)**

2B   Spider not as above. 3

3A (2)   Two small eyes at the front of the face, the other six eyes are arranged in a hexagon around the highest portion of the head (Fig. 34). The spiders have conspicuous spiny legs and the habit of jumping actively.
. . . . **lynx spiders (Oxyopidae; Plate 38)**

3B   Spider not as above. The remaining families in the foliage hunters group are relatively similar. There are a few characteristics that would separate the ghost spiders (Anyphaenidae), but these are not easily visible. The spiracular opening to the respiratory tubes (trachaea) is shifted forward toward the middle of the abdomen. In the other families this tiny slit is located just in front of the spinnerets. Also the claw tufts of the ghost spiders have unusual broad flat hairs, called lamelliform hairs. These can sometimes be seen as odd gray hairs in two groups at the tips of the tarsi. Otherwise the spiders of these families are remarkably similar in general form.
. . . . **ghost spiders (Anyphaenidae; Plate 36)** or **sac spiders (Clubionidae; Plate 37)**
or **prowling spiders (Miturgidae in part *Cheiracanthium*; Plate 37)**

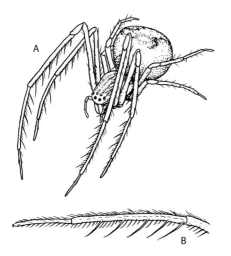

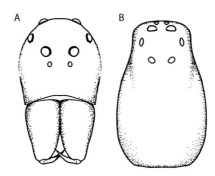

FIGURE 33. Specialized spine pattern in pirate spiders (Mimetidae). (A) Typical posture. (B) Metatarsus and tarsus of leg I.

FIGURE 34. Cephalothorax of a lynx spider (Oxyopidae). (A) Face. (B) Dorsal view showing the hexagonal arrangement of the eyes.

MYGALOMORPH GROUP

Plates 1–4

This group of spiders includes distinctive, usually large species, including the tarantulas. They represent an infraorder of spiders, the Mygalomorphae, which are thought to be an ancient lineage. The mygalomorph spiders represent only about 7 percent of spiders worldwide and only 3.5 percent of species in North America north of Mexico. One characteristic that they share are large, forward-oriented chelicerae (see Fig. 16A). Their jaws flex parallel to each other and in line with the long axis of the body. At rest the large cheliceral bases in mygalomorph spiders extend forward from the cephalothorax, and the fangs lie flat underneath, nearly parallel to each other (see Fig. 16B). When they strike, they raise their body, open the fangs, and strike downward. This method is quite different from that employed by the "typical" spiders in the infraorder Araneomorphae. In most araneomorph spiders the fangs fold facing each other (see Fig. 16D). When these spiders bite, the cheliceral fangs pinch the prey in an opposing motion (see Fig. 16C).

A few araneomorph spiders do have large forward-projecting cheliceral bases with fangs that fold back, nearly parallel, but they do not share the other characteristics listed here. In addition to the odd chelicerae, the mygalomorphs have their eight small eyes clustered in a group, usually on a slightly raised lobe at the front of the carapace. Most mygalomorph spiders in this region possess only two pairs of spinnerets; other spiders possess three pairs. In addition, all mygalomorphs possess at least one pair of spinnerets that are long and often extend beyond the end of the abdomen. The mygalomorph spiders possess two pairs of book lungs.

Mygalomorph spiders tend to be relatively long lived. The majority of these spiders live in burrows in the soil. Many of the burrowing forms have a patch of short stiff spines on their chelicerae called a rastellum that they use for digging. The anatomy and reproductive structures of these spiders show less modification than araneomorph spiders. Only now, with the benefit of modern molecular biology techniques, are we discovering the hidden diversity among these spiders. Many new species have been detected in the recent past, and there are undoubtedly more to be discovered. There are eight families of mygalomorph spiders found in North America north of Mexico.

For more information about mygalomorph spiders, consult *Spiders of North America: An Identification Guide* (Ubick et al. 2005).

Key to Families in the Mygalomorph Group

1A   Spider has a truncate abdomen with a hard sculptured disk at the back.
       . . . . **trapdoor spiders (Ctenizidae in part *Cyclocosmia*; Plate 2)**
1B   Spider not like this. 2

2A (1) Abdomen with one to four dorsal hard plates (tergites or sclerites), no claw tufts
       or scopulae. With two or three pairs of spinnerets. 3
2B   Abdomen without hard plates, with or without claw tufts. With only two pairs of
       spinnerets. 5

3A(2)   The sternum is fused to the labium so that it looks like one piece, slightly
       pointed at the front. The chelicerae are huge, equal to more than half of the
       length of the carapace. Usually found in a burrow with a closed silken tubular
       extension (purseweb). . . . . **purseweb spiders (Atypidae; Plate 1)**
3B   The sternum is separated from the labium by a visible joint. The chelicerae are
       large but not longer than half the length of the carapace. Usually found in a
       burrow without an above-ground tube or a silken retreat at the end of a funnel-
       shaped sheet. 4

4A (3) Posterior (longest) spinnerets with three segments. Last segment is shorter
       than the two basal segments combined. Usually found in a burrow with either a
       collapsible closure, turret, or thin door. Some with three pairs of spinnerets.
       . . . . **foldingdoor spiders (Antrodiaetidae; Plate 1)**
4B   Posterior (longest) spinnerets either three or four segmented with many pseudo-
       segments that render the last segments flexible and usually longer than the basal
       two segments. Usually found in a silken retreat at the end of a funnel-shaped
       sheet. Most have three pairs of spinnerets.
       . . . . **midget funnelweb tarantulas (Mecicobothriidae; Plate 3)**

5A (1) Very hairy spiders. The carapace is entirely clothed with dense hairs. With claw
       tufts and dense iridescent scopulae. Legs relatively thick, typically the tarsi are
       no more than twice as long as they are wide. Tarsi with two claws (claws may be
       hard to see). . . . . **tarantulas (Theraphosidae; Plate 4)**
5B   Typically less hairy. Legs lack claw tufts, scopulae (if present) not iridescent, tarsi
       with three claws (claws may be hard to see). 6

6A (5) Legs relatively thick, usually with many heavy spines on anterior legs. Spinnerets
       are short and blunt. Carapace with a U-shaped or inverted U-shaped thoracic
       groove (Fig. 35A, B). Usually found in burrow with a thick (cork-type) trapdoor.
       . . . . **trapdoor spiders (Ctenizidae; Plate 2)**
6B   Legs less thick, tibiae usually at least four times as long as they are wide.
       Spinnerets are long. May be found in burrow with a thin lid, no lid, or a funnel-
       shaped silken retreat. 7

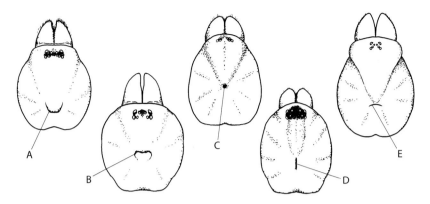

FIGURE 35. Fovea shape in mygalomorph spiders. (A) U-shape (*Ummidia*). (B) Inverted U-shape (*Bothriocyrtum*). (C) Pit (*Euagrus*). (D) Longitudinal (*Microhexura*). (E) Transverse (*Brachythele*).

7A (6)   Legs without claw tufts or scopulae. Cephalothorax has a thoracic pit (Fig. 35C) or longitudinal thoracic groove (Fig. 35D). Usually found in a silken retreat, often one with multiple entrances, sometimes hidden under debris.
. . . . **funnelweb spiders (Dipluridae; Plate 3)**

7B   The legs have scopulae on some tarsi. Cephalothorax has a transverse thoracic groove (Fig. 35E) or a shallow pit. . . . . **waferlid trapdoor spiders (Euctenizidae [formerly Cyrtaucheniidae]; Plate 3)** or **wishbone spiders (Nemesiidae; Plate 4)**

TINY SPIDERS GROUP • *Very Small, Often Cryptic Spiders*

Plates 79–82

The small spiders collected here represent a wide variety of families and many genera in the North American fauna. Because of their small size and typically reclusive habits, few naturalists take notice of spiders like these. The term "cryptozoic" describes them well; they remain hidden most of the time. The group as constituted here is a very diverse assemblage, perhaps including 920 species. Of these only 35 representative examples are included in this guide. There are a great many wonderful and fascinating spiders that inhabit the microworlds of leaf litter, soil debris, and mossy mats. There is so much to discover that careful study will surely be rewarded with new knowledge.

A few small species of spiders are not included here because they are usually associated with distinct webs, rendering them more conspicuous. These tiny spiders, including the members of the ground orbweb weavers (Anapidae) and the dwarf orbweavers (Symphytognathidae), are included with the orb-shaped web group. Likewise, the dwarf cobweb weavers (Mysmenidae) are illustrated with the space-filling web group.

Key to Families in the Tiny Spiders Group

1A   Either two or eight eyes grouped together near the center of the carapace. Carapace and legs typically orange or yellow-brown. Abdomen gray.
. . . . **bright lungless spiders (Caponiidae; Plate 80)**

1B   Spider not as above. 2

2A (1)	With only six eyes.	8
2B	With eight eyes.	3

3A (2) Cephalothorax is wider than long. Eyes in a group with the posterior median eyes kidney-shaped and shiny. Legs often held in an odd recurved posture. Sometimes found in flimsy star-shaped web but more often found away from the web. Very pale with dark spots . . . . **flatmesh weavers (Oecobiidae; Plate 82)**

3B Spider not as above. 4

4A (3) All six spinnerets in a transverse row.
. . . . **comb-tailed spiders (Hahniidae in part** *Hahnia*; **Plate 81)**

4B Spider not as above. 5

5A (4) With more than four pairs of strong spines on the ventral surfaces of the anterior tibiae . . . . **antmimic spiders (Corinnidae in part** *Phrurotimpus, Scotinella*; **Plate 80)**

5B Without paired ventral spines, or fewer than four pairs of ventral spines on the ventral surfaces of the anterior tibiae. 6

6A (5) Spider with a low clypeus. 7

6B With a high clypeus. A very diverse group of spiders in two large families; the sheetweb weavers (Linyphiidae) and the cobweb weavers (Theridiidae). Most of the spiders that key this way are members of the subfamily Erigoninae of the Linyphiidae (Plate 79). There are a few species of cobweb weavers that live near the ground and do not build conspicuous webs (*Crustulina, Dipoena, Robertus*; Plates 80–82) that would key out here.

7A (6) Posterior eye row clearly procurved (see Fig. 10).
. . . . **meshweavers (Dictynidae in part** *Argenna, Lathys*; **Plate 81)**

NOTE: Some species in the genus *Lathys* have only six eyes, the anterior median eyes are missing. In all of these spiders, the anterior median eyes are smaller than the anterior lateral eyes.

7B Posterior eye row straight (when viewed from above).
. . . . **comb-tailed spiders (Hahniidae in part** *Cryphoeca*; **Plate 81)**

8A (2) Yellow or orange spiders with six eyes usually tightly grouped on a dark spot. Make short, jerky movements, or hop. . . . . **goblin spiders (Oonopidae; Plates 80, 82)**

8B Spider not as above. 9

9A (9) Anterior spinnerets largest, with conspicuous colulus (see Fig. 12C).
. . . . **midget ground weavers (Ochyroceratidae, Plate 82)**
or **longlegged cave spiders (Telemidae; Plate 82)**

9B Spinnerets similar, grouped together, without a conspicuous colulus.
. . . . **midget cave spiders (Leptonetidae; Plate 82)**

COMMON SPIDERS OF NORTH AMERICA

PLATES AND GROUPS

Antrodiaetus riversi, turret spider, adult male, adult female
   Female TBL 17–18 mm, Male TBL 13–15 mm (see page 73)

Antrodiaetus unicolor, adult female, adult male
   Female TBL 20 mm, Male TBL 17 mm (see page 73)

Sphodros abboti, adult female, adult male, tube web against tree trunk
   Female TBL 13.0 mm, Male TBL 9.8 mm (see page 103)

Sphodros rufipes, redlegged purseweb spider, adult female, adult male
   Female TBL 24.0–25.0 mm, Male TBL 14.5 mm (see page 103)

redlegged
purseweb spider

S. rufipes

♀

♂

S. abboti
tubewebs

♀

♂

A. unicolor

♀

♂

S. abboti

turret
spider

A. riversi

♀

♂

Aliatypus californicus, adult female
    Female TBL 12.5–23.0 mm, Male TBL 3.5–12.0 mm (see page 72)

Bothriocyrtum californicum, California trapdoor spider, adult female and trapdoor
    Female TBL 25–33 mm, Male TBL 15–24 mm (see page 111)

Cyclocosmia truncata, adult female, adult male
    Female TBL 33 mm, Male TBL 19 mm (see page 112)

Ummidia audouini, adult female, adult male
    Female TBL 28 mm, Male TBL 15 mm (see page 112)

C. truncata

A. californicus

California
trapdoor spider

B. californicum

U. audouini

PLATE 3

MYGALOMORPH SPIDERS: FUNNELWEB SPIDERS, MIDGET FUNNELWEB SPIDERS, AND WAFERLID TRAPDOOR SPIDERS

Aptostichus simus, adult female, adult male
   Female TBL 12–18 mm, Male TBL 9–12 mm (see page 114)

Euagrus comstocki, adult female
   Female TBL 11–15 mm, Male TBL 8–13 mm (see page 122)

Hexura picea, adult female
   Female TBL 8.5–12.0 mm, Male TBL 8.0–8.5 mm (see page 164)

Megahexura fulva, adult female
   Female TBL 13–18 mm, Male TBL 8–13 mm (see page 164)

Microhexura idahoana, adult female
   Female TBL 2.9–5.6 mm, Male TBL 3.1–4.5 mm (see page 123)

Myrmekiaphila comstocki, adult female
   Female TBL 23.5–25.0 mm, Male TBL 16.5 mm (see page 114)

Microhexura
idahoana ♀

♀   E. comstocki

Megahexura fulva ♀

♀   H. picea

♀

Myrmekiaphila comstocki

♀

A. simus

♂

PLATE 4

MYGALOMORPH SPIDERS: TARANTULAS, WISHBONE SPIDERS, AND CREVICE WEAVERS

Aphonopelma chalcodes, desert blonde tarantula, adult female, adult male
  Female TBL 62 mm, Male TBL 55 mm (see page 222)

Aphonopelma moderatum, Rio Grande gold tarantula, adult female
  Female TBL 28–29 mm, Male TBL 29 mm (see page 222)

Aphonopelma hentzi, Texas brown tarantula, adult female
  Female TBL 44–58 mm, Male TBL 38–52 mm (see page 222)

Brachythele longitarsis, adult female, adult male
  Female TBL 18–46 mm, Male TBL 15–18 mm (see page 169)

Kukulcania hibernalis, southern house spider, adult female, adult male
  Female TBL 13–19 mm, Male TBL 9–10 mm (see page 124)

Rio Grande
gold tarantula

A. moderatum

♀

Texas brown
tranatula

A. hentzi

♀

desert blonde
tarantula

A. chalcodes

♂

♀

♂

♀

B. longitarsis

♀

K. hibernalis

♂

♀

♂

southern
house spider

PLATE 5

ORB-SHAPED WEB BUILDERS: ORBWEAVERS WITH SPINY ABDOMENS

Gasteracantha cancriformis, spinybacked spider, adult female black-and-white form,
adult female red-spined form, adult female yellow-bodied form, adult male
Female TBL 5.8–8.6 mm, Male TBL 1.9–2.7 mm (see page 90)

Kaira alba, frilled orbweaver, adult female, adult male below
Female TBL 6.5–7.3 mm, Male TBL 2.6–3.0 mm (see page 91)

Micrathena gracilis, spined micrathena, adult male, adult female black-and-white form,
adult female yellow form
Female TBL 7.0–11.0 mm, Male TBL 4.2–5.1 mm (see page 97)

Micrathena mitrata, white micrathena, adult female
Female TBL 4.7–6.0 mm, Male TBL 3.0–3.7 mm (see page 97)

Micrathena sagittata, arrowshaped micrathena, adult female
Female TBL 5.4–8.6 mm, Male TBL 4.2–5.9 mm (see page 98)

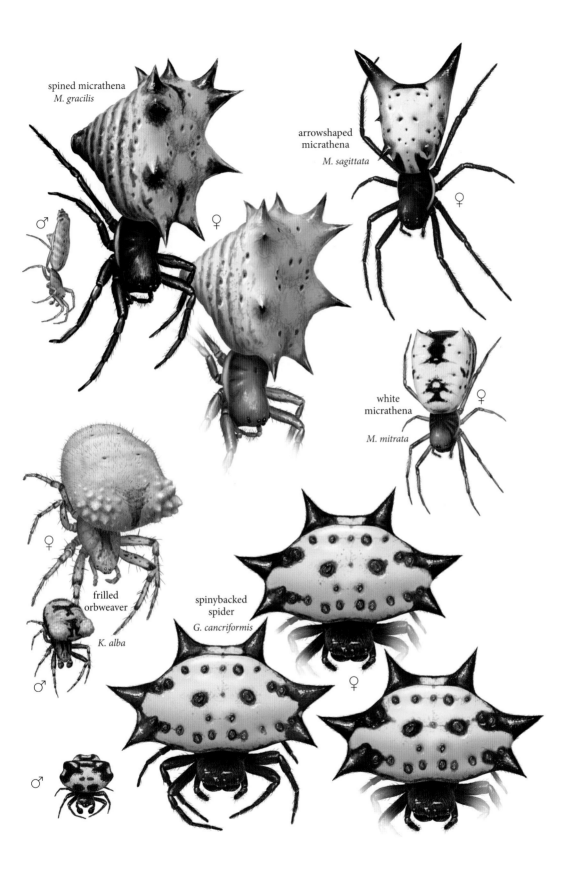

spined micrathena
M. gracilis

♂ ♀

arrowshaped
micrathena

M. sagittata

♀

white
micrathena

♀

M. mitrata

♀

frilled
orbweaver

spinybacked
spider

K. alba

G. cancriformis

♀

♂

♂

PLATE 6

ORB-SHAPED WEB BUILDERS: ORBWEAVERS WITH LUMPY ABDOMENS

Acanthepeira stellata, starbellied orbweaver, adult female
Female TBL 7.0–15.1 mm, Male TBL 5.1–8.1 mm (see page 77)

Colphepeira catawba, adult female
Female TBL 2.2–3.8 mm, Male TBL 1.6–2.2 mm (see page 88)

Ocrepeira ectypa, asterisk spider, adult female
Female TBL 5.2–9.4 mm, Male TBL 5.4–7.2 mm (see page 100)

Scoloderus nigriceps, ladderweb spider, adult female
Female TBL 3.2–4.3 mm, Male TBL 2.3–2.8 mm (see page 100)

Verrucosa arenata, triangulate orbweaver, adult female black-and-white form, adult male
orange form, adult female black-and-yellow form, adult female red-and-yellow form
Female TBL 5.0–9.5 mm, Male TBL 4.0–6.1 mm (see page 101)

Wagneriana tauricornis, adult female
Female TBL 4.3–6.1 mm, Male TBL 3.3–5.8 mm (see page 102)

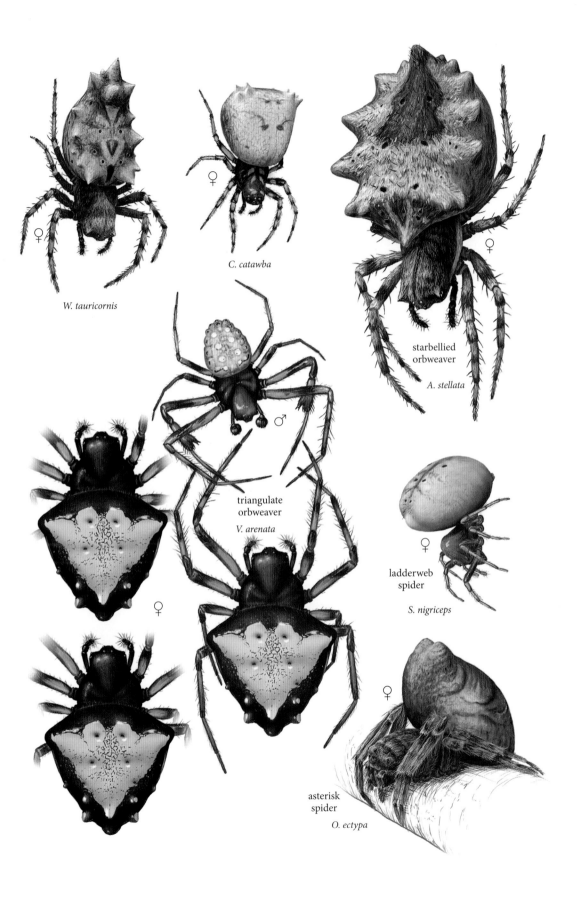

W. tauricornis

♀

C. catawba

♀

starbellied
orbweaver

A. stellata

♀

triangulate
orbweaver

V. arenata

♂

♀

ladderweb
spider

S. nigriceps

♀

♀

asterisk
spider

O. ectypa

♀

Allocyclosa bifurca, adult female
  Female TBL 5.1–9.0 mm, Male TBL 1.8 mm (see page 78)

Cyclosa conica, trashline orbweaver, adult female
  Female TBL 3.6–7.9 mm, Male TBL 3.5–4.9 mm (see page 88)

Cyclosa turbinata, humped trashline orbweaver, adult female
  Female TBL 3.3–5.2 mm, Male TBL 2.1–3.2 mm (see page 88)

Mastophora cornigera, southern bolas spider, adult female, egg case, adult male
  Female TBL 8.8–14.0 mm, Male TBL 1.6–1.7 mm (see page 94)

Mastophora hutchinsoni, cornfield bolas spider, adult female hunting, egg case,
  adult male, adult female
  Female TBL 6.2–10.4 mm, Male TBL 1.7 mm (see page 95)

Mastophora phrynosoma, toadlike bolas spider, adult female, egg case
  Female TBL 8.3–12.3 mm, Male TBL 1.5–1.7 mm (see page 96)

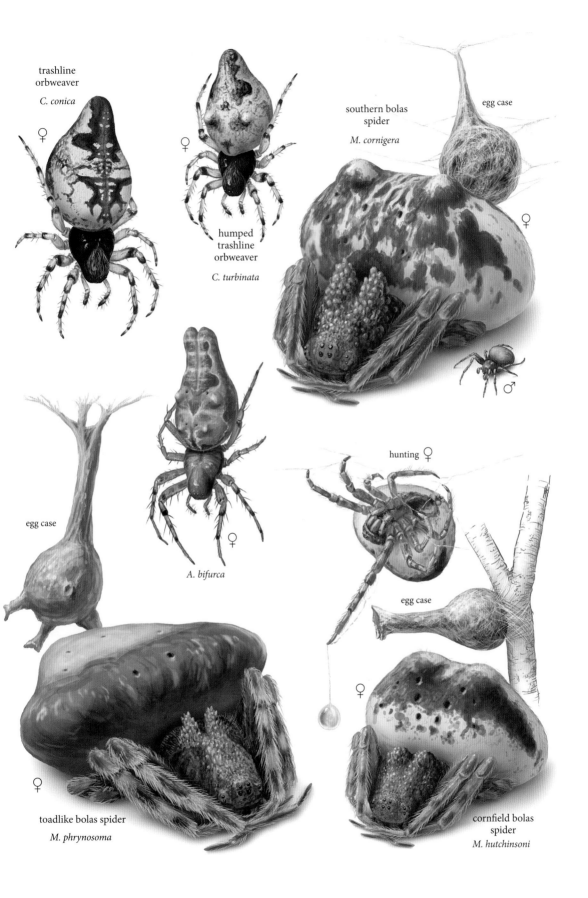

trashline
orbweaver

C. conica

♀

humped
trashline
orbweaver

C. turbinata

♀

southern bolas
spider

M. cornigera

egg case

♀

♂

egg case

A. bifurca

♀

hunting ♀

egg case

♀

toadlike bolas spider

M. phrynosoma

♀

cornfield bolas
spider

M. hutchinsoni

PLATE 8
ORB-SHAPED WEB BUILDERS: ORBWEAVERS WITH ANGULATE ABDOMENS

Araneus andrewsi, adult female
Female TBL 11–22 mm, Male TBL 8–11 mm (see page 78)

Araneus bicentenarius, lichenmarked orbweaver, adult female
Female TBL 21–28 mm, Male TBL 7 mm (see page 79)

Araneus cavaticus, barn orbweaver, adult female
Female TBL 13–22 mm, Male TBL 10–19 mm (see page 79)

Araneus diadematus, cross orbweaver, adult female orange form, adult male orange form
Female TBL 6.5–20.0 mm, Male TBL 5.7–13.0 mm (see page 80)

Araneus gemmoides, plains orbweaver, adult female plain form, adult female marked form
Female TBL 13.0–25.0 mm, Male TBL 5.4–7.9 mm (see page 81)

Araneus nordmanni, adult female brown form, adult black-and-white form
Female TBL 7–19 mm, Male TBL 6–10 mm (see page 83)

Araneus saevus, adult female
Female TBL 11–17 mm, Male TBL 9–11 mm (see page 85)

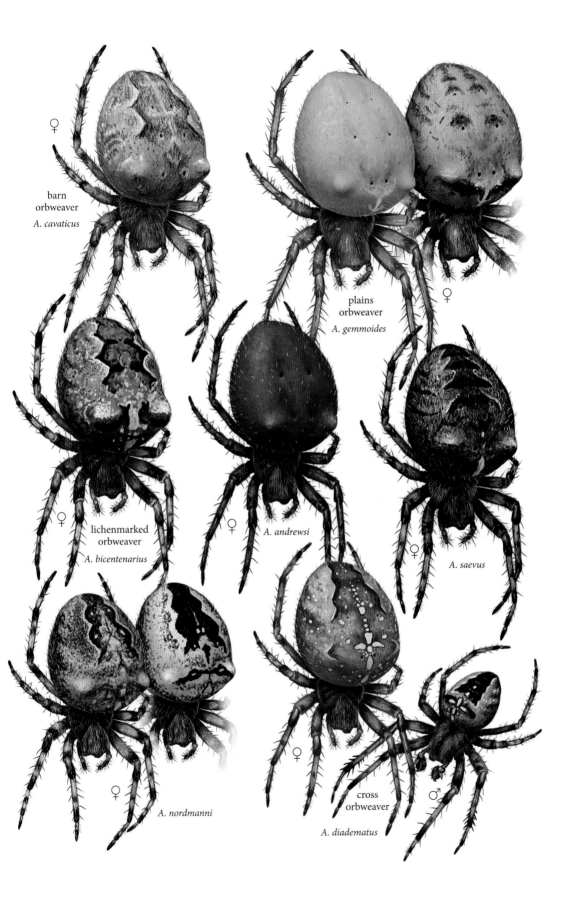

♀

barn
orbweaver
A. cavaticus

plains
orbweaver
A. gemmoides

♀

♀
lichenmarked
orbweaver

A. bicentenarius

♀
A. andrewsi

♀
A. saevus

♀
A. nordmanni

♀

cross
orbweaver
A. diadematus

♂

Araneus ivei, orange orbweaver, adult female
Female TBL 8.5–12.0 mm, Male TBL 5.0–7.0 mm (see page 81)

Araneus marmoreus, marbled orbweaver, adult female yellow form, adult female orange form,
adult female mottled form, adult female foliate form, adult female yellow form ventral view
Female TBL 9.0–18.0 mm, Male TBL 5.9 mm (see page 82)

Araneus trifolium, shamrock orbweaver, adult female red form, adult female pale form,
adult female yellow form, adult female green form, adult male
Female TBL 9–20 mm, Male TBL 5–8 mm (see page 85)

Larinioides cornutus, furrow orbweaver, adult male, adult female
Female TBL 6.5–14.0 mm, Male TBL 4.7–8.5 mm (see page 92)

Larinioides patagiatus, adult female
Female TBL 5.5–11.0 mm, Male TBL 5.8–6.5 mm (see page 92)

Larinioides sclopetarius, bridge orbweaver, adult female
Female TBL 8–14 mm, Male TBL 6–7 mm (see page 93)

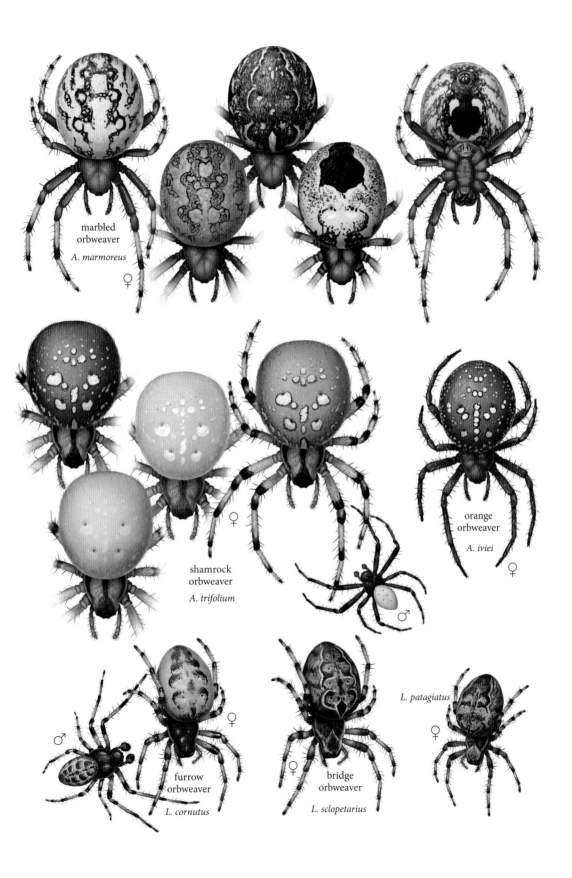

marbled
orbweaver

A. marmoreus

♀

shamrock
orbweaver

A. trifolium

♀

♂

orange
orbweaver

A. iviei

♀

♂

furrow
orbweaver

L. cornutus

♀

♀

bridge
orbweaver

L. sclopetarius

L. patagiatus

♀

PLATE 10

ORB-SHAPED WEB BUILDERS: SMALL GREEN ORBWEAVERS

Araneus cingulatus, adult female green form, adult female redmask form
Female TBL 4.6–6.0 mm, Male TBL 2.7–3.5 mm (see page 80)

Araneus detrimentosus, adult female
Female TBL 4.0–6.0 mm, Male TBL 2.5–4.2 mm (see page 80)

Araneus guttulatus, adult female
Female TBL 3.8–6.0 mm, Male TBL 3.9–4.8 mm (see page 81)

Araneus juniperi, adult female
Female TBL 2.5–5.2 mm, Male TBL 3.2–4.6 mm (see page 82)

Araneus niveus, adult female
Female TBL 3.2–5.0 mm, Male TBL 2.9–4.3 mm (see page 83)

A. cingulatus ♀

A. juniperi ♀

A. guttulatus ♀

A. detrimentosus ♀

A. niveus ♀

PLATE 11

ORB-SHAPED WEB BUILDERS: SMALL ORBWEAVERS
WITH DISTINCTIVELY MARKED ABDOMENS

Araneus bispinosus, adult female
 Female TBL 4.1–6.8 mm, Male TBL 3.2–5.0 mm (see page 79)

Araneus miniatus, adult female
 Female TBL 3.0–4.7 mm, Male TBL 2.5–3.7 mm (see page 82)

Araneus montereyensis, adult female reddish form, adult female brown form,
 adult female bicolored form
 Female TBL 3.2–5.5 mm, Male TBL 3.5–3.6 mm (see page 83)

Araneus partitus, adult female
 Female TBL 3.3–4.3 mm, Male TBL 2.7–3.3 mm (see page 84)

Araneus pegnia, adult female
 Female TBL 3.5–8.2 mm, Male TBL 2.5–4.9 mm (see page 84)

Araneus thaddeus, lattice orbweaver, adult female
 Female TBL 5.9–8.0 mm, Male TBL 3.7–5.7 mm (see page 85)

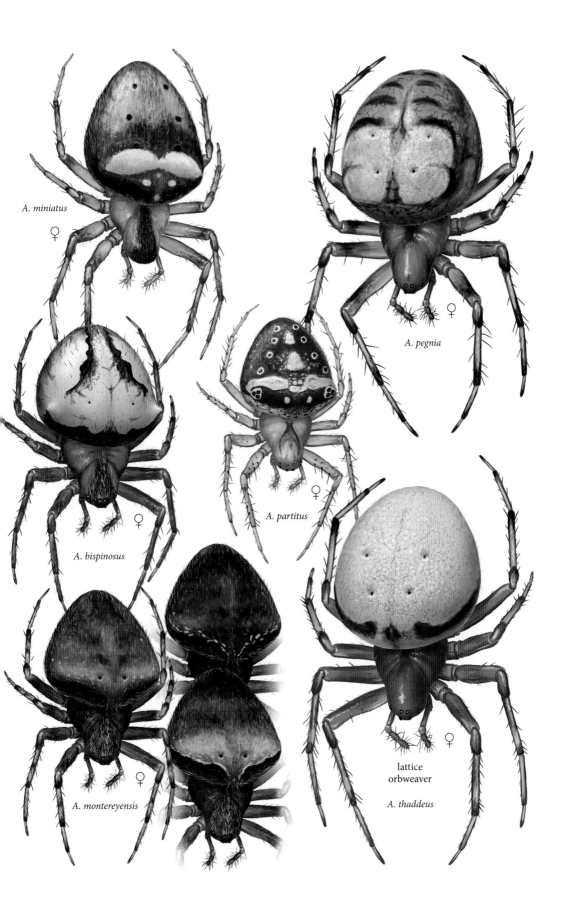

A. miniatus
♀

A. pegnia
♀

A. bispinosus
♀

A. partitus
♀

A. montereyensis
♀

lattice
orbweaver
A. thaddeus
♀

PLATE 12
ORB-SHAPED WEB BUILDERS: SMALL ORBWEAVERS WITH SPOTTED ABDOMENS

Araniella displicata, sixspotted orbweaver, adult female pale form, adult female red form
Female TBL 4.8–7.2 mm, Male TBL 4.0–5.0 mm (see page 86)

Glenognatha foxi, adult male, adult female
Female TBL 1.6–2.6 mm, Male TBL 1.4–2.2 mm (see page 217)

Genus *Mangora*: The spiders of the genus *Mangora* are small spiny orbweavers with tufts of long, thin sensory hairs (trichobothria) on the front edge of the tibiae of leg III.

Mangora gibberosa, lined orbweaver, adult female
Female TBL 3.4–4.8 mm, Male TBL 2.6–3.2 mm (see page 93)

Mangora maculata, greenlegged orbweaver, adult female
Female TBL 3.6–5.5 mm, Male TBL 2.7–4.0 mm (see page 93)

Mangora placida, tuftlegged orbweaver, adult female
Female TBL 2.3–4.5 mm, Male TBL 2.0–2.8 mm (see page 94)

Mangora spiculata, adult female
Female TBL 2.4–4.3 mm, Male TBL 1.9–2.2 mm (see page 94)

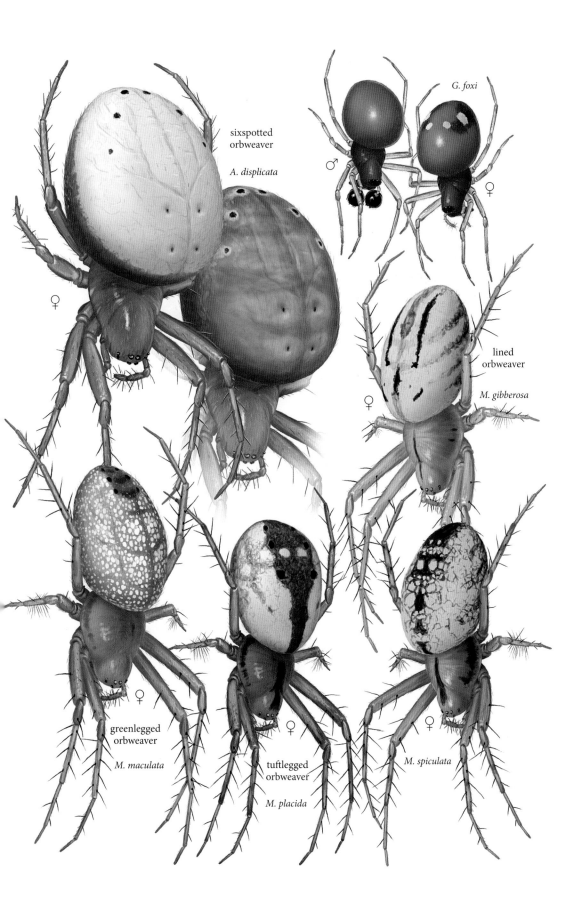

sixspotted
orbweaver

A. displicata

♀

G. foxi

♂

♀

lined
orbweaver

M. gibberosa

♀

greenlegged
orbweaver

M. maculata

♀

tuftlegged
orbweaver

M. placida

♀

M. spiculata

♀

PLATE 13

ORB-SHAPED WEB BUILDERS: ORBWEAVERS WITH A DISTINCT FOLIUM ON THE ABDOMEN

Acacesia hamata, difoliate orbweaver, adult female
Female TBL 4.7–9.1 mm, Male TBL 3.6–4.8 mm (see page 77)

Azilia affinis, adult female Florida form
Female TBL 6.9–9.9 mm, Male TBL 5.2–6.9 mm (see page 216)

Eustala anastera, humpbacked orbweaver, adult female
Female TBL 5.4–10.0 mm, Male TBL 3.9–9.5 mm (see page 89)

Metazygia calix, adult female
Female TBL 3.9–6.0 mm, Male TBL 3.8–4.0 mm (see page 96)

Metepeira labyrinthea, labyrinth spider, adult female
Female TBL 4.0–8.6 mm, Male TBL 3.0–6.8 mm (see page 97)

Zygiella x-notata, opensector orbweaver, adult female
Female TBL 7.4–8.7 mm, Male TBL 6.0–6.5 mm (see page 102)

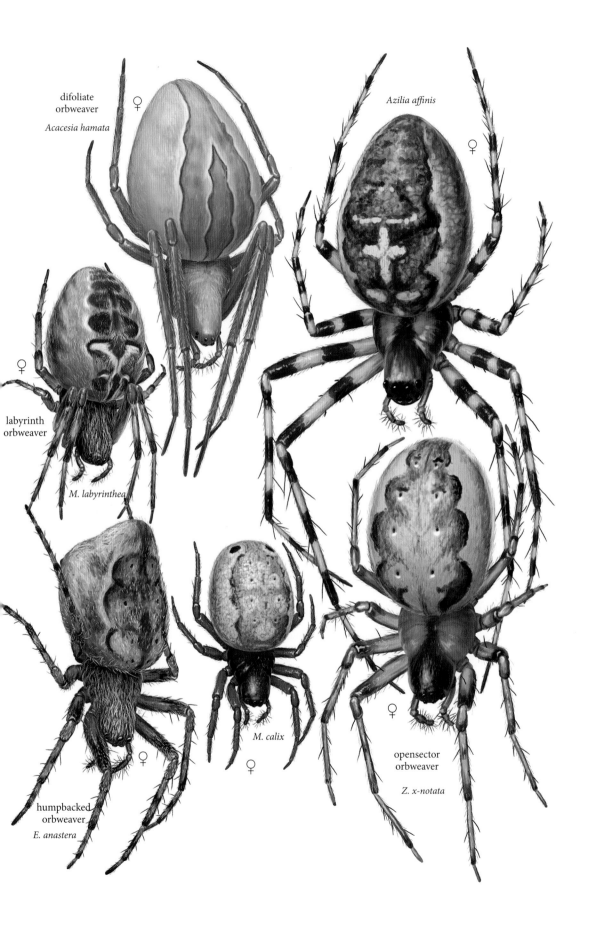

difoliate
orbweaver
♀

Acacesia hamata

Azilia affinis
♀

labyrinth
orbweaver
♀

M. labyrinthea

humpbacked
orbweaver
♀

E. anastera

M. calix
♀

opensector
orbweaver
♀

Z. x-notata

Araneus pratensis, openfield orbweaver, adult female
Female TBL 3.8–5.0 mm, Male TBL 3.0–3.5 mm (see page 84)

Hypsosinga pygmaea, adult female
Female TBL 2.9–3.9 mm, Male TBL 2.2–2.6 mm (see page 91)

Hypsosinga rubens, adult male, adult female
Female TBL 7.4–8.7 mm, Male TBL 6.0–6.5 mm (see page 91)

Pachygnatha furcillata, thickjawed orbweaver, adult female, adult male
Female TBL 5.2–6.2 mm, Male TBL 4.6–5.9 mm (see page 219)

Singa eugeni, adult female
Female TBL 4.3–6.5 mm, Male TBL 3.6–5.4 mm (see page 101)

Singa keyserlingi, adult female
Female TBL 5.1–6.0 mm, Male TBL 2.3–4.0 mm (see page 101)

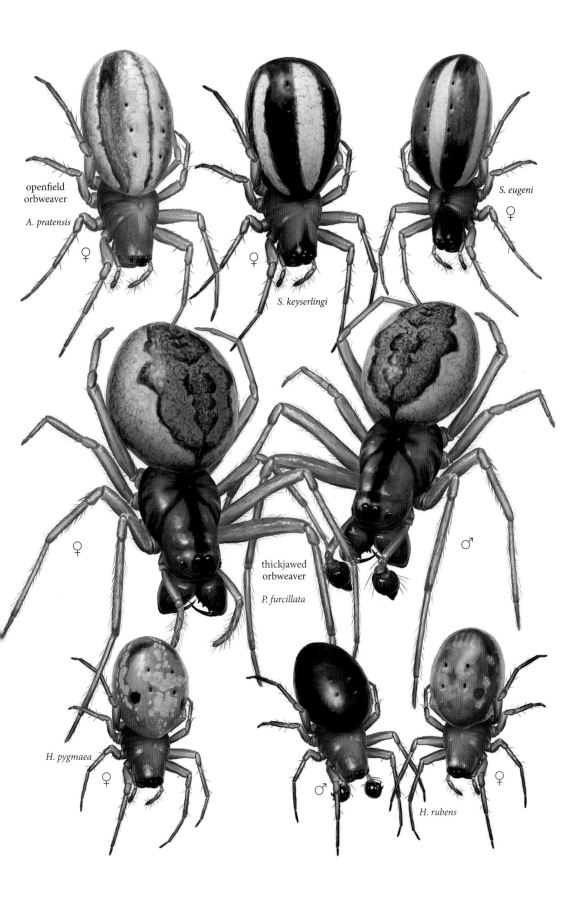

openfield
orbweaver

A. pratensis

♀

♀

S. keyserlingi

S. eugeni

♀

♀

thickjawed
orbweaver

P. furcillata

♂

H. pygmaea

♀

♂

H. rubens

♀

PLATE 15

ORB-SHAPED WEB BUILDERS: ORBWEAVERS; *NEOSCONA* AND *ERIOPHORA*

Eriophora ravilla, tropical orbweaver, adult male, adult female spotted form, adult female plain form
  Female TBL 12–24 mm, Male TBL 9–13 mm (see page 89)

Neoscona arabesca, arabesque orbweaver, adult female
  Female TBL 5.2–7.7 mm, Male TBL 4.2–5.9 mm (see page 98)

Neoscona crucifera, arboreal orbweaver, adult female bold form, adult female plain form, adult male
  Female TBL 8.5–19.7 mm, Male TBL 4.5–15.0 mm (see page 99)

Neoscona domiciliorum, adult female
  Female TBL 7.2–16.2 mm, Male TBL 8.0–9.0 mm (see page 99)

Neoscona oaxacensis, western spotted orbweaver, adult female
  Female TBL 8.9–18.0 mm, Male TBL 6.3–12.7 mm (see page 99)

Neoscona pratensis, marsh orbweaver, adult female
  Female TBL 6.5–10.2 mm, Male TBL 6.7–7.9 mm (see page 100)

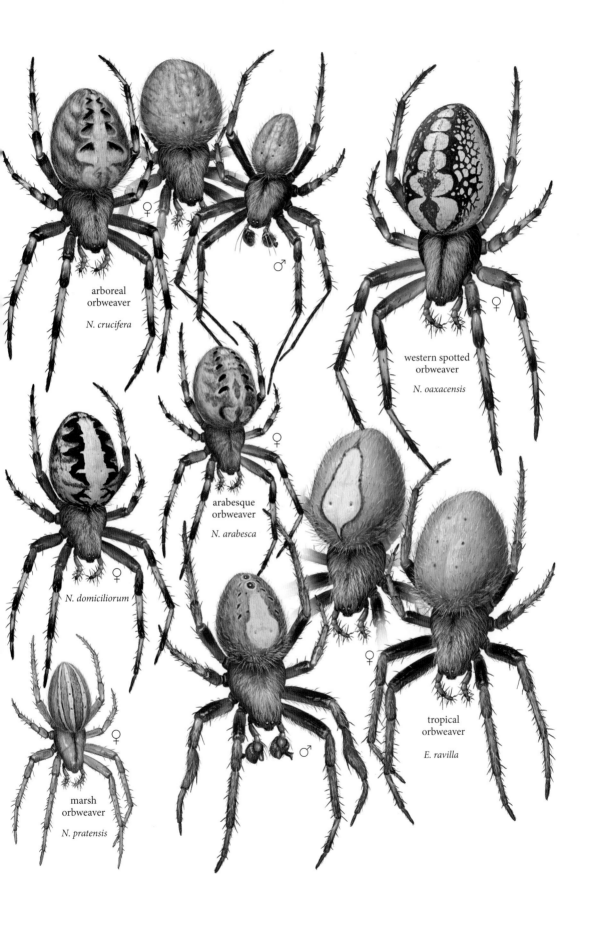

arboreal
orbweaver

N. crucifera

♀

♂

western spotted
orbweaver

N. oaxacensis

♀

arabesque
orbweaver

N. arabesca

♀

N. domiciliorum

♀

♀

tropical
orbweaver

E. ravilla

marsh
orbweaver

N. pratensis

♀

♂

PLATE 16

ORB-SHAPED WEB BUILDERS: ORBWEAVERS WITH LONG ABDOMENS; *NEPHILA* AND *LEUCAUGE*

Aculepeira packardi, adult female
   Female TBL 5.6–16.5 mm, Male TBL 5.0–8.9 mm (see page 78)

Larinia directa, adult female
   Female TBL 4.8–11.7 mm, Male TBL 4.5–6.5 mm (see page 91)

Leucauge argyra, adult female
   Female TBL 4.5–10 mm, Male TBL 4.1–6.3 mm (see page 217)

Leucauge venusta, orchard spider, adult female
   Female TBL 3.7–8.0 mm, Male TBL 3.2–5.1 mm (see page 217)

Mecynogea lemniscata, basilica orbweaver, adult female
   Female TBL 6.0–9.0 mm, Male TBL 5.0–6.9 mm (see page 96)

Nephila clavipes, golden silk orbweaver, adult female, adult male
   Female TBL 19–34 mm, Male TBL 5–8 mm (see page 219)

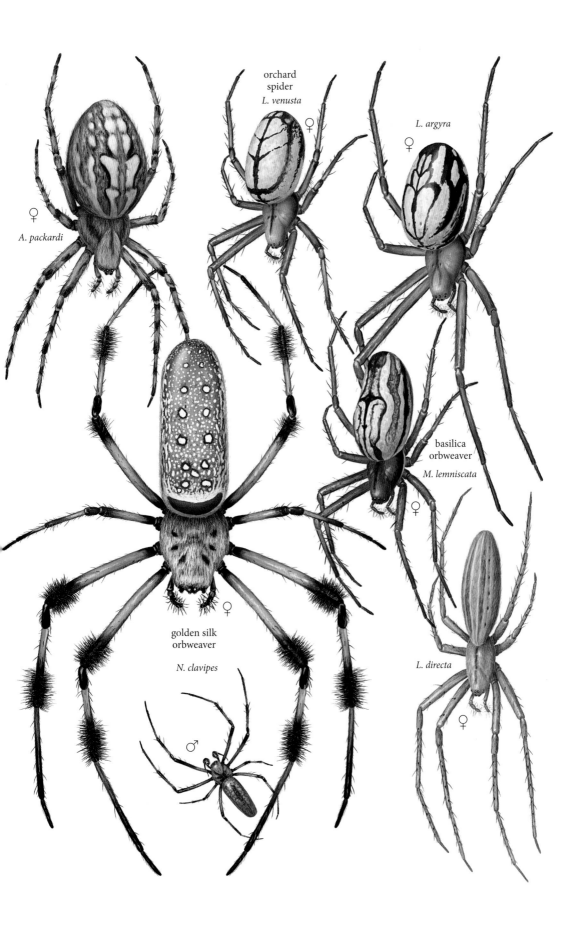

orchard
spider
L. venusta
♀

L. argyra
♀

♀
A. packardi

basilica
orbweaver

M. lemniscata
♀

golden silk
orbweaver

N. clavipes
♀

♂

L. directa
♀

Argiope argentata, silver garden spider, adult female
    Female TBL 12.0–16.0 mm, Male TBL 3.7–4.7 mm (see page 86)

Argiope aurantia, black-and-yellow garden spider, adult male, adult female
    Female TBL 19–28 mm, Male TBL 5–8 mm (see page 86)

Argiope florida, Florida garden spider, adult female
    Female TBL 12.5–18.0 mm, Male TBL 5.1–6.7 mm (see page 87)

Argiope trifasciata, banded garden spider, adult female
    Female TBL 15.0–25.0 mm, Male TBL 4.0–5.5 mm (see page 87)

Dolichognatha pentagona, adult female
    Female TBL 2.6–4.0 mm, Male TBL 2.6–3.2 mm (see page 216)

Gea heptagon, adult female
    Female TBL 4.5–5.8 mm, Male TBL 2.6–4.3 mm (see page 90)

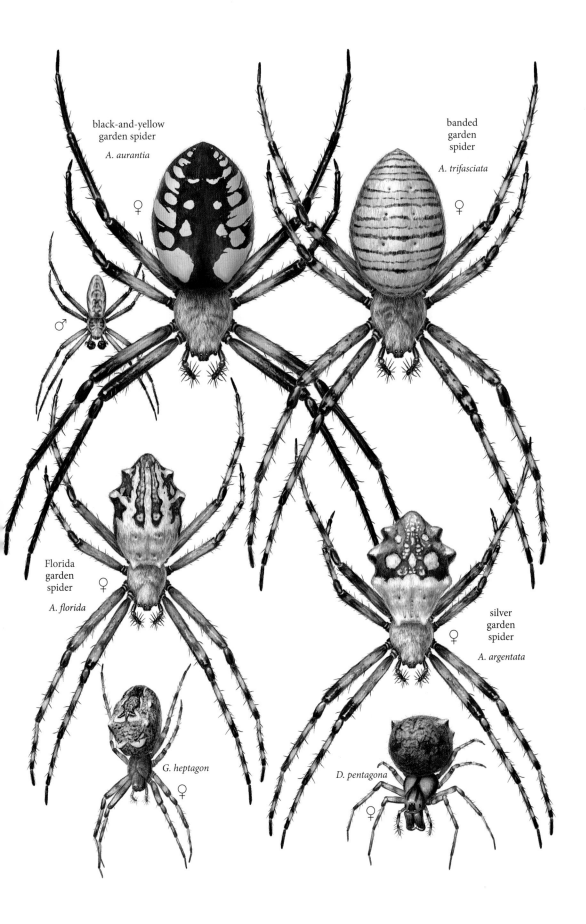

black-and-yellow
garden spider

A. aurantia

♀

♂

banded
garden
spider

A. trifasciata

♀

Florida
garden
spider

♀

A. florida

silver
garden
spider

♀

A. argentata

G. heptagon

♀

D. pentagona

♀

PLATE 18

ORB-SHAPED WEB BUILDERS: LONGJAWED ORBWEAVERS

Metellina mimetoides, adult female
Female TBL 3.3–6.0 mm, Male TBL 3.1–4.9 mm (see page 218)

Metleucauge eldorado, adult female
Female TBL 8.8–11.3 mm, Male TBL 7.8–11.7 mm (see page 219)

Tetragnatha elongata, adult female, adult male
Female TBL 8.2–13.2 mm, Male TBL 4.8–10.5 mm (see page 220)

Tetragnatha guatemalensis, adult female
Female TBL 5.4–11.5 mm, Male TBL 5.2–10.2 mm (see page 220)

Tetragnatha laboriosa, silver longjawed orbweaver, adult female
Female TBL 5.2–9.0 mm, Male TBL 3.8–7.4 mm (see page 220)

Tetragnatha versicolor, adult female
Female TBL 5.4–13.3 mm, Male TBL 4.3–9.2 mm (see page 221)

Tetragnatha viridis, green longjawed orbweaver, adult female
Female TBL 5.7–7.4 mm, Male TBL 4.4–6.7 mm (see page 221)

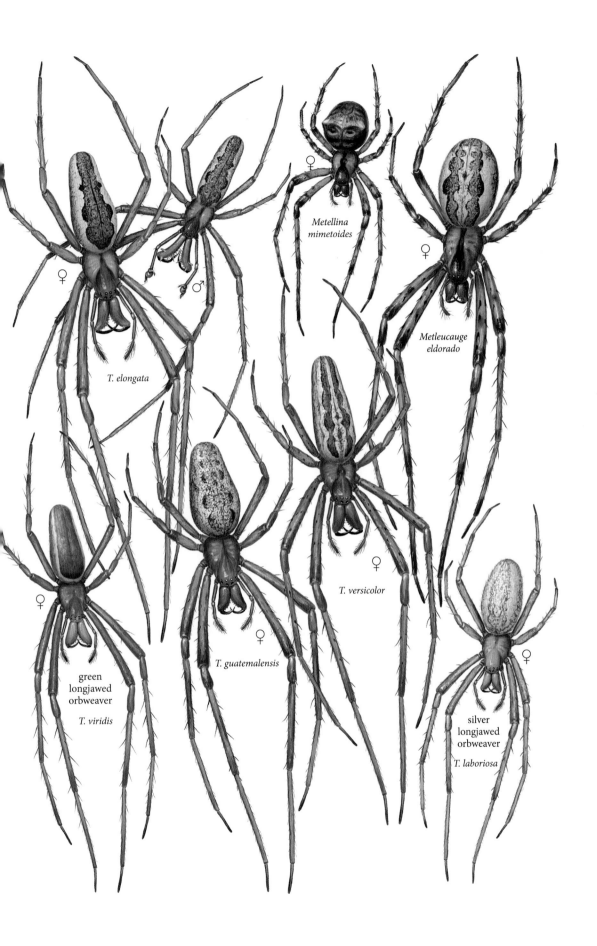

T. elongata ♀

T. elongata ♂

Metellina mimetoides ♀

Metleucauge eldorado ♀

green
longjawed
orbweaver

T. viridis ♀

T. guatemalensis ♀

T. versicolor ♀

silver
longjawed
orbweaver

T. laboriosa ♀

PLATE 19

ORB-SHAPED WEB BUILDERS: HACKLED ORBWEAVERS

Hyptiotes cavatus, triangle weaver, adult female, adult male
 Female TBL 2.3–3.8 mm, Male TBL 2.0–2.6 mm (see page 245)

Miagrammopes mexicanus, Mexican stickspider, adult female
 Female TBL 9 mm, Male TBL 7 mm (see page 246)

Octonoba sinensis, adult female
 Female TBL 3.0–5.8 mm, Male TBL 2.8–3.3 mm (see page 246)

Philoponella oweni, adult female
 Female TBL 10.3 mm, Male TBL 7.7 mm (see page 246)

Siratoba referens, adult female
 Female TBL 2.8 mm, Male TBL 2.5 mm (see page 247)

Uloborus glomosus, featherlegged orbweaver, adult female, adult male
 Female TBL 2.8–4.3 mm, Male TBL 2.3–3.2 mm (see page 247)

Zosis geniculata, adult female
 Female TBL 5.2–8.3 mm, Male TBL 4.2–5.5 mm (see page 247)

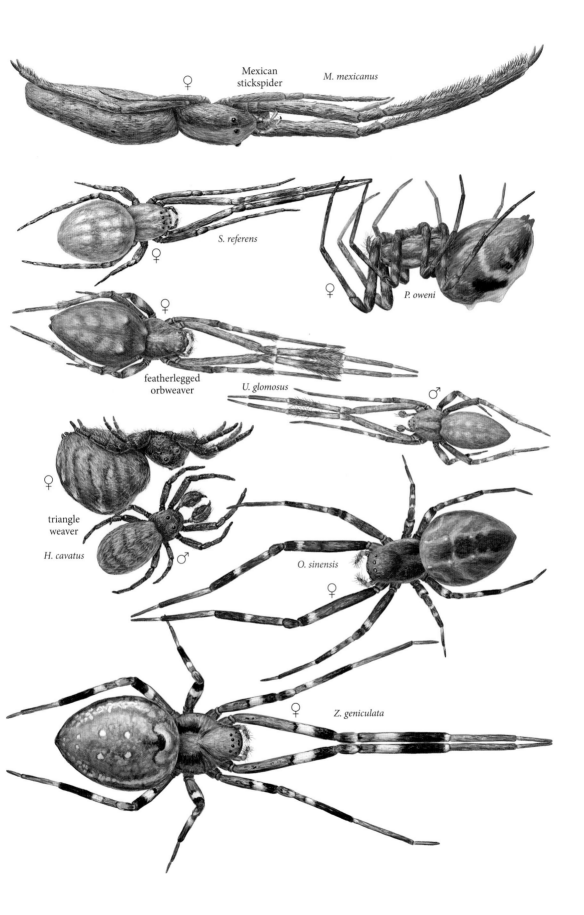

♀ Mexican stickspider *M. mexicanus*

S. referens ♀

♀ *P. oweni*

♀ featherlegged orbweaver *U. glomosus* ♂

♀ triangle weaver

H. cavatus ♂

O. sinensis ♀

♀ *Z. geniculata*

PLATE 20

ORB-SHAPED WEB BUILDERS: TINY ORBWEAVERS; *META*

Anapistula secreta, adult female
   Female TBL 0.2–0.5 mm, Male TBL 0.2–0.4 mm (see page 213)

Gertschanapis shantzi, adult female
   Female TBL 1.0–1.5 mm, Male TBL 1.0–1.5 mm (see page 72)

Meta ovalis, cave orbweaver, adult female
   Female TBL 8.0–10.0 mm, Male TBL 8.0–9.5 mm (see page 218)

Microdipoena guttata, adult female
   Female TBL 0.9–1.1 mm, Male TBL 0.7–0.9 mm (see page 168)

Theridiosoma gemmosum, ray orbweaver, egg case, adult female
   Female TBL 2.7 mm, Male TBL 1.6 mm (see page 237)

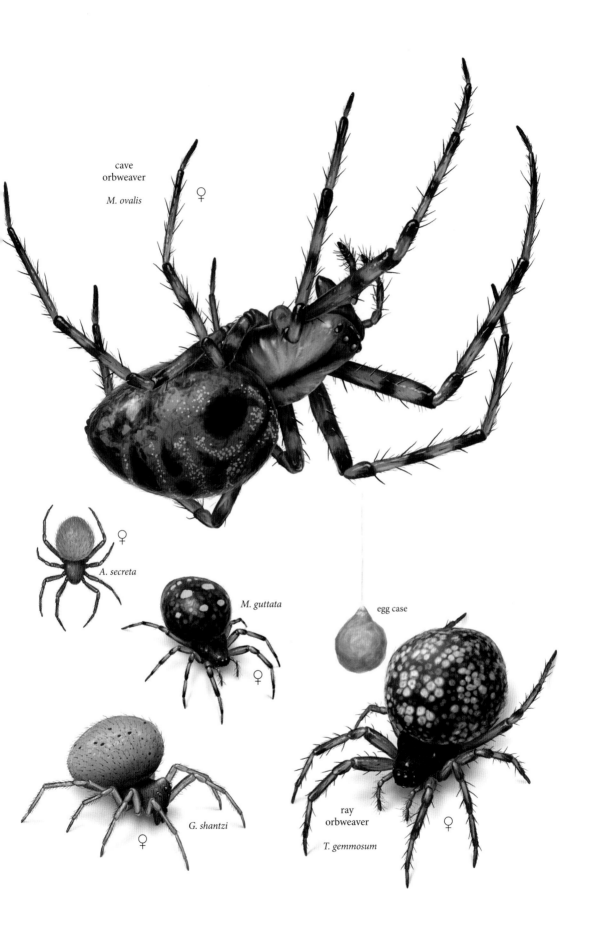

cave
orbweaver

M. ovalis

♀

♀

A. secreta

M. guttata

♀

egg case

♀

G. shantzi

♀

ray
orbweaver

♀

T. gemmosum

PLATE 21

SPACE-FILLING WEB BUILDERS: MESHWEAVERS; *DICTYNA* AND *EMBLYNA*

Dictyna bostoniensis, adult female
   Female TBL 3.1–3.8 mm, Male TBL 2.2–2.7 mm (see page 117)

Dictyna coloradensis, adult female
   Female TBL 3.8 mm, Male TBL 3.2 mm (see page 118)

Dictyna foliacea, adult female, adult male with chelicerae partly spread
   Female TBL 2.0–2.7 mm, Male TBL 1.7–2.1 mm (see page 118)

Dictyna volucripes, adult female
   Female TBL 3.4–4.2 mm, Male TBL 2.8–3.3 mm (see page 118)

Emblyna annulipes, adult female
   Female TBL 2.9–4.4 mm, Male TBL 2.4–3.8 mm (see page 119)

Emblyna sublata, adult female
   Female TBL 2.3–3.7 mm, Male TBL 2.0–2.5 mm (see page 119)

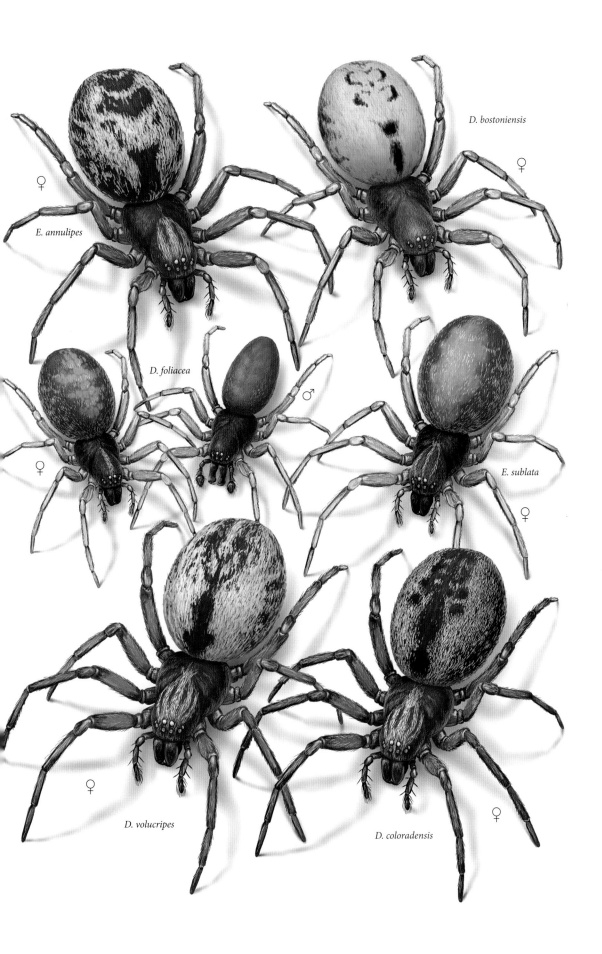

E. annulipes ♀

D. bostoniensis ♀

D. foliacea ♀ ♂

E. sublata ♀

D. volucripes ♀

D. coloradensis ♀

PLATE 22

SPACE-FILLING WEB BUILDERS: MESHWEAVERS; *DIGUETIA*

Diguetia canities, adult female
  Female TBL 5.0–10.0 mm, Male TBL 5.0–6.8 mm (see page 122)

Mallos pallidus, adult female
  Female TBL 2.3–5.0 mm, Male TBL 3.0 mm (see page 120)

Mexitlia trivittata, adult female
  Female TBL 4.8–8.0 mm, Male TBL 4.0–5.7 mm (see page 120)

Nigma linsdalei, adult male, adult female
  Female TBL 3.0 mm, Male TBL 2.8 mm (see page 120)

Phantyna bicornis, adult female
  Female TBL 2.6–3.0 mm, Male TBL 2.3–2.5 mm (see page 121)

Saltonia incerta, adult female
  Female TBL 2.5–4.5 mm, Male TBL 3.3 mm (see page 121)

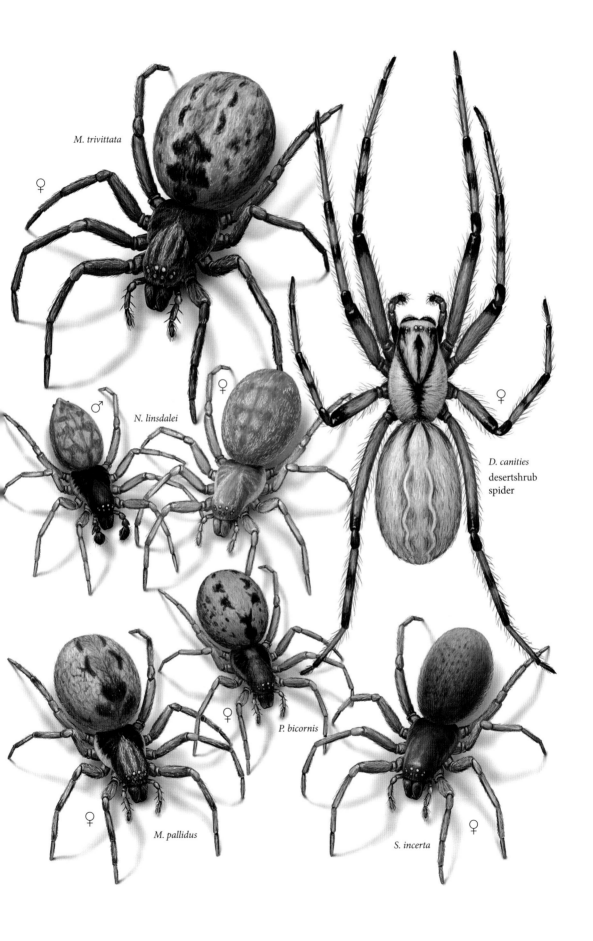

M. trivittata

♀

N. linsdalei

♂

♀

D. canities
desertshrub
spider

♀

P. bicornis

♀

♀

M. pallidus

♀

S. incerta

PLATE 23

SPACE-FILLING WEB BUILDERS: CELLAR SPIDERS

Some of these are illustrated with abbreviated legs to permit room for larger view of the body. All of the spiders in this family have long thin legs, similar to those shown for *Pholcus*.

Artema atlanta, adult female
   Female TBL 5.0–10.0 mm, Male TBL 9.5 mm (see page 176)

Crossopriza lyoni, adult female
   Female TBL 5–9 mm, Male TBL 4–8 mm (see page 177)

Holocnemus pluchei, marbled cellar spider, adult female
   Female TBL 5–9 mm, Male TBL 5–7 mm (see page 177)

Pholcophora americana, adult female
   Female TBL 2.1–2.6 mm, Male TBL 2.2 mm (see page 177)

Pholcus phalangioides, longbodied cellar spider, adult male, adult female
   Female TBL 7.0–9.0 mm, Male TBL 6.0–7.5 mm (see page 178)

Physocyclus californicus, adult female
   Female TBL 5.3–7.0 mm, Male TBL 4.8–6.0 mm (see page 178)

Psilochorus hesperus, adult female
   Female TBL 2.9–3.7 mm, Male TBL 1.8–3.1 mm (see page 178)

Spermophora senoculata, shortbodied cellar spider, adult female
   Female TBL 2.0 mm, Male TBL 1.6 mm (see page 179)

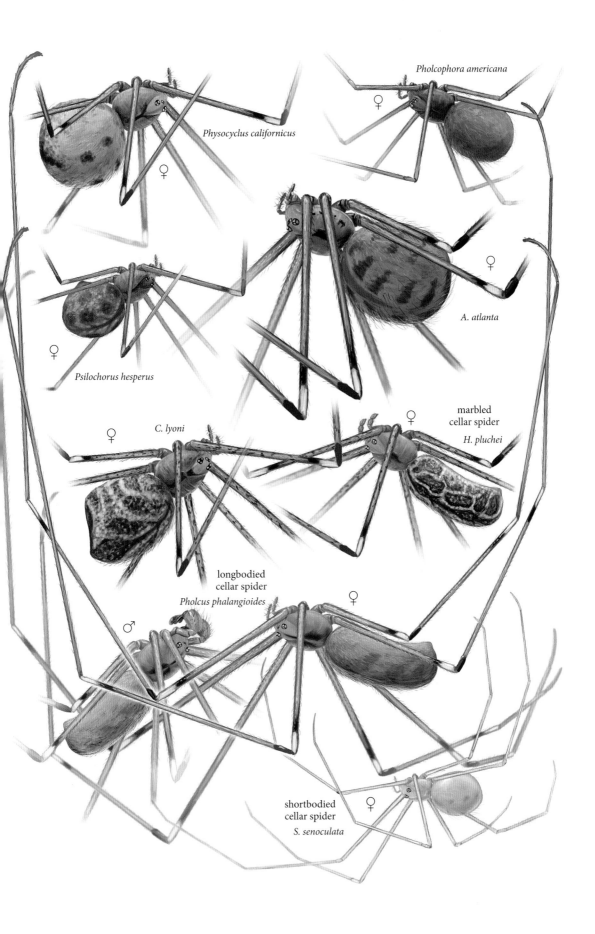

Physocyclus californicus

♀

Pholcophora americana

♀

♀

A. atlanta

♀

Psilochorus hesperus

♀

marbled
cellar spider

H. pluchei

♀

C. lyoni

♀

longbodied
cellar spider

Pholcus phalangioides

♂

♀

shortbodied
cellar spider

♀

S. senoculata

PLATE 24

SPACE-FILLING WEB BUILDERS: COBWEB WEAVERS WITH TAN OR BROWN TEARDROP-SHAPED ABDOMENS

Cryptachaea porteri, adult female
Female TBL 2.2–4.9 mm, Male TBL 1.6–2.8 mm (see page 225)

Eidmannella pallida, adult female
Female TBL 3.0–4.0 mm, Male TBL 3.5 mm (see page 169)

Hentziectypus globosus, adult female, adult male
Female TBL 1.6–2.0 mm, Male TBL 1.2–1.8 mm (see page 227)

Nesticus silvestrii, adult female
Female TBL 3.7–5.0 mm, Male TBL 4.0 mm (see page 170)

Parasteatoda tabulata, adult female
Female TBL 5.9 mm, Male TBL 4.1 mm (see page 229)

Parasteatoda tepidariorum, common house spider, adult female, adult male
Female TBL 5.0–7.0 mm, Male TBL 3.8–4.7 mm (see page 230)

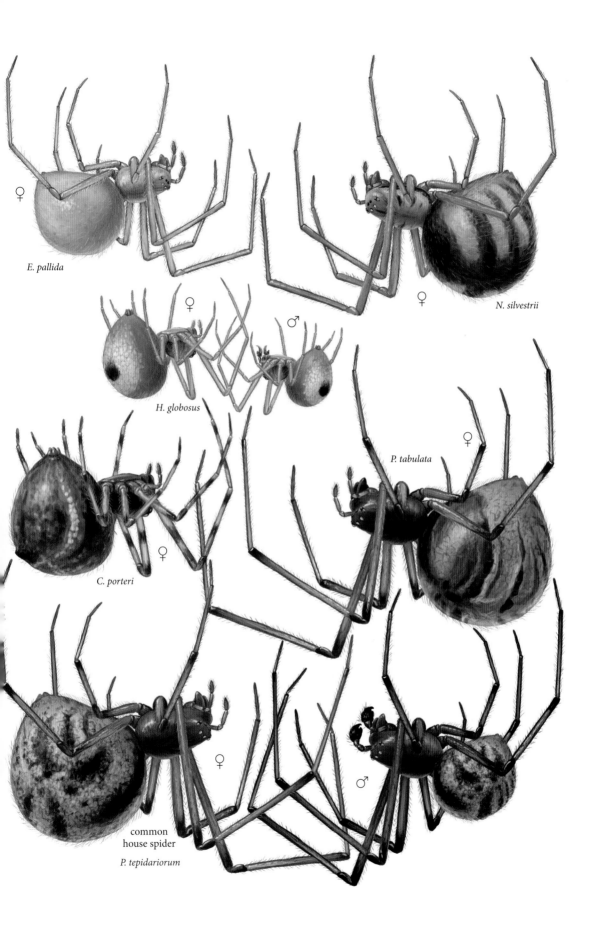

E. pallida

♀

N. silvestrii

♀

♀

♂

H. globosus

P. tabulata

♀

C. porteri

♀

common
house spider

P. tepidariorum

♀

♂

PLATE 25

SPACE-FILLING WEB BUILDERS: COBWEB WEAVERS
WITH UNUSUALLY SHAPED ABDOMENS

Argyrodes elevatus, dewdrop spider, adult female
   Female TBL 3.4–5.2 mm, Male TBL 3.3–4.0 mm (see page 224)

Episinus amoenus, adult female
   Female TBL 3.0–4.5 mm, Male TBL 3.0 mm (see page 226)

Euryopis funebris, adult female
   Female TBL 3–4 mm, Male TBL 3 mm (see page 227)

Neospintharus trigonum, adult female, adult male
   Female TBL 3.7–4.2 mm, Male TBL 2.5–3.3 mm (see page 229)

Rhomphaea fictilium, adult female
   Female TBL 5.0–10.5 mm, Male TBL 4.0–7.0 mm (see page 231)

Spintharus flavidus, adult female
   Female TBL 4.0–4.5 mm, Male TBL 2.8 mm (see page 232)

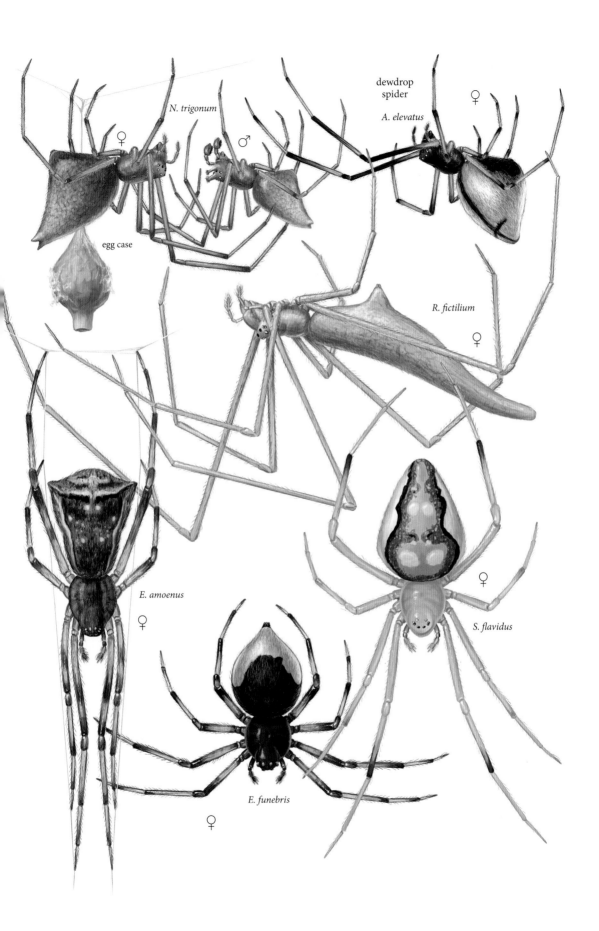

N. trigonum
♀
♂

dewdrop
spider
A. elevatus
♀

egg case

R. fictilium
♀

E. amoenus
♀

S. flavidus
♀

E. funebris
♀

PLATE 26

SPACE-FILLING WEB BUILDERS: COBWEB WEAVERS
WITH OVAL ABDOMENS AND WIDOWS

Enoplognatha marmorata, marbled cobweb spider, adult female
   Female TBL 6–7 mm, Male TBL 5–6 mm (see page 226)

Enoplognatha ovata, adult females, plain, banded, and pink–backed forms
   Female TBL 4.3–7.0 mm, Male TBL 3.5–5.2 mm (see page 226)

Latrodectus bishopi, red widow, adult female
   Female TBL 7.8–9.0 mm, Male TBL 3.0–3.8 mm (see page 228)

Latrodectus geometricus, brown widow, adult female, adult male
   Female TBL 5–10 mm, Male TBL 3–5 mm (see page 228)

Latrodectus mactans, southern black widow, adult male, adult female typical form
   and spotted form (abdomen from above)
   Female TBL 8.0–10.0 mm, Male TBL 3.2–4.0 mm (see page 228)

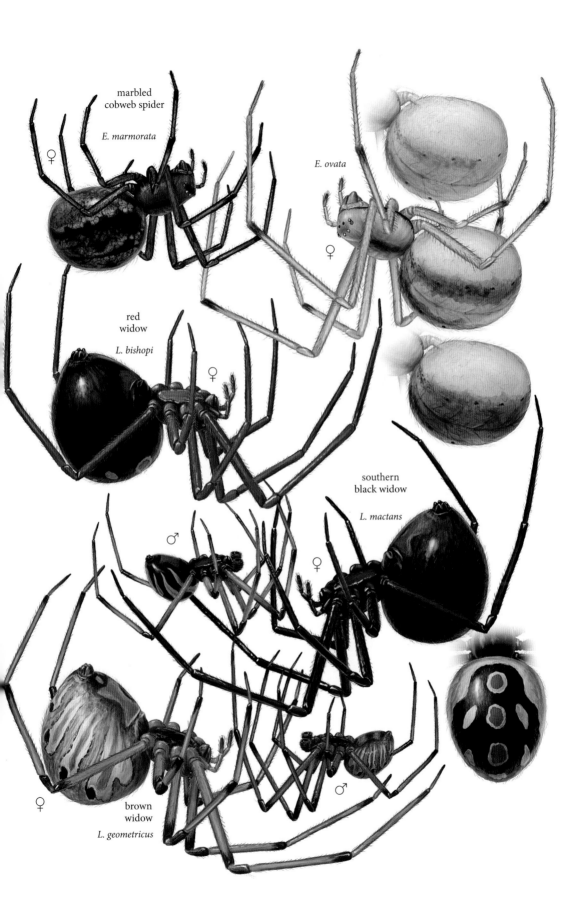

marbled
cobweb spider

E. marmorata

♀

E. ovata

♀

red
widow

L. bishopi

♀

southern
black widow

L. mactans

♂

♀

♀

brown
widow

L. geometricus

♂

Asagena americana, twospotted cobweb weaver, adult female, adult male
Female TBL 3.5–4.7 mm, Male TBL 3.2–4.4 mm (see page 224)

Asagena fulva, adult female
Female TBL 3.0–5.9 mm, Male TBL 2.4–5.0 mm (see page 224)

Steatoda albomaculata, adult female
Female TBL 4.0–8.0 mm, Male TBL 4.3–6.8 mm (see page 232)

Steatoda borealis, adult female, adult male
Female TBL 6.0–7.0 mm, Male TBL 4.7–6.0 mm (see page 233)

Steatoda grossa, false black widow, adult female
Female TBL 5.9–10.5 mm, Male TBL 4.1–7.2 mm (see page 233)

Steatoda triangulosa, checkered cobweb weaver, adult female
Female TBL 3.6–5.9 mm, Male TBL 3.5–4.7 mm (see page 233)

false black
widow

♀ *S. grossa*

A. fulva

♀

S. albomaculata

♀

checkered
cobweb weaver

S. triangulosa

♀

twospotted
cobweb
weaver

A. americana

♀

♂

♀

S. borealis

♂

PLATE 28

SPACE-FILLING WEB BUILDERS: COBWEB WEAVERS WITH ABDOMENS MARKED WITH A FOLIUM OR LINES

Anelosimus studiosus, adult female
Female TBL 3.2–4.7 mm, Male TBL 2.1–2.3 mm (see page 223)

Canalidion montanum, adult female
Female TBL 2.7–4.4 mm, Male TBL 2.2–3.8 mm (see page 225)

Platnickina alabamensis, adult female
Female TBL 1.9–3.7 mm, Male TBL 1.8–2.7 mm (see page 231)

Theridion differens, adult female
Female TBL 1.8–3.5 mm, Male TBL 1.6–2.5 mm (see page 234)

Theridion murarium, adult female
Female TBL 2.8–4.0 mm, Male TBL 2.7–3.0 mm (see page 234)

Tidarren sisyphoides, adult male, adult female
Female TBL 5.8–8.6 mm, Male TBL 1.2–1.4 mm (see page 236)

Yunohamella lyrica, adult female
Female TBL 3.0–3.5 mm, Male TBL 2.1–2.8 mm (see page 236)

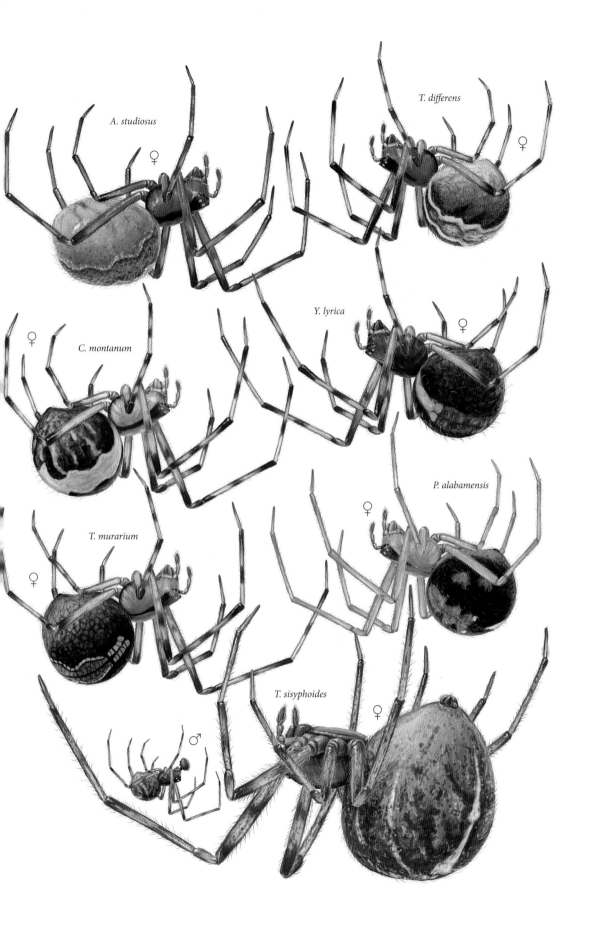

A. studiosus ♀

T. differens ♀

C. montanum ♀

Y. lyrica ♀

T. murarium ♀

P. alabamensis ♀

T. sisyphoides ♀ ♂

Phoroncidia americana, adult female
  Female TBL 1.5–1.8 mm, Male TBL 1.5–1.6 mm (see page 230)

Theridion frondeum, adult male, adult females, three color forms
  Female TBL 3.0–4.2 mm, Male TBL 3.0–3.5 mm (see page 234)

Theridula emertoni, adult female, adult male
  Female TBL 2.3–2.8 mm, Male TBL 2.0–2.3 mm (see page 235)

Theridula gonygaster, adult female
  Female TBL 1.7–2.5 mm, Male TBL 1.3–1.7 mm (see page 235)

Theridula opulenta, adult female
  Female TBL 2.3–2.8 mm, Male TBL 2.0–2.3 mm (see page 235)

T. frondeum

♂

♀

T. gonygaster

♀

T. emertoni

♀

♂

T. opulenta

♀

P. americana

♀

PLATE 30

FUNNEL-SHAPED WEB BUILDERS: FUNNEL WEAVERS

Agelenopsis aperta, adult female
  Female TBL 11–19 mm, Male TBL 10–19 mm (see page 65)

Agelenopsis pennsylvanica, grass spider, adult male, adult female
  Female TBL 10–17 mm, Male TBL 9–12 mm (see page 65)

Calilena arizonica, adult female
  Female TBL 7.3–11.3 mm, Male TBL 7.0–9.7 mm (see page 66)

Hololena curta, adult female
  Female TBL 10–12 mm, Male TBL 9–10 mm (see page 67)

Melpomene rita, adult female
  Female TBL 7.0–8.5 mm, Male TBL 7.0–8.5 mm (see page 67)

Novalena intermedia, adult female
  Female TBL 9.0–10.0 mm, Male TBL 7.5–9.0 mm (see page 67)

Rualena cockerelli, adult female
  Female TBL 6.5–9.3 mm, Male TBL 6.0 mm (see page 68)

R. cockerelli

♀

A. aperta

♀

M. rita

♀

N. intermedia

♀

A. pennsylvanica
grass
spider

♀

♂

C. arizonica

♀

H. curta

♀

PLATE 31

FUNNEL-SHAPED WEB BUILDERS: FUNNEL WEAVERS
AND HACKLEDMESH WEAVERS

Coras juvenilis, adult female, adult male
  Female TBL 6.7–11.0 mm, Male TBL 6.1–9.2 mm (see page 66)

Coras montanus, adult female
  Female TBL 9.7–13.1 mm, Male TBL 8.2–9.7 mm (see page 66)

Cybaeopsis tibialis, adult female
  Female TBL 5.0–9.0 mm, Male TBL 5.5–6.5 mm (see page 70)

Pimus fractus, adult female
  Female TBL 5.0–6.5 mm, Male TBL 5.0–5.5 mm (see page 71)

Tegenaria agrestis, hobo spider, adult female, adult male
  Female TBL 10–15 mm, Male TBL 7–10 mm (see page 68)

Tegenaria domestica, barn funnel weaver, adult female
  Female TBL 7.5–11.5 mm, Male TBL 6.0–9.0 mm (see page 68)

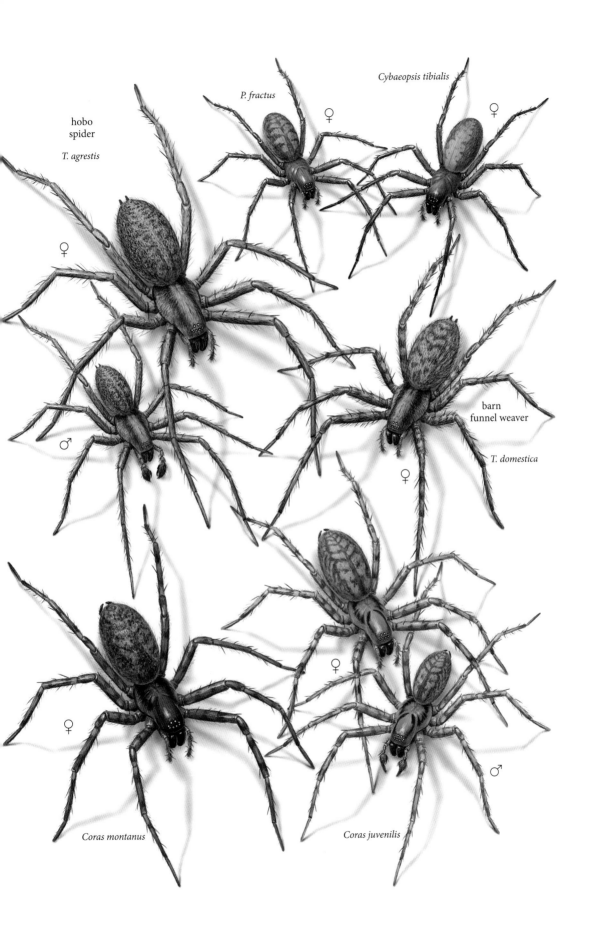

hobo
spider

T. agrestis

P. fractus

♀

Cybaeopsis tibialis

♀

♀

♂

barn
funnel weaver

♀

T. domestica

♀

Coras montanus

♀

♂

Coras juvenilis

PLATE 32

FUNNEL-SHAPED WEB BUILDERS: HACKLEDMESH WEAVERS, *SOSIPPUS*, AND *WADOTES*

Amaurobius borealis, adult female, adult male
  Female TBL 4.5–6.0 mm, Male TBL 3.5–5.0 mm (see page 69)

Amaurobius ferox, adult male
  Female TBL 8.5–14.0 mm, Male TBL 8.0–12.5 mm (see page 69)

Arctobius agelenoides, adult female
  Female TBL 7–10 mm, Male TBL 6–8 mm (see page 70)

Callobius bennetti, adult female
  Female TBL 5–12 mm, Male TBL 5–9 mm (see page 70)

Sosippus is a wolf spider that spins a funnel-shaped web.

Sosippus mimus, adult female, adult male
  Female TBL 12.9–18.2 mm, Male TBL 13.1–14.2 mm (see page 162)

Wadotes calcaratus, adult female, adult male
  Female TBL 6.0–11.0 mm, Male TBL 6.0–8.7 mm (see page 69)

C. bennetti

♀

Acrtobius agelenoides

♀

♂

Amaurobius ferox

♂

W. calcaratus

♀

♀

♂

Amaurobius borealis

♀

S. mimus

♂

These spiders build atypical funnel-shaped webs.

Ariadna bicolor, tubeweb spider, adult female
Female TBL 6.1–15.0 mm, Male TBL 5.4–7.2 mm (see page 207)

Badumna longinqua, adult female
Female TBL 10–18 mm, Male TBL 8–10 mm (see page 115)

Calymmaria persica, adult female
Female TBL 4.0–9.7 mm, Male TBL 6.1–7.3 mm (see page 133)

Cybaeus giganteus, adult female
Female TBL 9–12 mm, Male TBL 8–10 mm (see page 113)

Cybaeus reticulatus, adult female, adult male
Female TBL 9–10 mm, Male TBL 7–8 mm (see page 113)

Segestria pacifica, western tubeweb spider, adult female
Female TBL 8–16 mm, Male TBL 7–10 mm (see page 208)

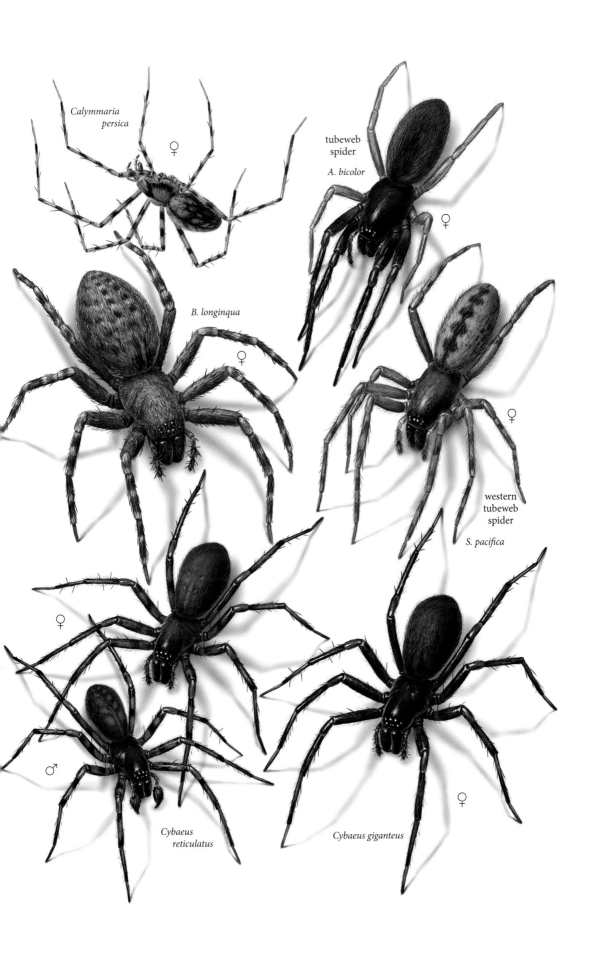

Calymmaria persica ♀

tubeweb spider
A. bicolor ♀

B. longinqua ♀

western tubeweb spider

S. pacifica ♀

♀

Cybaeus reticulatus ♂

Cybaeus giganteus ♀

Kibramoa madrona, adult female
    Female TBL 10 mm, Male TBL 8 mm (see page 183)

Metaltella simoni, adult male, adult female
    Female TBL 7–10 mm, Male TBL 6–9 mm (see page 71)

Plectreurys tristis, adult female, adult male
    Female TBL 12.5–17.0 mm, Male TBL 11.5–17.0 mm (see page 183)

Titanoeca nigrella, adult female
    Female TBL 5.0–8.0 mm, Male TBL 3.5–7.0 mm (see page 244)

T. nigrella
♀

♂

M. simoni
♀

K. madrona
♀

♀

P. tristis
♂

PLATE 35

SPIDERS FOUND UNDER DEBRIS: PALE OR VERY SMALL SPIDERS FOUND UNDER DEBRIS; *ZOROCRATES*

Blabomma californicum, adult female
   Female TBL 7.0 mm, Male TBL 5.5 mm (see page 117)

Cicurina arcuata, adult female, adult male
   Female TBL 4.7–7.0 mm, Male TBL 4.3–5.7 mm (see page 117)

Neoantistea agilis, adult female
   Female TBL 2.6–3.2 mm, Male TBL 2.5–3.0 mm (see page 134)

Yorima angelica, adult female
   Female TBL 3.3–5.5 mm, Male TBL 2.8–4.1 mm (see page 121)

Zorocrates unicolor, adult female
   Female TBL 15 mm, Male TBL 12 mm (see page 250)

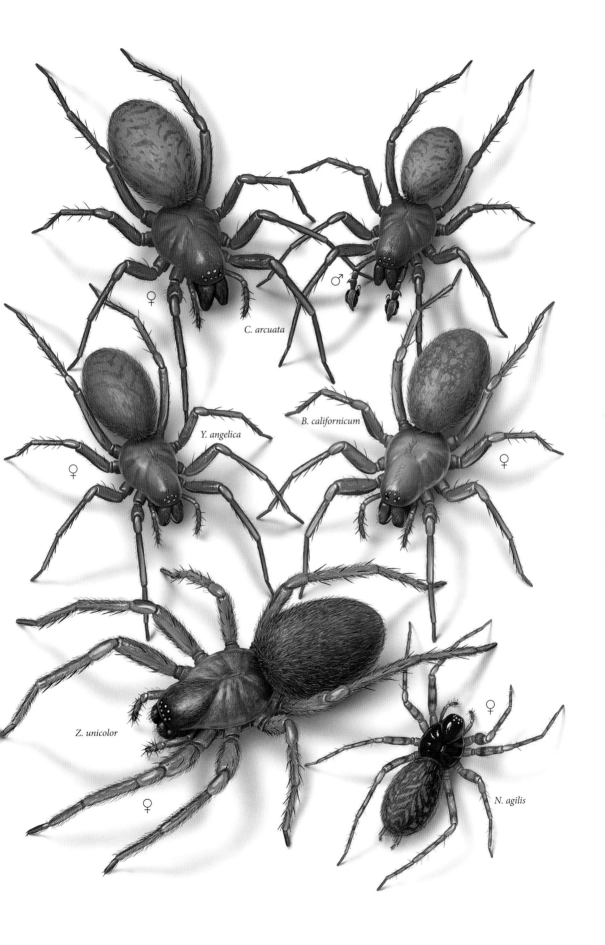

C. arcuata

Y. angelica

B. californicum

Z. unicolor

N. agilis

PLATE 36

FOLIAGE HUNTERS: GHOST SPIDERS

Anyphaena maculata, adult female
Female TBL 4.5–5.5 mm, Male TBL 3.8–4.7 mm (see page 74)

Anyphaena pectorosa, adult female, adult male
Female TBL 5.0–5.5 mm, Male TBL 4.7–5.0 mm (see page 74)

Arachosia cubana, adult female
Female TBL 5.9 mm, Male TBL 5.2 mm (see page 75)

Hibana gracilis, garden ghost spider, adult female
Female TBL 6.4–7.0 mm, Male TBL 5.7–6.5 mm (see page 75)

Hibana velox, adult female
Female TBL 8.4 mm, Male TBL 7.3 mm (see page 75)

Lupettiana mordax, adult female
Female TBL 3.0–5.5 mm, Male TBL 3.7–5.0 mm (see page 75)

Pippuhana calcar, adult male
Female TBL 5.0 mm, Male TBL 3.2 mm (see page 76)

Wulfila saltabundus, adult female
Female TBL 3.7–4.2 mm, Male TBL 2.9–3.5 mm (see page 76)

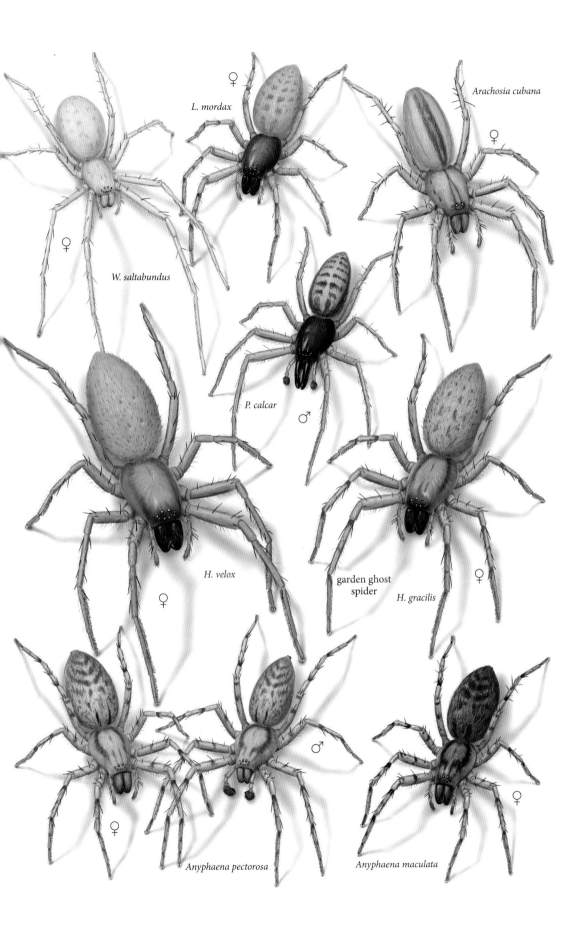

♀

L. mordax

♀

Arachosia cubana

♀

W. saltabundus

P. calcar

♂

H. velox

♀

garden ghost
spider

H. gracilis

♀

♀

♂

Anyphaena pectorosa

Anyphaena maculata

♀

PLATE 37

FOLIAGE HUNTERS: SAC SPIDERS; *CHEIRACANTHIUM* AND *ZORA*

Cheiracanthium inclusum, agrarian sac spider, adult female
   Female TBL 4.9–9.7 mm, Male TBL 4–7.7 mm (see page 166)

Clubiona abboti, adult female, adult male
   Female TBL 4.0–5.4 mm, Male TBL 3.7–4.4 mm (see page 105)

Clubiona canadensis, adult female
   Female TBL 6.0–11.5 mm, Male TBL 4.8–8.0 mm (see page 105)

Clubiona kastoni, adult female
   Female TBL 4.3–5.4 mm, Male TBL 3.3–4.4 mm (see page 105)

Clubiona riparia, adult female, adult male
   Female TBL 5.4–8.7 mm, Male TBL 4.4–7.4 mm (see page 106)

Elaver excepta, adult female
   Female TBL 6.7–7.4 mm, Male TBL 4.5–6.5 mm (see page 106)

Zora pumila, adult female
   Female TBL 5.0–6.0 mm, Male TBL 3.5–4.0 mm (see page 249)

E. excepta ♀

♀

agrarian
sac spider

*Cheiracanthium
inclusum*

Z. pumila ♀

*Clubiona
kastoni* ♀

Clubiona riparia ♀

♂

♀

*Clubiona
abboti* ♂

♀

Clubiona canadensis

PLATE 38

FOLIAGE HUNTERS: LYNX SPIDERS AND PIRATE SPIDERS

Ero canionis, adult female
    Female TBL 2.7–3.4 mm, Male TBL 2.2–2.5 mm (see page 165)

Hamataliwa grisea, bark lynx spider, adult female
    Female TBL 8.7–10.9 mm, Male TBL 8.4–11.1 mm (see page 172)

Mimetus puritanus, adult female
    Female TBL 5.0–5.6 mm, Male TBL 4.0–4.5 mm (see page 165)

Oxyopes salticus, striped lynx spider, adult female, adult male
    Female TBL 5.7–6.7 mm, Male TBL 4.0–4.5 mm (see page 173)

Oxyopes scalaris, western lynx spider, adult female, adult male
    Female TBL 7–8 mm, Male TBL 5 mm (see page 173)

Peucetia viridans, green lynx spider, adult male, adult female
    Female TBL 14–16 mm, Male TBL 12–13 mm (see page 173)

Reo eutypus, adult female
    Female TBL 4–5 mm, Male TBL 3–4 mm (see page 166)

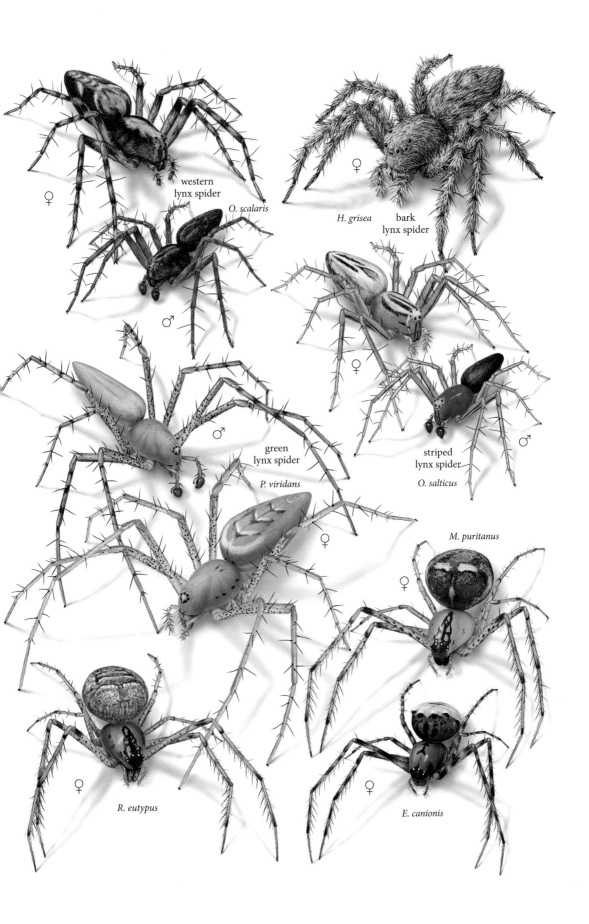

western
lynx spider

O. scalaris

♀

♂

♀

H. grisea　bark
lynx spider

♀

green
lynx spider

P. viridans

♂

striped
lynx spider

O. salticus

♂

♀

M. puritanus

♀

R. eutypus

♀

E. canionis

♀

PLATE 39

GROUND HUNTERS: WANDERING SPIDERS; *MERIOLA*, AND *TRACHELAS*

Anahita punctulata, adult female, adult male
    Female TBL 7–9 mm, Male TBL 6–8 mm (see page 111)

Ctenus hibernalis, adult female
    Female TBL 16–18 mm, Male TBL 13 mm (see page 111)

Meriola decepta, adult female, adult male
    Female TBL 3.4–4.2 mm, Male TBL 3.1–4.1 mm (see page 109)

Trachelas tranquillus, bullheaded sac spider, adult female, adult male
    Female TBL 6.0–7.7 mm, Male TBL 5.3–6.5 mm (see page 110)

M. decepta ♀ ♂

bullheaded
sac spider

T. tranquillus ♂

♀

A. punctulata ♂

♀

C. hibernalis ♀

PLATE 40
GROUND HUNTERS: ANTMIMIC SPIDERS

Castianeira amoena, orange antmimic, adult female, adult male
Female TBL 7.0–8.8 mm, Male TBL 5.7–6.8 mm (see page 107)

Castianeira cingulata, twobanded antmimic, adult female, adult male
Female TBL 6.7–8.0 mm, Male TBL 6.4–7.0 mm (see page 107)

Castianeira descripta, redspotted antmimic, adult female
Female TBL 8.0–10.0 mm, Male TBL 6.2–7.6 mm (see page 107)

Castianeira longipalpata, manybanded antmimic, adult female
Female TBL 7.0–9.0 mm, Male TBL 5.5–6.0 mm (see page 108)

Castianeira occidens, adult female
Female TBL 7.1–10.4 mm, Male TBL 6.7–7.8 mm (see page 108)

Castianeira variata, adult female
Female TBL 9.3 mm, Male TBL 5.7–5.8 mm (see page 108)

Drassinella gertschi, adult female
Female TBL 3.2 mm, Male TBL 3.4 mm (see page 108)

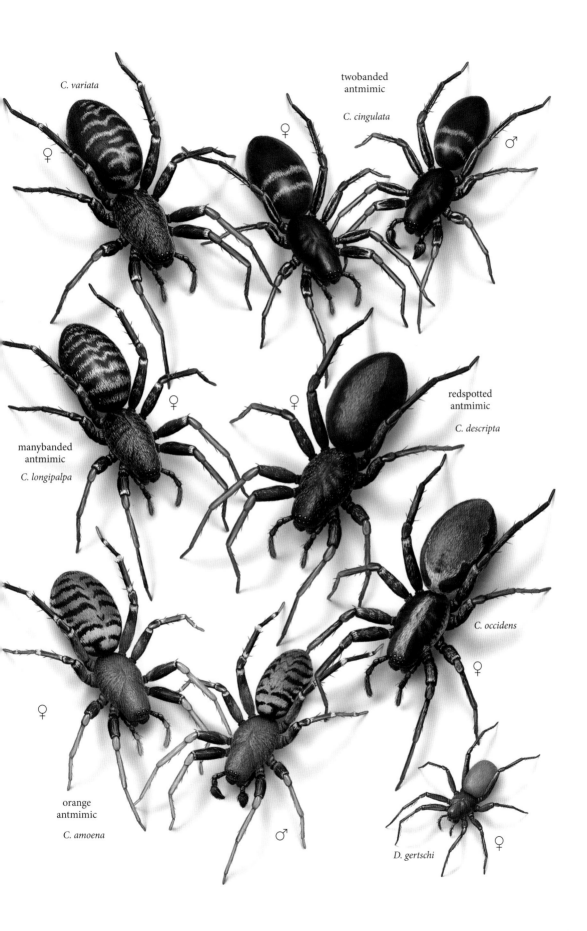

C. variata ♀

twobanded
antmimic

C. cingulata ♀ ♂

manybanded
antmimic

C. longipalpa ♀ ♀

redspotted
antmimic

C. descripta

C. occidens ♀

orange
antmimic

C. amoena ♀ ♂

D. gertschi ♀

PLATE 41

GROUND HUNTERS: DARK GROUND SPIDERS
WITH A TAN OR RED CEPHALOTHORAX

Callilepis imbecilla, adult female
   Female TBL 3.7–5.6 mm, Male TBL 3.0–3.6 mm (see page 125)

Drassyllus depressus, adult female, adult male
   Female TBL 4.7–6.0 mm, Male TBL 5.0 mm (see page 126)

Drassyllus insularis, adult male
   Female TBL 5.0–6.5 mm, Male TBL 4.8–5.0 mm (see page 127)

Gnaphosa fontinalis, adult female orange form, adult female dark form
   Female TBL 6–10 mm, Male TBL 6–7 mm (see page 127)

Haplodrassus signifer, adult female
   Female TBL 7.5–11.0 mm, Male TBL 7.0–9.0 mm (see page 128)

Scotophaeus blackwalli, mouse spider, adult female
   Female TBL 8.9–12.0 mm, Male TBL 8.0–9.0 mm (see page 130)

♀

G. fontinalis

H. signifer

♀

C. imbecilla

♀

♀

♂

D. depressus

mouse
spider

S. blackwalli

♂

♀

D. insularis

PLATE 42

GROUND HUNTERS: BOLD-PATTERNED GROUND SPIDERS

Cesonia bilineata, adult female
 Female TBL 4.3–7.0 mm, Male TBL 3.5–4.4 mm (see page 125)

Cesonia josephus, adult female
 Female TBL 5.2–7.5 mm, Male TBL 4.0–4.8 mm (see page 126)

Micaria longipes, adult female
 Female TBL 5–6 mm, Male TBL 5 mm (see page 129)

Micaria pulicaria, adult female, adult male
 Female TBL 3.4–4.0 mm, Male TBL 2.4–4.0 mm (see page 129)

Sergiolus capulatus, adult female
 Female TBL 6.0–10.0 mm, Male TBL 5.5–7.0 mm (see page 131)

Sergiolus montanus, adult female
 Female TBL 7.3–8.6 mm, Male TBL 5.8 mm (see page 131)

Sosticus insularis, adult female
 Female TBL 5.7–7.7 mm, Male TBL 4.8–6.0 mm (see page 131)

M. longipes ♀

M. pulicaria ♀

♂

*Sosticus
insularis* ♀

*Sergiolus
capulatus* ♀

Sergiolus montanus ♀

C. bilineata ♀

C. josephus ♀

PLATE 43

GROUND HUNTERS: PALE GROUND SPIDERS

Drassodes auriculoides, adult female
Female TBL 8.2–11.0 mm, Male TBL 7.3–11.0 mm (see page 126)

Gnaphosa muscorum, adult female
Female TBL 7.9–15.0 mm, Male TBL 6.8–12.0 mm (see page 127)

Litopyllus temporarius, adult female
Female TBL 5.0–7.1 mm, Male TBL 4.0–5.5 mm (see page 129)

Orodrassus coloradensis, adult female
Female TBL 8.1–9.8 mm, Male TBL 7.5–8.6 mm (see page 130)

Scopoides catharius, adult female
Female TBL 5.7–7.7 mm, Male TBL 4.7–6.5 mm (see page 130)

Talanites echinus, adult female, adult male
Female TBL 4.8–5.9 mm, Male TBL 4.5–5.2 mm (see page 132)

♂ *T. echinus* ♀

D. auriculoides ♀

G. muscorum ♀

L. temporarius ♀

O. coloradensis ♀

S. catharius ♀

PLATE 44

GROUND HUNTERS: RED AND GRAY OR BLACK GROUND SPIDERS

Herpyllus ecclesiasticus, parson spider, adult female
  Female TBL 6.4–9.0 mm, Male TBL 4.8–6.1 mm (see page 128)

Herpyllus hesperolus, adult female
  Female TBL 7.4–10.2 mm, Male TBL 5.9–7.2 mm (see page 128)

Nodocion voluntarius, adult female
  Female TBL 6.0–7.8 mm, Male TBL 5.1–6.5 mm (see page 130)

Trachyzelotes lyonneti, adult female
  Female TBL 6.3–7.9 mm, Male TBL 4.3–5.0 mm (see page 132)

Urozelotes rusticus, adult female
  Female TBL 5.9–7.4 mm, Male TBL 5.8–7.3 mm (see page 132)

Zelotes fratris, adult female
  Female TBL 6.3–7.7 mm, Male TBL 5.5–6.6 mm (see page 132)

parson
spider
H. ecclesiasticus   ♀

H. hesperolus   ♀

N. voluntarius

♀   *T. lyonneti*

♀

U. rusticus   ♀   *Z. fratris*

Agroeca pratensis, adult female
  Female TBL 7 mm, Male TBL 5.5 mm (see page 146)

Neoanagraphis chamberlini, adult female
  Female TBL 5.5–9.7 mm, Male TBL 3.8–9.1 mm (see page 147)

Neozimiris pubescens, adult female
  Female TBL 1.7–4.0 mm, Male TBL 1.7–4.0 mm (see page 184)

Prodidomus rufus, adult female
  Female TBL 3.0–5.5 mm, Male TBL 3.0–5.5 mm (see page 184)

Strotarchus piscatorius, adult female, adult male
  Female TBL 7.6–9.2 mm, Male TBL 7.2–8.7 mm (see page 167)

Syspira longipes, adult female
  Female TBL 8.5–16.5 mm, Male TBL 7.4–12.5 mm (see page 167)

Teminius affinis, adult female
  Female TBL 10 mm, Male TBL 9 mm (see page 168)

S. longipes ♀

S. piscatorius ♀

T. affinis ♀

♂

P. rufus ♀

N. pubescens ♀

N. chamberlini ♀

A. pratensis ♀

PLATE 46

GROUND HUNTERS: TENGELIDS, DUSTY DESERT SPIDER, AND *ZOROPSIS*

Anachemmis sober, adult female
    Female TBL 10.3 mm, Male TBL 8.0 mm (see page 214)

Homalonychus selenopoides, adult female
    Female TBL 8–9 mm, Male TBL 7 mm (see page 135)

Lauricius hooki, adult female, adult male
    Female TBL 10.7–18.0 mm, Male TBL 10.1–10.8 mm (see page 214)

Liocranoides tennesseensis, adult female
    Female TBL 9.1 mm, Male TBL 8.0 mm (see page 214)

Socalchemmis dolichopus, adult female
    Female TBL 7.5 mm, Male TBL 8.1 mm (see page 215)

Titiotus californicus, adult female
    Female TBL 12 mm, Male TBL 10 mm (see page 215)

Zoropsis spinimana, adult female
    Female TBL 15–17 mm, Male TBL 10 mm (see page 250)

L. tennesseensis
♀

Z. spinimana
♀

S. dolichopus
♀

A. sober
♀

T. californicus
♀

H. selenopoides
♀

L. hooki
♀

♂

Deinopis spinosa, netcasting spider, adult female
   Female TBL 12–17 mm, Male TBL 10–14 mm (see page 115)

Lutica maculata, adult female
   Female TBL 10–14 mm, Male TBL 7–10 mm (see page 248)

Neotama mexicana, adult female
   Female TBL 6.5–11.8 mm, Male TBL 6.1–8.2 mm; lengths excluding spinnerets (see page 135)

Paratheuma insulana, adult male, adult female
   Female TBL 3.3–5.9 mm, Male TBL 3.5–5.0 mm (see page 116)

Zodarion rubidum, adult female
   Female TBL 2.3–2.9 mm, Male TBL 2.1–2.6 mm (see page 248)

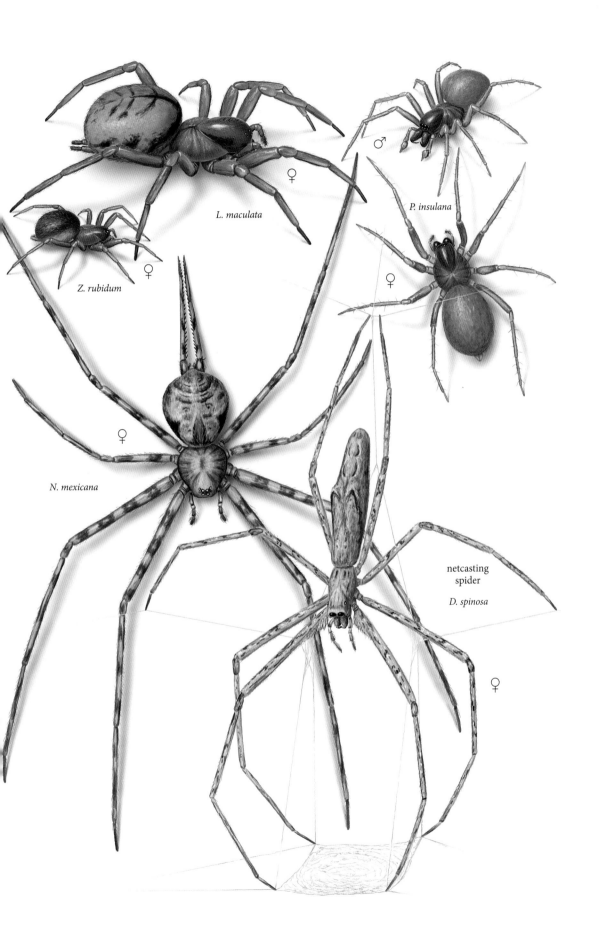

L. maculata ♀

P. insulana ♂

Z. rubidum ♀

♀

N. mexicana ♀

netcasting
spider

D. spinosa

♀

PLATE 48

GROUND HUNTERS: BURROWING WOLF SPIDERS

Geolycosa missouriensis, adult female
    Female TBL 15.8–21.0 mm, Male TBL 15.0–16.3 mm (see page 150)

Geolycosa pikei, adult female
    Female TBL 18–22 mm, Male TBL 14 mm (see page 150)

Geolycosa rafaelana, adult female, adult male
    Female TBL 17 mm, Male TBL 15 mm (see page 151)

Geolycosa turricola, adult female
    Female TBL 22 mm, Male TBL 21 mm (see page 151)

Geolycosa wrighti, adult female
    Female TBL 15.0–17.4 mm, Male TBL 14–16 mm (see page 151)

Geolycosa xera, adult female, adult male
    Female TBL 17–20 mm, Male TBL 15 mm (see page 151)

G. rafaelana ♀

G. pikei ♀

♂

G. missouriensis ♀

G. wrighti ♀

G. xera ♂

G. turricola ♀

♀

PLATE 49

GROUND HUNTERS: SMALL AND TINY WOLF SPIDERS

Allocosa funerea, adult male, adult female
Female TBL 5.0–6.0 mm, Male TBL 3.5–4.5 mm (see page 148)

Hesperocosa unica, adult female
Female TBL 5.9 mm, Male TBL 4.0 mm (see page 153)

Pirata insularis, adult male, adult female
Female TBL 4.0–5.3 mm, Male TBL 4.5–5.0 mm (see page 158)

Pirata minutus, adult male, adult female
Female TBL 3.3–3.7 mm, Male TBL 2.7–3.0 mm (see page 159)

Pirata piraticus, adult female, adult male
Female TBL 5.8–7.5 mm, Male TBL 5.5–6.2 mm (see page 159)

Trabeops aurantiacus, adult male, adult female
Female TBL 3.0–3.5 mm, Male TBL 2.5–3.0 mm (see page 163)

H. unica

♀

A. funerea

♂

♀

T. aurantiacus

♀

♂

P. minutus

♂

♀

P. insularis

♂

♀

♀

P. piraticus

♂

Pardosa distincta, adult female, adult male
  Female TBL 5–7 mm, Male TBL 4–6 mm (see page 156)

Pardosa sternalis, adult female, adult male
  Female TBL 6.0–7.0 mm, Male TBL 4.6–5.5 mm (see page 158)

Rabidosa punctulata, dotted wolf spider, adult male, female underside of abdomen, adult female
  Female TBL 11–17 mm, Male TBL 13–15 mm (see page 159)

Rabidosa rabida, adult male, adult female
  Female TBL 16–21 mm, Male TBL 11–12 mm (see page 160)

Varacosa avara, adult female, adult male
  Female TBL 8.9–10.3 mm, Male TBL 7.7–8.2 mm (see page 164)

P. sternalis

♀

♂

P. distincta

♀

♂

R. rabida

♂

♀

♀

♂

V. avara

♀

♂

dotted
wolf spider

R. punctulata

PLATE 51

GROUND HUNTERS: WOLF SPIDERS; *PARDOSA*

Pardosa lapidicina, adult male, adult female
Female TBL 7.7–9.3 mm, Male TBL 6.0–7.0 mm (see page 156)

Pardosa milvina, adult female, adult male
Female TBL 5.2–6.5 mm, Male TBL 4.0–4.7 mm (see page 157)

Pardosa moesta, adult male, adult female
Female TBL 5.1–6.1 mm, Male TBL 4.4–5.3 mm (see page 157)

Pardosa saxatilis, adult female, adult male
Female TBL 4.7–5.7 mm, Male TBL 3.8–4.5 mm (see page 157)

Pardosa xerampelina, adult female brown form, adult male gray form
Female TBL 7.4–9.5 mm, Male TBL 5.0–6.0 mm (see page 158)

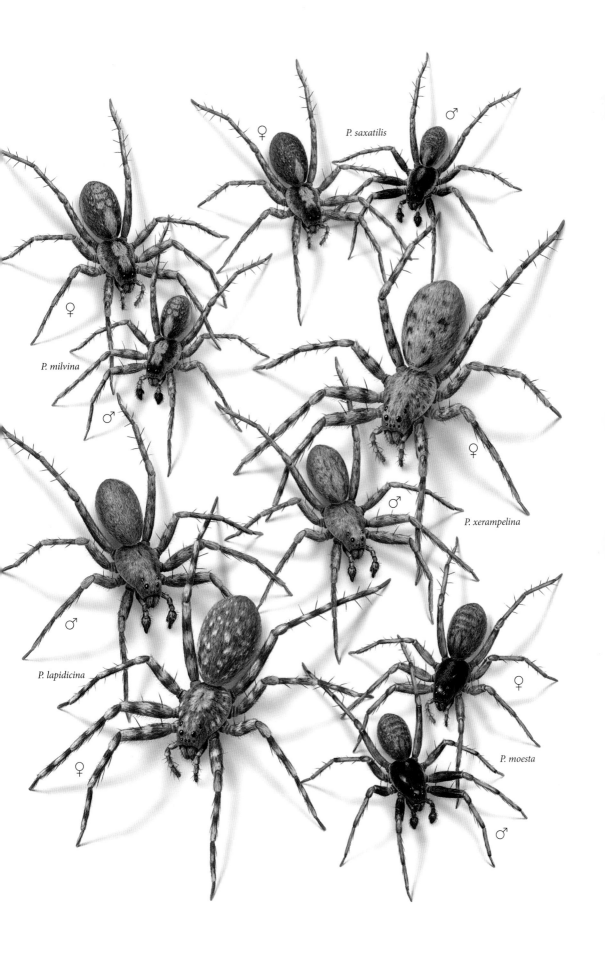

♀

P. saxatilis

♂

♀

P. milvina

♂

♀

P. xerampelina

♂

♂

P. lapidicina

♀

♀

P. moesta

♀

♂

Alopecosa aculeata, adult male, adult female
  Female TBL 8.1–11.3 mm, Male TBL 7.2–9.5 mm (see page 148)

Alopecosa kochi, adult male, adult female
  Female TBL 9.0–16.0 mm, Male TBL 6.6–11.0 mm (see page 149)

Gladicosa gulosa, adult female, adult male
  Female TBL 11–14 mm, Male TBL 10–11 mm (see page 152)

Gladicosa pulchra, adult female, brown and gray forms
  Female TBL 15.3–16.5 mm, Male TBL 12–12.7 mm (see page 152)

Trochosa terricola, adult female, adult male
  Female TBL 9–14 mm, Male TBL 9–12 mm (see page 163)

♂ A. aculeata

♀

A. kochi

♀

♀ T. terricola

♂

♀ G. gulosa

♂

♀ G. pulchra

PLATE 53

GROUND HUNTERS: WOLF SPIDERS; *SCHIZOCOSA*

Schizocosa avida, adult female, adult male
Female TBL 6.6–14.7 mm, Male TBL 6.3–9.8 mm (see page 160)

Schizocosa bilineata, adult male, adult female
Female TBL 5.7–8.7 mm, Male TBL 5.0–6.3 mm (see page 161)

Schizocosa mccooki, adult female, adult male
Female TBL 9.6–22.7 mm, Male TBL 9.1–15.5 mm (see page 161)

Schizocosa ocreata, brushlegged wolf spider, adult female, adult male
Female TBL 7.3–10.4 mm, Male TBL 6.0–10.0 mm (see page 161)

Schizocosa rovneri, adult male
Female TBL 6.0–8.0 mm, Male TBL 6.5–8.1 mm (see page 162)

Schizocosa saltatrix, adult male, adult female
Female TBL 7.0–11.8 mm, Male TBL 6.5–8.2 mm (see page 162)

♂ *S. saltatrix*

♀

♀

S. mccooki

brushlegged
wolf spider

S. ocreata

♂

♂

♀

S. rovneri

♀

S. avida

♂

♂

♂

♀

S. bilineata

PLATE 54

GROUND HUNTERS: WOLF SPIDERS; *ARCTOSA* AND *HOGNA*

Arctosa littoralis, shoreline wolf spider, adult male, adult female
  Female TBL 11.2–14.7 mm, Male TBL 9.6–12.8 mm (see page 149)

Arctosa rubicunda, adult male, adult female
  Female TBL 8.0–12.0 mm, Male TBL 6.6–9.3 mm (see page 149)

Arctosa virgo, adult female
  Female TBL 5.0–6.6 mm, Male TBL 5.1–6.9 mm (see page 150)

Hogna antelucana, adult female
  Female TBL 13.5–19.0 mm, Male TBL 13.0–18.0 mm (see page 153)

Hogna georgicola, adult male, adult female
  Female TBL 18–24 mm, Male TBL 18–22 mm (see page 155)

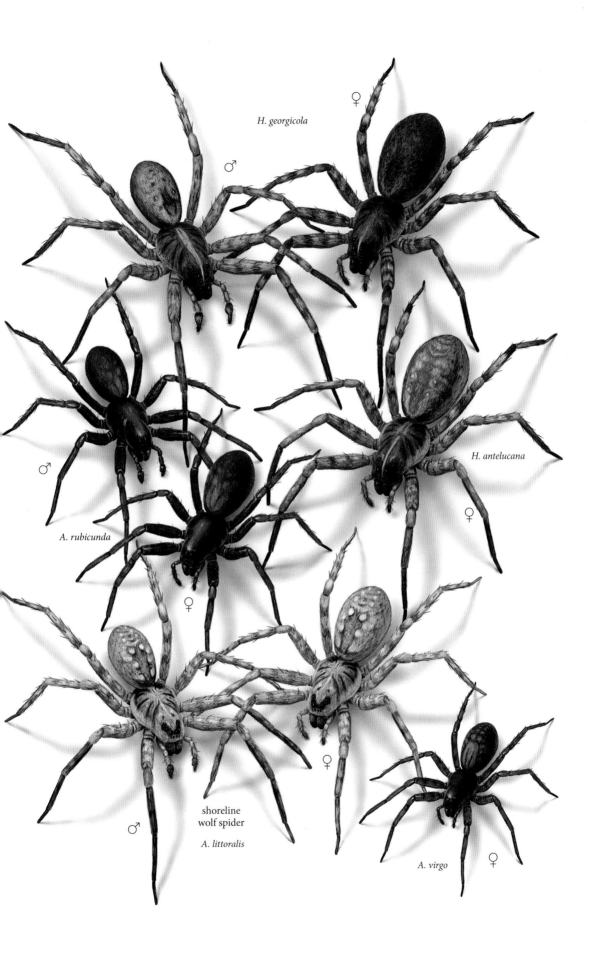

H. georgicola

♀

♂

♂

A. rubicunda

H. antelucana

♀

♀

shoreline
wolf spider

A. littoralis

♂

♀

A. virgo

♀

PLATE 55

GROUND HUNTERS: WOLF SPIDERS; *HOGNA*

Hogna aspersa, tiger wolf spider, adult female
Female TBL 18–25 mm, Male TBL 16–18 mm (see page 153)

Hogna baltimoriana, adult female, adult male
Female TBL 15.0–18.0 mm, Male TBL 14.0–17.5 mm (see page 154)

Hogna carolinensis, Carolina wolf spider, adult female, adult male
Female TBL 22–35 mm, Male TBL 18–20 mm (see page 154)

Hogna frondicola, adult female
Female TBL 11–14 mm, Male TBL 9–12 mm (see page 155)

Hogna helluo, adult male, adult female
Female TBL 18–21 mm, Male TBL 10–12 mm (see page 155)

Hogna lenta, adult female
Female TBL 17.0–22.0 mm, Male TBL 13.0–20.5 mm (see page 156)

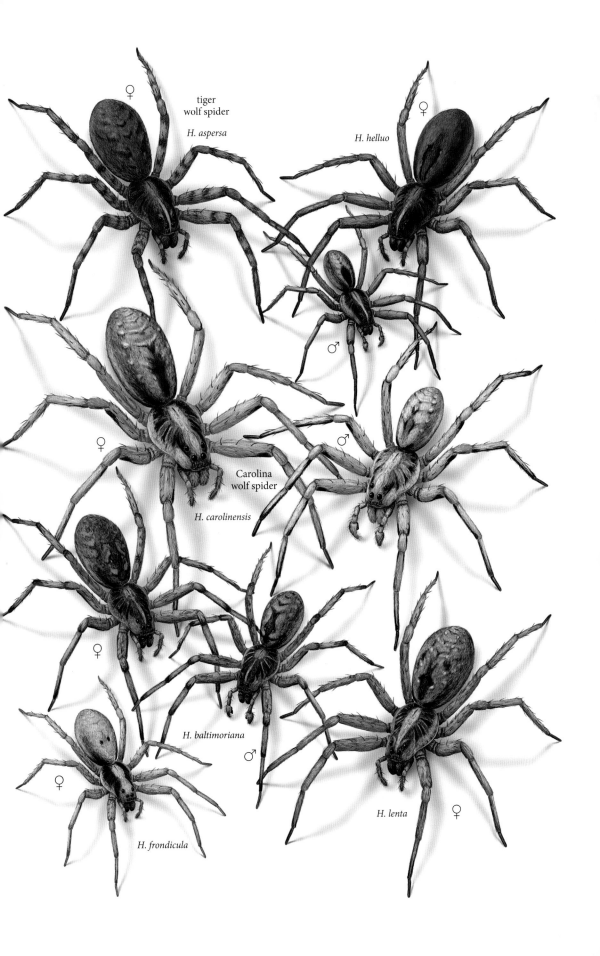

♀
tiger
wolf spider
H. aspersa

♀
H. helluo

♂

♀

♂

Carolina
wolf spider

H. carolinensis

♀

H. baltimoriana

♂

♀
H. frondicula

H. lenta ♀

PLATE 56

SURFACE HUNTERS: FISHING SPIDERS

Dolomedes scriptus, adult female, adult male
  Female TBL 17–24 mm, Male TBL 13–16 mm (see page 180)

Dolomedes tenebrosus, adult female, adult male
  Female TBL 15–26 mm, Male TBL 7–13 mm (see page 181)

Dolomedes triton, sixspotted fishing spider, adult female, adult male
  Female TBL 17–20 mm, Male TBL 9–13 mm (see page 181)

Dolomedes vittatus, adult female, adult male
  Female TBL 19–28 mm, Male TBL 10–25 mm (see page 182)

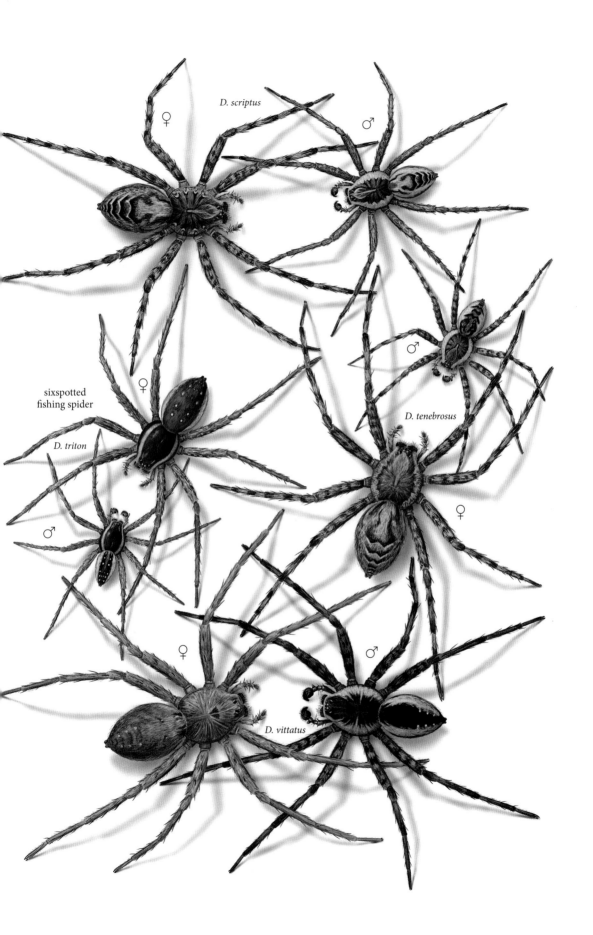

D. scriptus

♀

♂

sixspotted
fishing spider

♀

D. triton

♂

D. tenebrosus

♂

♀

♀

D. vittatus

♂

PLATE 57

SURFACE HUNTERS: FISHING SPIDER, NURSERY WEB SPIDER, AND *HYPOCHILUS*

Hypochilus builds a lampshade-shaped web but also clings to rock faces.

Dolomedes albineus, adult female, adult male
    Female TBL 23 mm, Male TBL 18 mm (see page 180)

Hypochilus thorelli, lampshade weaver, adult female
    Female TBL 14.0–15.5 mm, Male TBL 10.0–11.0 mm (see page 136)

Pisaurina mira, nursery web spider, adult female, adult male
    Female TBL 12.5–16.5 mm, Male TBL 10.5–15.0 mm (see page 182)

Tinus peregrinus, adult female
    Female TBL 5.0–10.0 mm, Male TBL 4.3–9.0 mm (see page 182)

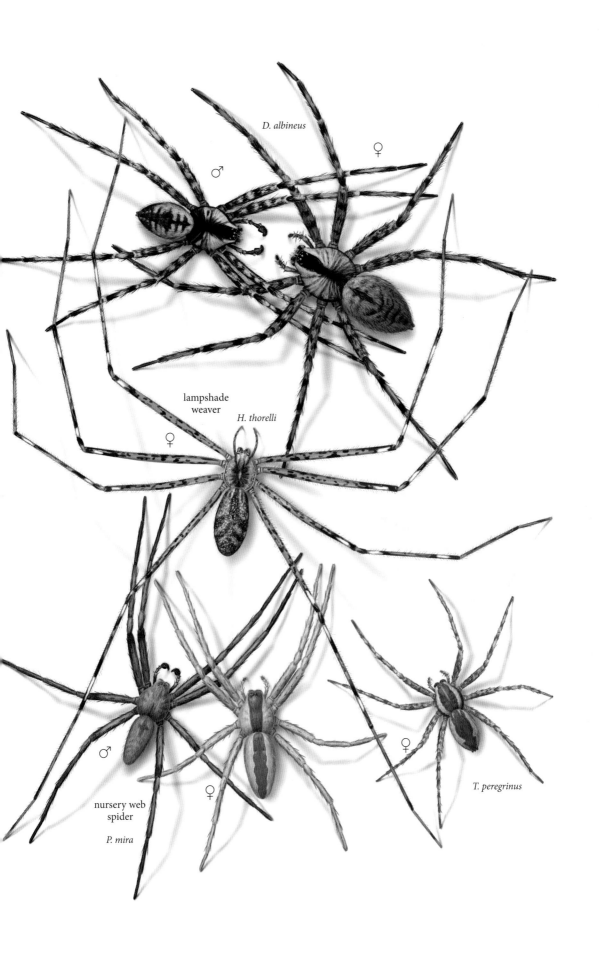

D. albineus

♂   ♀

lampshade
weaver

H. thorelli

♀

nursery web
spider

P. mira

♂   ♀

T. peregrinus

♀

PLATE 58

UNUSUAL JUMPING SPIDERS

Anasaitis canosa, twinflagged jumper, adult female, adult male
Female TBL 5–6 mm, Male TBL 4–5 mm (see page 185)

Bagheera prosper, adult male, adult female
Female TBL 6.0–8.0 mm, Male TBL 5.0–6.5 mm (see page 186)

Euophrys monadnock, adult female, adult male
Female TBL 4.0–5.0 mm, Male TBL 3.6–4.0 mm (see page 186)

Neon nelli, adult male, adult female
Female TBL 2.3–3.0 mm, Male TBL 2.0–2.5 mm (see page 194)

Paramarpissa albopilosa, adult female, adult male
Female TBL 6.5 mm, Male TBL 6.5 mm (see page 194)

A. canosa

twinflagged
jumper

♂

♀

B. prosper

♂

♀

P. albopilosa

♂

E. monadnock

♀

♀

♂

N. nelli

♀

♂

PLATE 59

SMALL BROWN OR GRAY JUMPING SPIDERS WITH A BAND
AROUND THE ABDOMEN

Ghelna canadensis, adult female, adult male
  Female TBL 4.6–6.4 mm, Male TBL 3.9–5.0 mm (see page 187)

Metaphidippus manni, adult male, adult female
  Female TBL 4.3–4.9 mm, Male TBL 3.3–4.8 mm (see page 193)

Pelegrina aeneola, adult male, adult female
  Female TBL 4.5–5.5 mm, Male TBL 4.0–5.0 mm (see page 195)

Pelegrina exigua, adult female, adult male striped form, adult male dull form
  Female TBL 4.0–5.6 mm, Male TBL 3.3–5.1 mm (see page 196)

Pelegrina galathea, peppered jumper, adult female, adult male
  Female TBL 3.6–5.4 mm, Male TBL 2.7–4.4 mm (see page 196)

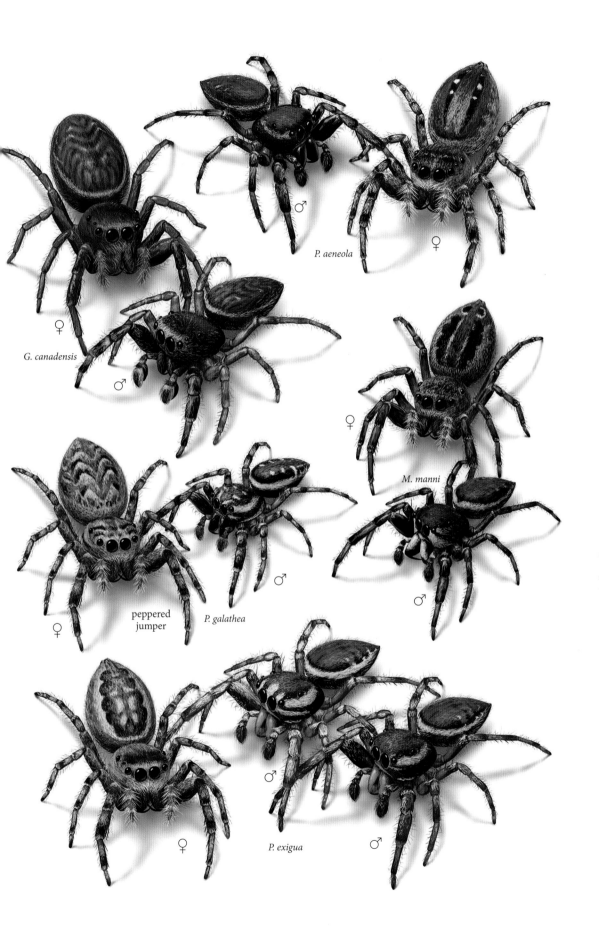

P. aeneola

♀

G. canadensis

♂

♀

M. manni

peppered
jumper

P. galathea

♂

♂

♀

P. exigua

♂

PLATE 60

BOLDLY PATTERNED JUMPING SPIDERS

Eris militaris, bronze jumper, adult female, adult male
  Female TBL 6.0–8.0 mm, Male TBL 4.7–6.7 mm (see page 186)

Paraphidippus aurantius, emerald jumper, adult female, adult male
  Female TBL 8–12 mm, Male TBL 7–10 mm (see page 195)

Plexippus paykulli, pantropical jumper, adult female, adult male
  Female TBL 10.0 12.0 mm, Male TBL 9.5 mm (see page 202)

Salticus scenicus, zebra jumper, adult male, adult female
  Female TBL 4.3 6.4 mm, Male TBL 4.0 5.5 mm (see page 202)

Zygoballus rufipes, hammerjawed jumper, adult male, adult female
  Female TBL 4.3 6.0 mm, Male TBL 3.0 4.0 mm (see page 206)

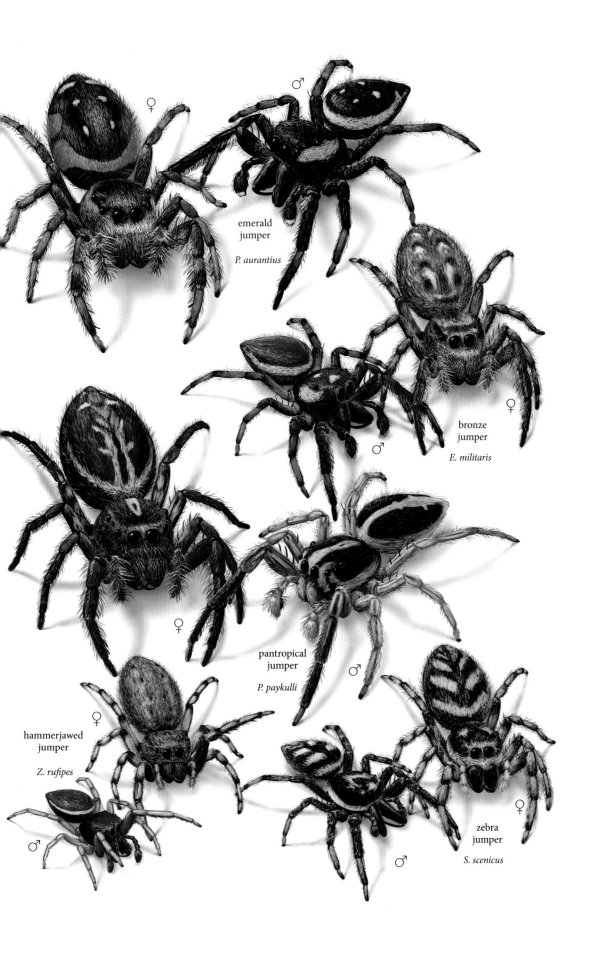

♀
♂

emerald
jumper

P. aurantius

♂

bronze
jumper

E. militaris

♀

♀

pantropical
jumper

P. paykulli

♂

hammerjawed
jumper

Z. rufipes

♀

♂

zebra
jumper

S. scenicus

♂

PLATE 61

SMALL BROWN OR GRAY JUMPING SPIDERS

Evarcha hoyi, adult female, adult male
Female TBL 4.6–6.3 mm, Male TBL 4.3–5.5 mm (see page 187)

Habronattus agilis, adult female, adult male
Female TBL 5.2–6.4 mm, Male TBL 5.0–5.5 mm (see page 187)

Naphrys pulex, adult male, adult female
Female TBL 4.5–5.5 mm, Male TBL 4.0–5.5 mm (see page 194)

Sitticus concolor, adult female, adult male
Female TBL 3.0–3.5 mm, Male TBL 2.5–3.0 mm (see page 204)

Sitticus palustris, adult female, adult male
Female TBL 5.0–6.0 mm, Male TBL 3.5–5.0 mm (see page 204)

E. hoyi ♀ ♂

N. pulex ♀ ♂

S. palustris ♀ ♂

H. agilis ♀ ♂

S. concolor ♀ ♂

PLATE 62

JUMPING SPIDERS: *HABRONATTUS*

Habronattus americanus, adult male, adult female
Female TBL 5.5 mm, Male TBL 4.2–4.8 mm (see page 188)

Habronattus coecatus, adult male, adult female
Female TBL 5.5 mm, Male TBL 4.3–4.7 mm (see page 188)

Habronattus decorus, adult male, adult female
Female TBL 5–6 mm, Male TBL 4–5 mm (see page 188)

Habronattus oregonensis, adult female, adult male
Female TBL 4.8–7.6 mm, Male TBL 4.3–6.0 mm (see page 189)

Habronattus viridipes, adult male displaying, adult female
Female TBL 5.5–7.0 mm, Male TBL 4.3–5.7 mm (see page 189)

♂ H. americanus ♀

♀ H. oregonensis

♂

H. coecatus ♀

♂

H. viridipes ♂

♀ H. decorus ♀

♂

PLATE 63

JUMPING SPIDERS: *HENTZIA, SASSACUS,* AND *TUTELINA*

Hentzia mitrata, adult female, adult male
Female TBL 2.9–4.5 mm, Male TBL 3.5–4.1 mm (see page 189)

Hentzia palmarum, adult female, adult male
Female TBL 4.7–6.1 mm, Male TBL 4.0–5.3 mm (see page 190)

Sassacus cyaneus, adult male, adult female
Female TBL 3.3–4.6 mm, Male TBL 3.1–4.0 mm (see page 203)

Sassacus papenhoei, adult male, adult female
Female TBL 4.4–5.5 mm, Male TBL 2.8–4.8 mm (see page 203)

Tutelina elegans, adult male, adult female
Female TBL 5.5–7.0 mm, Male TBL 4.0–4.5 mm (see page 205)

T. elegans

♀

♂

H. mitrata

♀

♂

H. palmarum

♀

♀

♂

S. papenhoei

♂

♀

S. cyaneus

♀

♂

PLATE 64

JUMPING SPIDERS: *LYSSOMANES, METACYRBA,* AND *THIODINA*

Lyssomanes viridis, magnolia green jumper, adult female, adult male
Female TBL 7–8 mm, Male TBL 5–6 mm (see page 190)

Maevia inclemens, dimorphic jumper, adult male tufted form, adult male plain form, adult female
Female TBL 6.5–10.0 mm, Male TBL 4.8–7.0 mm (see page 191)

Metacyrba taeniola, adult female, adult male
Female TBL 6–7 mm, Male TBL 5–6 mm (see page 193)

Thiodina sylvana, adult female, adult male red form, adult male black form
Female TBL 8–10 mm, Male TBL 7–9 mm (see page 205)

magnolia
green jumper   *L. viridis*

♀

♂

♀

T. sylvana

♂

♂

♂

♀

dimorphic
jumper

♂

Maevia inclemens

♀

♂

Metacyrba taeniola

♂

PLATE 65

JUMPING SPIDERS THAT MIMIC ANTS

Messua limbata, adult female, adult male
  Female TBL 5.5 mm, Male TBL 5.0 mm (see page 192)

Myrmarachne formicaria, adult female, adult male
  Female TBL 5.0–6.0 mm, Male TBL 5.5–6.5 mm (including chelicerae) (see page 193)

Peckhamia americana, adult female
  Female TBL 5 mm, Male TBL 4 mm (see page 195)

Sarinda hentzi, adult female, adult male
  Female TBL 5–7 mm, Male TBL 5–7 mm (see page 203)

Synageles noxiosus, adult female
  Female TBL 3.0–3.5 mm, Male TBL 4.2–5.0 mm (see page 204)

Synemosyna formica, adult male
  Female TBL 4.7–5.7 mm, Male TBL 4.2–5.0 mm (see page 205)

♀

♂ *Messua limbata*

Myrmarachne formicaria

♀

♂

Synemosyna formica

Sarinda hentzi

♀

♀

♂

Synageles noxiosus

♀

♀

P. americana

PLATE 66

JUMPING SPIDERS WITH ELONGATED ABDOMENS

Marpissa formosa, adult female, adult male
Female TBL 6.5–9.0 mm, Male TBL 5.2–6.0 mm (see page 191)

Marpissa lineata, adult male, adult female
Female TBL 4–5.3 mm, Male TBL 3.0–4.0 mm (see page 191)

Marpissa pikei, Pike slender jumper, adult male, adult female
Female TBL 6.5–9.5 mm, Male TBL 6.0–8.2 mm (see page 192)

Menemerus bivittatus, gray wall jumper, adult female, adult male
Female TBL 7.0–10.2 mm, Male TBL 5.5–7.3 mm (see page 192)

Platycryptus undatus, adult female, adult male
Female TBL 10.0–13.0 mm, Male TBL 8.5–9.5 mm (see page 201)

♀ P. undatus ♂

gray wall
jumper

Menemerus bivittatus

♂

♀

Marpissa lineata

Marpissa formosa

♂

♀

♂

Pike slender
jumper

Marpissa pikei

♀

PLATE 67

JUMPING SPIDERS: *PELLENES, PHIDIPPUS,* AND *PHLEGRA*

Pellenes wrighti, adult male, adult female
   Female TBL 5.3 mm, Male TBL 5.1 mm (see page 196)

Phidippus audax, bold jumper, adult female, adult male var. *bryanti,* adult male typical
   Female TBL 8–15 mm, Male TBL 6–13 mm (see page 197)

Phidippus clarus, adult male, adult female tan form, adult female red form
   Female TBL 8–10 mm, Male TBL 5–7 mm (see page 199)

Phlegra hentzi, adult female, adult male
   Female TBL 7–8 mm, Male TBL 6–7 mm (see page 201)

bold
jumper

Phidippus audax

♂

♂

♀

Phidippus clarus

♀

♀

♂

♀

Phlegra hentzi

♂

Pellenes wrighti

♀

♂

PLATE 68

JUMPING SPIDERS: *PHIDIPPUS*

Phidippus apacheanus, adult male, adult female
  Female TBL 7.1–13.4 mm, Male TBL 5.2–10.6 mm (see page 197)

Phidippus cardinalis, cardinal jumper, adult female, adult male
  Female TBL 6.0–14.3 mm, Male TBL 4.4–9.5 mm (see page 198)

Phidippus johnsoni, Johnson jumper, adult female, adult male
  Female TBL 9.0–14.2 mm, Male TBL 6.2–10.7 mm (see page 199)

Phidippus regius, regal jumper, adult female red form, adult female gray form, adult male
  Female TBL 14–23 mm, Male TBL 10–18 mm (see page 200)

P. apacheanus

♀

♂

♀

cardinal
jumper

P. cardinalis

♀

♂

♂

Johnson
jumper

P. johnsoni

♀

♀

regal
jumper

P. regius

♂

PLATE 69

JUMPING SPIDERS: *PHIDIPPUS*

Phidippus ardens, adult female, adult male
    Female TBL 11.5–15.2 mm, Male TBL 8.3–12.7 mm (see page 197)

Phidippus carneus, adult female, adult male banded form, adult male plain form
    Female TBL 8.4–15.2 mm, Male TBL 7.2–10.4 mm (see page 198)

Phidippus otiosus, adult male plain form, adult female
    Female TBL 7.6–17.1 mm, Male TBL 6.3–13.7 mm (see page 200)

Phidippus princeps, adult male, adult female
    Female TBL 6.0–11.5 mm, Male TBL 4.4–8.5 mm (see page 200)

Phidippus whitmani, adult female, adult male
    Female TBL 5.2–11.6 mm, Male TBL 4.1–9.6 mm (see page 201)

♀ *P. whitmani* ♂

P. princeps ♂ ♀

P. ardens ♀ ♂

P. otiosus ♀

♂

♀ *P. carneus* ♂ ♂

PLATE 70

CRABLIKE SPIDERS: RUNNING CRAB SPIDERS

Apollophanes margareta, adult female
Female TBL 5.6–10.0 mm, Male TBL 4.8–8.8 mm (see page 174)

Philodromus cespitum, adult male, adult female
Female TBL 4.5–6.1 mm, Male TBL 4.0–5.0 mm (see page 174)

Philodromus rufus, adult female
Female TBL 2.7–4.5 mm, Male TBL 2.5–3.5 mm (see page 174)

Philodromus vulgaris, adult female
Female TBL 4.5–8.0 mm, Male TBL 4.5–6.3 mm (see page 175)

Thanatus vulgaris, adult female
Female TBL 5.0–10.0 mm, Male TBL 4.5–6.0 mm (see page 175)

Tibellus oblongus, adult female
Female TBL 6–10 mm, Male TBL 5–7 mm (see page 175)

Titanebo albocaudatus, adult female, adult male
Female TBL 4.0–5.0 mm, Male TBL 2.8–3.9 mm (see page 176)

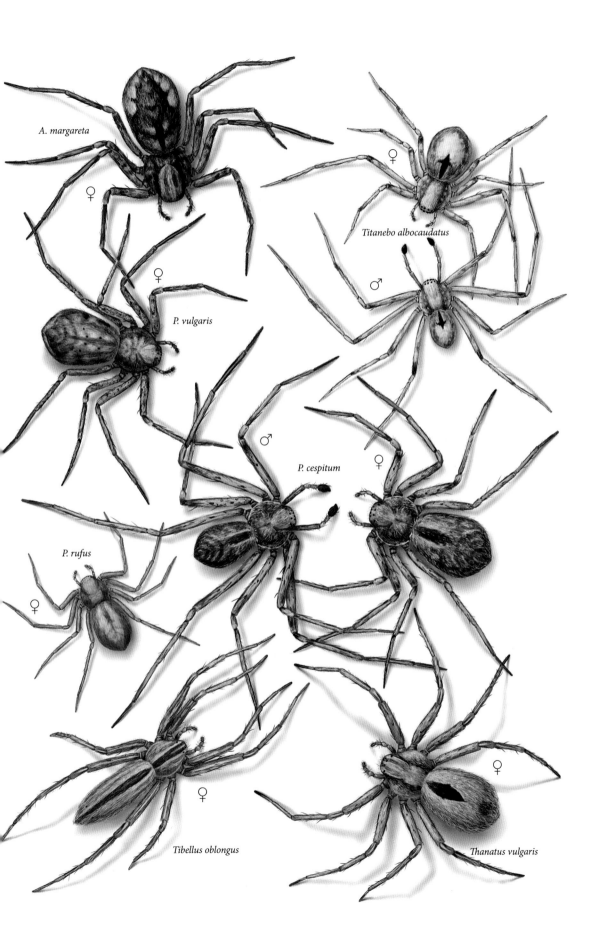

A. margareta

♀

♀

P. vulgaris

Titanebo albocaudatus

♀

♂

♂

P. cespitum

♀

P. rufus

♀

♀

Tibellus oblongus

♀

♀

Thanatus vulgaris

PLATE 71

CRABLIKE SPIDERS: HUNTSMAN SPIDERS, FLATTIES, AND LONGLEGGED WATER SPIDER

Heteropoda venatoria, huntsman spider, adult male, adult female
Female TBL 20–30 mm, Male TBL 15–25 mm (see page 212)

Olios giganteus, golden huntsman spider, adult female
Female TBL 14.6–48.0 mm, Male TBL 11.3–29.4 mm (see page 212)

Selenops actophilus, adult female
Female TBL 11.0–12.0 mm, Male TBL 9.0–9.5 mm (see page 208)

Trechalea gertschi, adult female
Female TBL 16.4–20.0 mm, Male TBL 13.0–15.0 mm (see page 244)

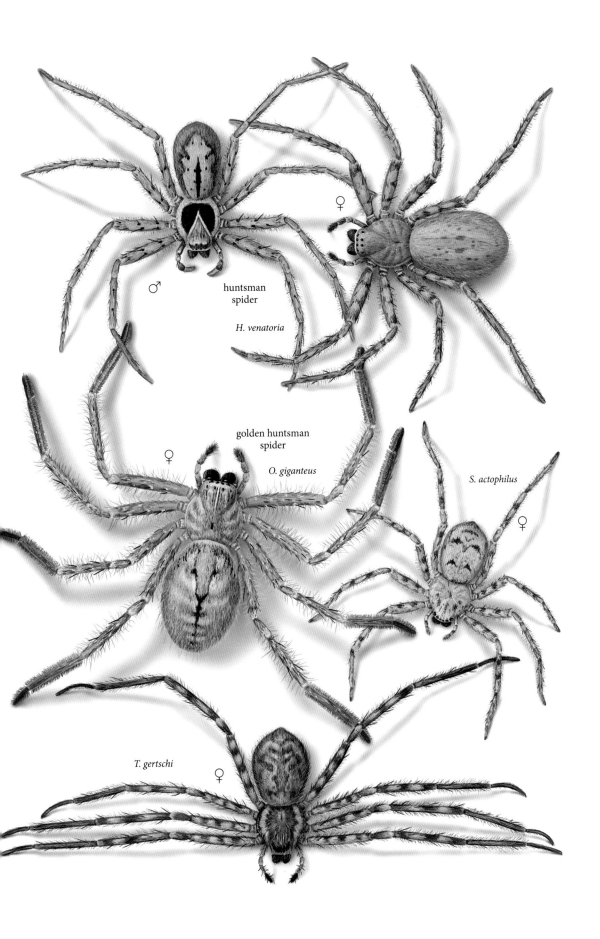

♀

♂

huntsman
spider

H. venatoria

golden huntsman
spider

♀

O. giganteus

S. actophilus

♀

T. gertschi   ♀

PLATE 72

CRABLIKE SPIDERS: BOLDLY MARKED CRAB SPIDERS AND *TMARUS*

Bassaniana versicolor, bark crab spider, adult female, adult male
Female TBL 4.4–7.7 mm, Male TBL 3.9–5.7 mm (see page 238)

Ozyptila americana, leaflitter crab spider, adult female
Female TBL 3.5–4.0 mm, Male TBL 3.0–3.5 mm (see page 241)

Synema parvulum, adult female, adult male
Female TBL 2.5–2.8 mm, Male TBL 2.5 mm (see page 241)

Tmarus angulatus, adult female
Female TBL 4.5–7.0 mm, Male TBL 3.0–5.0 mm (see page 241)

Xysticus alboniger, adult female
Female TBL 4.7 mm, Male TBL 3.6 mm (see page 242)

bark
crab spider
♀

B. versicolor
♂

♂
S. parvulum

♀

X. alboniger

leaflitter
crab spider

♀
O. americana

♀

♀

T. angulatus   ♀

PLATE 73

CRABLIKE SPIDERS: CRAB SPIDERS; BROWN-COLORED *XYSTICUS*

Xysticus auctificus, adult female
   Female TBL 4.4 mm, Male TBL 3.5 mm (see page 242)

Xysticus californicus, adult female
   Female TBL 6.5–7.0 mm, Male TBL 4.0–5.0 mm (see page 242)

Xysticus elegans, elegant crab spider, adult male, adult female
   Female TBL 8–10 mm, Male TBL 6–7 mm (see page 242)

Xysticus ferox, adult female, adult male
   Female TBL 6–7 mm, Male TBL 5–6 mm (see page 243)

Xysticus pretiosus, adult female
   Female TBL 4.8–5.5 mm, Male TBL 3.3–4.0 mm (see page 243)

♂

X. ferox

♀

X. californicus

♀

X. pretiosus

♀

X. auctificus

♀

♂

elegant
crab spider

X. elegans

♀

PLATE 74

CRABLIKE SPIDERS: COLORFUL CRAB SPIDERS

Diaea livens, adult female
Female TBL 6.0–6.5 mm, Male TBL 4.0–5.0 mm (see page 238)

Misumena vatia, goldenrod crab spider, adult female yellow form, red-banded form, pale form, adult male
Female TBL 6.0–9.0 mm, Male TBL 2.9–4.0 mm (see page 239)

Misumenoides formosipes, whitebanded crab spider, adult male, adult female pale form, yellow form, pink form, face with carina
Female TBL 5.0–11.3 mm, Male TBL 2.5–3.2 mm (see page 240)

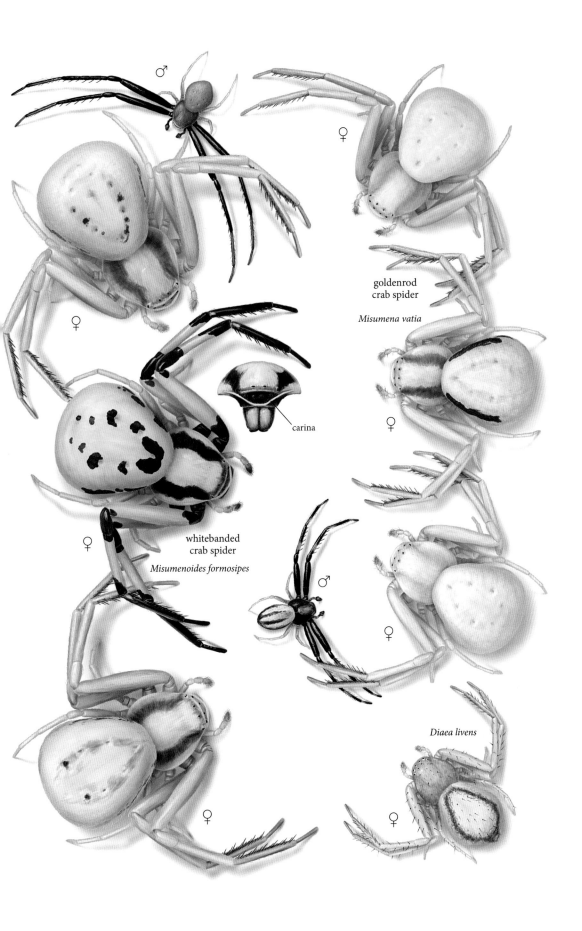

♂

♀

goldenrod
crab spider

Misumena vatia

♀

carina

♀

whitebanded
crab spider

Misumenoides formosipes

♂

♀

♀

Diaea livens

♀

PLATE 75

CRABLIKE SPIDERS: CRAB SPIDERS; *MECAPHESA* AND *MISUMESSUS*

Mecaphesa asperata, northern crab spider, adult female, adult male
Female TBL 4.4–6.0 mm, Male TBL 3.0–4.0 mm (see page 238)

Mecaphesa celer, celer crab spider, adult female, adult male
Female TBL 5.0–6.7 mm, Male TBL 2.5–3.2 mm (see page 239)

Mecaphesa lepida, adult female, adult male
Female TBL 5 mm, Male TBL 3 mm (see page 239)

Misumessus oblongus, adult female
Female TBL 4.9–6.2 mm, Male TBL 1.5–2.6 mm (see page 240)

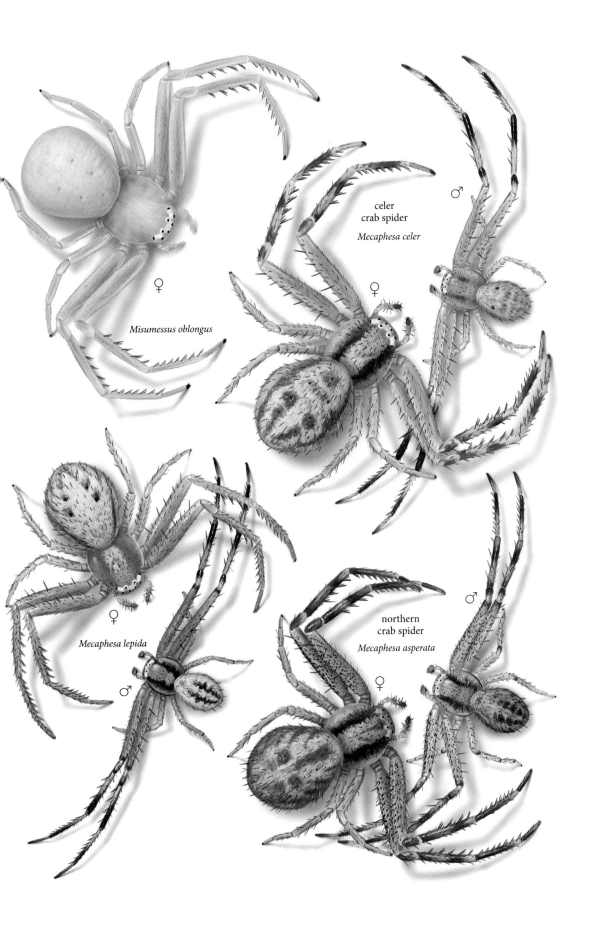

Misumessus oblongus
♀

celer
crab spider
Mecaphesa celer
♀
♂

Mecaphesa lepida
♀
♂

northern
crab spider
Mecaphesa asperata
♂
♀

Dysdera crocata, woodlouse spider, adult male, adult female
Female TBL 11–15 mm, Male TBL 9–10 mm (see page 123)

Loxosceles deserta, desert recluse, adult female
Female TBL 7.5 mm, Male TBL 6.0 mm (see page 209)

Loxosceles laeta, Chilean recluse, adult female
Female TBL 12.0 mm, Male TBL 9.4 mm (see page 209)

Loxosceles reclusa, brown recluse, adult male, adult female
Female TBL 9 mm, Male TBL 8 mm (see page 210)

Loxosceles rufescens, Mediterranean recluse, adult female
Female TBL 7.5 mm, Male TBL 7.0 mm (see page 211)

Scytodes thoracica, spitting spider, adult female
Female TBL 4.0–6.0 mm, Male TBL 3.5–4.0 mm (see page 207)

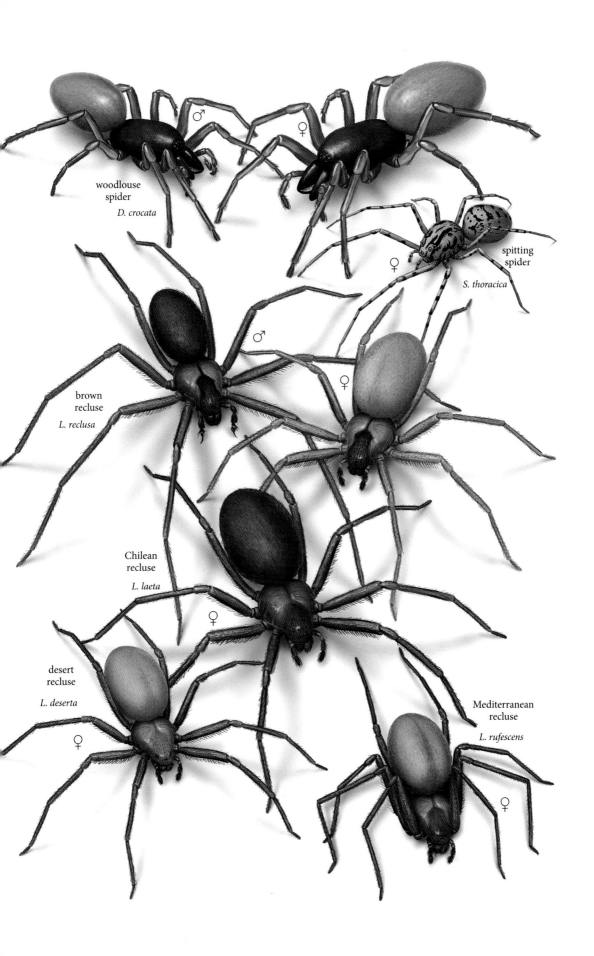

woodlouse
spider

D. crocata

♂ ♀

spitting
spider

S. thoracica

♀

brown
recluse

L. reclusa

♂ ♀

Chilean
recluse

L. laeta

♀

desert
recluse

L. deserta

♀

Mediterranean
recluse

L. rufescens

♀

PLATE 77

SHEET WEAVERS: SHEETWEB WEAVERS, LARGE HAMMOCKWEB SPIDER

Frontinella pyramitela, bowl and doily spider, adult female
Female TBL 3.0–4.0 mm, Male TBL 3.0–3.3 mm (see page 141)

Megalepthyphantes nebulosus, adult female
Female TBL 4.0–4.5 mm, Male TBL 3.5–4.0 mm (see page 141)

Microlinyphia mandibulata, platform spider, adult female
Female TBL 4.0–5.0 mm, Male TBL 3.3–4.6 mm (see page 142)

Neriene clathrata, adult male, adult female, mating pair
Female TBL 3.0–5.2 mm, Male TBL 3.6–4.8 mm (see page 143)

Neriene litigiosa, Sierra dome spider, adult female
Female TBL 5.2–8.5 mm, Male TBL 5.1–6.8 mm (see page 143)

Neriene radiata, filmy dome spider, adult female
Female TBL 4.0–6.5 mm, Male TBL 3.4–5.3 mm (see page 143)

Neriene variabilis, variable sheetweaver, adult female
Female TBL 3.4–5.4 mm, Male TBL 3.8–4.7 mm (see page 144)

Pimoa altioculata, large hammockweb spider, adult female
Female TBL 8.8 mm, Male TBL 6.5 mm (see page 179)

Tenuiphantes tenuis, adult female
Female TBL 2.1–3.2 mm, Male TBL 2.0–2.6 mm (see page 145)

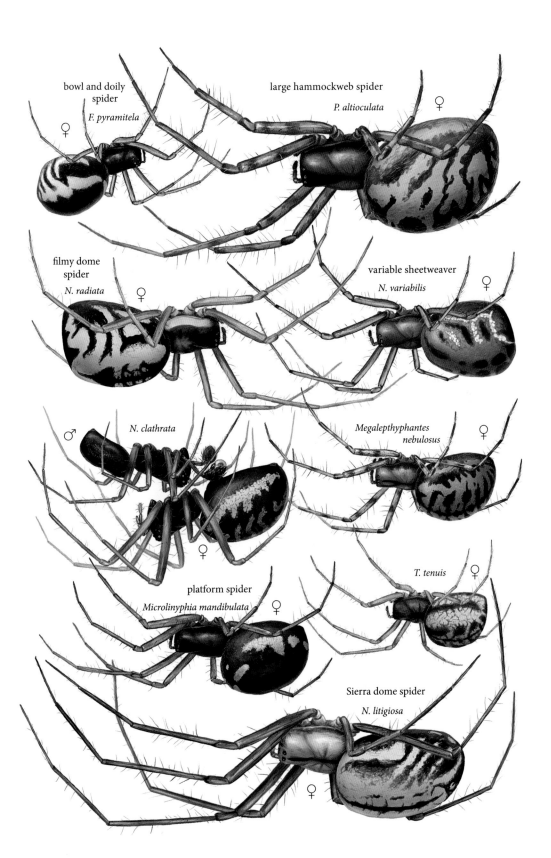

bowl and doily spider

F. pyramitela ♀

large hammockweb spider

P. altioculata ♀

filmy dome spider

N. radiata ♀

variable sheetweaver

N. variabilis ♀

N. clathrata ♂

♀

Megalepthyphantes nebulosus ♀

platform spider

Microlinyphia mandibulata ♀

T. tenuis ♀

Sierra dome spider

N. litigiosa

♀

PLATE 78

SHEET WEAVERS: SHEETWEB WEAVERS

Bathyphantes alboventris, adult female
  Female TBL 2.5 mm, Male TBL 2.5 mm (see page 138)

Bathyphantes pallidus, adult female
  Female TBL 2.3–2.8 mm, Male TBL 1.9–2.1 mm (see page 139)

Drapetisca alteranda, adult female
  Female TBL 4.0–4.5 mm, Male TBL 3.2–3.8 mm (see page 140)

Estrandia grandaeva, conifer sheetweaver, adult female
  Female TBL 2.4–2.8 mm, Male TBL 2.5 mm (see page 140)

Helophora insignis, adult female
  Female TBL 3.2–4.4 mm, Male TBL 2.8–3.7 mm (see page 141)

Pityohyphantes costatus, hammock spider, adult female
  Female TBL 5.0–7.0 mm, Male TBL 4.5–6.0 mm (see page 144)

Pityohyphantes rubrofasciatus, red hammock spider, adult female
  Female TBL 5.0–7.0 mm, Male TBL 4.5–6.0 mm (see page 144)

Stemonyphantes blauveltae, adult female
  Female TBL 5–6 mm, Male TBL 4 mm (see page 145)

Tapinopa bilineata, adult female
  Female TBL 5 mm, Male TBL 4–5 mm (see page 145)

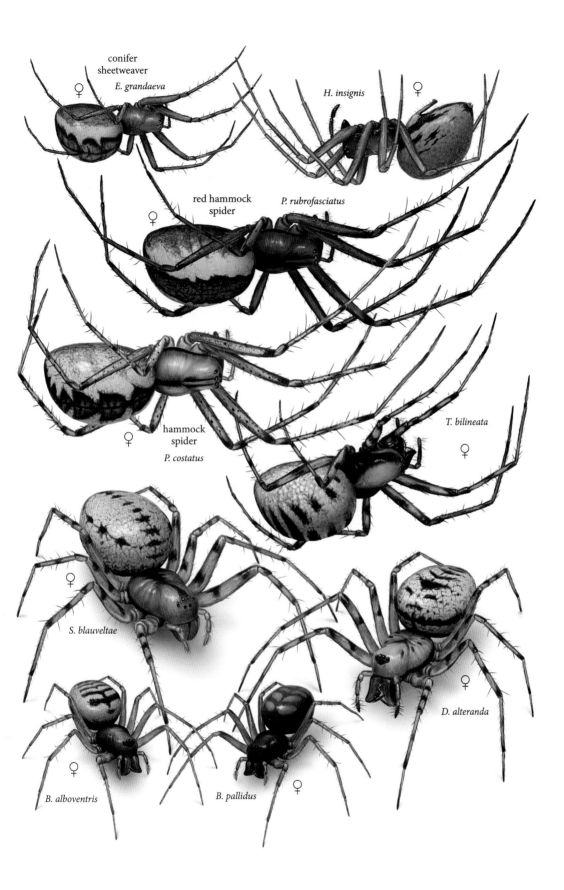

conifer
sheetweaver
E. grandaeva ♀

H. insignis ♀

red hammock
spider ♀   *P. rubrofasciatus*

hammock
spider ♀
P. costatus

T. bilineata ♀

S. blauveltae ♀

D. alteranda ♀

B. alboventris ♀

B. pallidus ♀

PLATE 79

SHEET WEAVERS: SHEETWEB WEAVERS, DWARF SHEETWEAVERS

Centromerus cornupalpis, adult female
Female TBL 2.0–2.6 mm, Male TBL 2.0–2.8 mm (see page 139)

Ceratinopsidis formosa, adult female
Female TBL 2.2–3.0 mm, Male TBL 2.1–2.3 mm (see page 139)

Diplostyla concolor, adult female, adult male
Female TBL 2.3–2.8 mm, Male TBL 2.0–2.3 mm (see page 139)

Florinda coccinea, scarlet sheetweaver, adult female
Female TBL 3.5 mm, Male TBL 3.0 mm (see page 140)

Hypselistes florens, splendid dwarf spider, adult female
Female TBL 2.5–3.0 mm, Male TBL 2.3–2.5 mm (see page 141)

Meioneta fabra, adult male, adult female
Female TBL 2.0–2.5 mm, Male TBL 1.8–2.4 mm (see page 142)

Microneta viaria, adult female
Female TBL 2.0–2.5 mm, Male TBL 2.0–2.5 mm (see page 142)

Tennesseellum formica, adult female
Female TBL 1.8–2.5 mm, Male TBL 1.8–2.4 mm (see page 145)

Walckenaeria directa, money spider, adult female, adult male, male head magnified
Female TBL 2.5 mm, Male TBL 2.5 mm (see page 146)

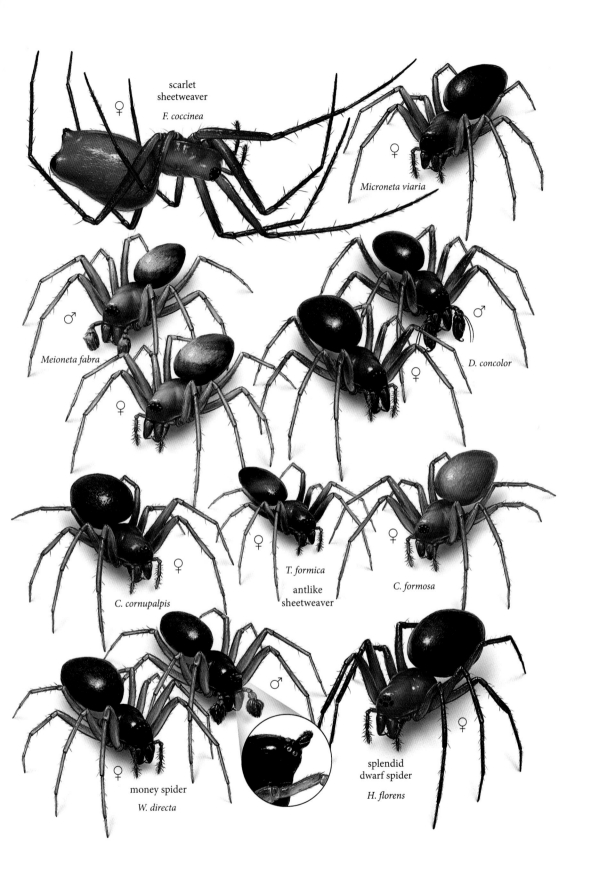

scarlet
sheetweaver

F. coccinea

♀

Microneta viaria

♀

♂

Meioneta fabra

♀

D. concolor

♂

♀

C. cornupalpis

♀

♀

T. formica

antlike
sheetweaver

C. formosa

♀

♂

money spider

W. directa

♀

splendid
dwarf spider

H. florens

♀

PLATE 80
TINY SPIDERS: TINY ORANGE OR BROWN SPIDERS

Calponia harrisonfordi, adult female
Female TBL 5.2 mm, Male TBL 5.1 mm (see page 104)

Escaphiella hespera, adult female
Female TBL 1.5 mm, Male TBL 1.6 mm (see page 171)

Hesperocranum rothi, adult female
Female TBL 2.7–4.1 mm, Male TBL 2.3–3.2 mm (see page 147)

Orthonops gertschi, adult female
Female TBL 4.0–5.4 mm, Male TBL 3.0 mm (see page 104)

Phrurotimpus alarius, adult female
Female TBL 2.9 mm, Male TBL 2.2 mm (see page 109)

Robertus riparius, adult female, adult male
Female TBL 2.8–4.1 mm, Male TBL 2.5–3.7 mm (see page 231)

Scotinella fratrella, adult female, adult male
Female TBL 2.0 mm, Male TBL 1.8 mm (see page 109)

S. fratrella

♀

♂

♀

H. rothi

P. alarius

♀

C. harrisonfordi

♀

O. gertschi

♀

♀

R. riparius

♂

E. hespera

♀

PLATE 81

TINY SPIDERS: TINY SPIDERS WITH PATTERNED ABDOMENS AND *CRUSTULINA*

Argenna obesa, adult male, adult female
  Female TBL 2.5–3.1 mm, Male TBL 2.3 mm (see page 116)

Crustulina sticta, adult female, adult male
  Female TBL 2.3–2.7 mm, Male TBL 2.7 mm (see page 225)

Cryphoeca montana, adult female
  Female TBL 3.0–3.5 mm, Male TBL 2.5–3.0 mm (see page 133)

Hahnia cinerea, adult male, adult female
  Female TBL 1.5–2.0 mm, Male TBL 1.7 mm (see page 134)

Lathys maculina, adult female
  Female TBL 1.4–1.7 mm, Male TBL 1.4 mm (see page 119)

Crustulina sticta

♀

♂

♀

L. maculina

♀

A. obesa

♂

♀

♂

H. cinerea

♀

♀

Cryphoeca montana

PLATE 82

TINY SPIDERS: FLATMESH WEAVER, LONGLEGGED CAVE SPIDER, MIDGET CAVE SPIDERS, MIDGET GROUND WEAVER, *ORCHESTINA*, AND *DIPOENA*

Archoleptoneta schusteri, adult female
Female TBL 1.3 mm, Male TBL 1.3 mm (see page 137)

Dipoena nigra, adult male, adult female
Female TBL 2.8–4.0 mm, Male TBL 1.5–2.0 mm (see page 225)

Neoleptoneta myopica, Tooth Cave spider, adult female
Female TBL 1.6 mm, Male TBL 1.6 mm (see page 137)

Oecobius navus, adult female
Female TBL 2.5–2.9 mm, Male TBL 2.0–2.6 mm (see page 171)

Orchestina saltitans, adult female
Female TBL 1 mm, Male TBL 1 mm (see page 172)

Theotima minutissima, adult female
Female TBL 1 mm, Male unknown (see page 170)

Usofila pacifica, adult female
Female TBL 1.5 mm, Male TBL 1.5 mm (see page 213)

U. pacifica ♀

A. schusteri ♀

Tooth Cave
spider
N. myopica ♀

O. navus ♀

T. minutissima ♀

O. saltitans ♀

D. nigra ♂ ♀

A female *Theridion frondeum* with her egg case and recently emerged spiderlings.

A female *Trachelas tranquillus* (bullheaded sac spider) with her egg case. The egg case was attached to the leaf while the spider was in captivity.

A female *Parasteatoda tepidariorum* (common house spider) and her teardrop-shaped egg case.

A female *Steatoda triangulosa* (checkered cobweb weaver) and several egg cases. There are at least three small males visible between the egg cases.

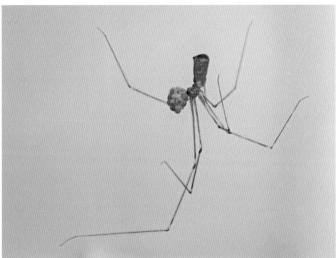

A female *Pholcus phalangioides* (longbodied cellar spider) carrying her loosely secured eggs with her chelicerae.

A female *Peucetia viridans* (green lynx spider) with her distinctive lumpy green egg case laid in captivity.

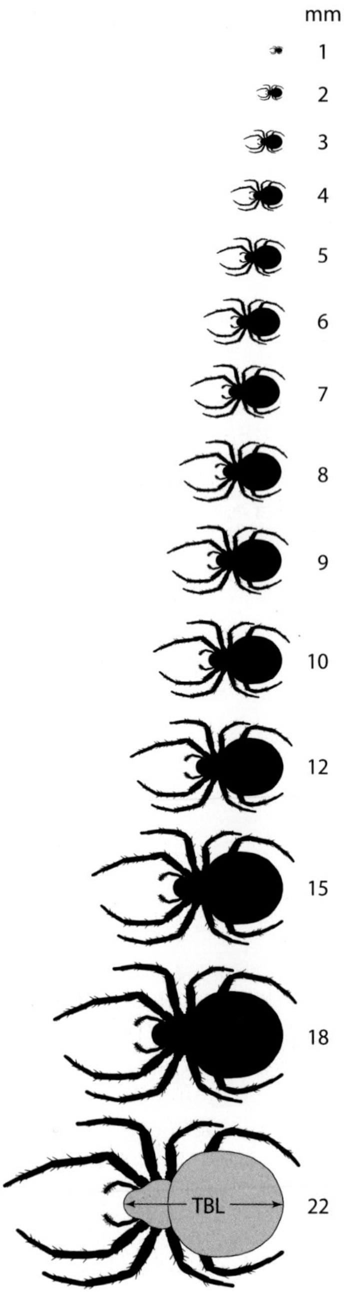

mm

1

2

3

4

5

6

7

8

9

10

12

15

18

TBL

22

Spider Accounts

What follow are short descriptions of the 68 families of spiders included in this book as well as brief introductions to the 469 species that are illustrated. They were chosen to represent the spider fauna of North America north of Mexico. There are nearly 4,000 described species known from this region. The species described in this guide are the ones most likely to be encountered by an interested naturalist.

Some of the spiders illustrated represent large numbers of similar species in the same, or closely related, genera. Many of these would be indistinguishable without careful examination with a microscope. The distinctive features are usually small details of the reproductive parts or genitalia. In the females these are parts of the epigynum, or the internal reproductive ducts and spermathecae. For the males the key features are usually parts of the palps. Technical identification is beyond the scope of this general guide. If you wish to make a correct technical identification of a spider, you should use *Spiders of North America: An Identification Manual* (Ubick et al. 2005). Using that manual, you will be able to identify the genus to which the spider belongs. In addition, you will find references to the scientific literature reference for each genus. These generic reviews will provide the detailed information necessary to distinguish among the species.

The following accounts are organized in alphabetical order by family, and by genus and species within families. This should make finding a species account straightforward. At this time there is no agreed-upon taxonomic sequence for spiders. So in addition to being convenient, this is the same arrangement used in published checklists as well as the aforementioned manual.

FAMILY AGELENIDAE • *Funnel Weavers, Agelenids*

The funnel weavers occur worldwide, with 1,148 described species. There are 111 species in North America north of Mexico. Some members of this family in the genera *Coras* and *Wadotes* have been placed in the Amaurobiidae in some recent publications. Current genetic evidence confirms that these spiders are agelenids (Miller et al. 2010). Funnel weavers are named for constructing large, relatively flat sheets of nonsticky silk connected to a funnel-shaped tube used as a retreat. The spider spends much time in the tube, or waiting at the entrance. When the web is disturbed, the spider may actually escape out a rear opening in the tubular retreat. Most species are active at night, but the spider may rush out onto the sheet to capture prey in the daytime.

The members of this family have eight eyes in two procurved rows. Many, but not all, funnel weavers have long and conspicuous posterior spinnerets that extend well beyond the end of the abdomen. They have relatively long legs and are capable of remarkable speed. They rely on this speed, rather than sticky silk, to capture prey that wander onto, or land on, the web. Many funnel weavers have light brown bodies with paired darker longitudinal stripes on the cephalothorax and sometimes also the abdomen. Members of the genus *Tegenaria* sometimes take up residence in buildings. Their characteristic webs may be found in a disused corner or basement. A number of the species of *Tegenaria* that are found in North America north of Mexico have been introduced from Eurasia.

Agelenopsis aperta (Gertsch, 1934)

Plate 30

IDENTIFICATION This is a large, grayish-brown agelenid. The cephalothorax is grayish tan with two dark brown stripes. The abdomen is tan with two parallel black lines with light centers. The lines break into spots toward the back half of the abdomen. The ventral dark markings are not as distinct as those of *Agelenopsis pennsylvanica*. The spinnerets extend well beyond the end of the abdomen.

OCCURRENCE This species is widespread throughout the arid Southwest, from Texas to California. The large webs often have the funnel extending into a shrub, rock crevice, or debris on the ground. Wandering adult males may stray into buildings.

SEASONALITY Small individuals appear in their webs in the spring. Adults: July through autumn.

REMARKS Susan Reichert (1988) has conducted extensive research on the territorial behavior of this species.

Agelenopsis pennsylvanica (C. L. Koch, 1843) • *Grass spider*

Plate 30

IDENTIFICATION This is a large, grayish-brown agelenid. The cephalothorax is tan with two dark brown stripes. There is a distinct V-shaped mark on the sternum. The abdomen is variegated brown with a lighter central region that may have a rusty tinge. The underside of the abdomen has a dark median band. The spinnerets extend well beyond the end of the abdomen.

OCCURRENCE This species is widespread throughout the Northern states. It is abundant in the Northeast, and its funnel-shaped webs have been found in many habitats. The webs are most frequently seen near the ground in open grassy areas. This species can also be found in woods, but it is not as common there. Wandering adult males often stray into buildings.

SEASONALITY Small individuals appear in their webs in the late spring. Adults: July through autumn.

REMARKS The females lay their eggs in a flat, white, silken case that is about 2 cm in diameter, usually under bark or debris near the ground. The female stays close by the egg case until she dies with the first hard frost.

Calilena arizonica Chamberlin and Ivie, 1941

Plate 30

IDENTIFICATION This is a medium-sized grayish agelenid. The cephalothorax is light grayish tan with two dark lines of variable width extending back from the lateral eyes. The light central area between the dark stripes is widest in the head region. There is a somewhat darker median band on the abdomen that is darkest at the front. The legs are distinctly ringed with gray or black. The posterior spinnerets are long, their last segment being almost twice as long as the basal one.

OCCURRENCE This species has been found in Arizona; similar members of the genus are western from Utah to the Pacific Coast and south to Mexico. These spiders have been found in small funnel-shaped webs near the ground. The funnel retreat is often under a rock, log, cow droppings, or other debris on the ground.

SEASONALITY Males: late summer or autumn. Females: all year.

Coras juvenilis (Keyserling, 1881)

Plate 31

IDENTIFICATION This is a medium-sized agelenid. It has a light orange or tan color with paired dark markings on either side of the cervical grooves, widest at the back. There is a light marginal band on the carapace and an irregular band of dark markings at each radial groove. The legs have conspicuous dark rings. The abdomen has a dark heart mark tapering to a dark line at the back. There are light areas on either side, forming paired chevrons at the back.

OCCURRENCE This species occurs from New England to Michigan and south to Virginia and Kentucky. Others in this genus are found throughout the eastern United States and southeastern Canada. This species inhabits loose bark, logs, rock walls, and cliffside crevices.

SEASONALITY Adults: April through late November.

REMARKS This species builds a fine-meshed funnel web, often with two or more openings (see Fig. 27, upper left).

Coras montanus (Emerton, 1889)

Plate 31

IDENTIFICATION This is a medium-sized agelenid. It has a light brown or tan color with paired dark markings on either side of the cervical grooves, widest at the back. There is a light marginal band on the carapace and an irregular band of dark markings at each radial groove. The legs are less obviously banded than in *Coras juvenilis*. The abdomen has a dark heart mark tapering to a dark line at the back. There may be lighter areas on the abdomen but not as clearly defined as in *Coras juvenilis*.

OCCURRENCE This species occurs in the Northeast from southeastern Canada to New England and Minnesota. Others in this genus are found throughout the eastern United States and southeastern Canada. This species inhabits loose bark, rock walls, and crevices.

SEASONALITY Adults: May through September.

REMARKS This spider builds a fine-meshed funnel web, often with two or more openings.

Hololena curta (McCook, 1894)

Plate 30

IDENTIFICATION This is a medium-sized to large agelenid. There are two dark bands extending from the lateral eyes. The light area enclosed by the dark bands is widest in the head region. The median band of the abdomen is often lighter than the lateral areas. It is separated from the dark sides by a light line that becomes a series of spots at the back. The legs are indistinctly ringed. The spinnerets are shorter than in *Agelenopsis* or *Calilena*, barely visible from above. The last segment of the long posterior spinnerets is shorter than the basal one.

OCCURRENCE This species has been found along the Pacific Coast. It is the most often reported member of the genus in southern California. They are common in funnel-shaped webs in bushes, trees, fences, or near the ground around buildings.

SEASONALITY Adults: March through August.

Melpomene rita (Chamberlin and Ivie, 1941)

Plate 30

IDENTIFICATION This is a medium-sized agelenid. The cephalothorax has two dark longitudinal bands and a light central stripe. The abdomen is variegated brown with two light lines on either side of a light central band. The posterior spinnerets are long and visible from above. The legs are only indistinctly banded.

OCCURRENCE This spider has been found in Arizona. It builds its funnel web in bushes or near the ground.

SEASONALITY Adults: summer.

REMARKS The funnel retreat may incorporate a leaf that has been folded into a tubular shape.

Novalena intermedia (Chamberlin and Gertsch, 1930)

Plate 30

IDENTIFICATION This is a medium-sized agelenid. The cephalothorax is grayish with two variable brown stripes extending back from the lateral eyes. The abdomen is dusky with a darker central band bounded by lighter lines. The posterior spinnerets are long and visible from above. The legs are only indistinctly banded on the femora and tibiae.

OCCURRENCE This species has been found from southwestern Washington south to central California. The funnel webs have been found extending from under the bark of dead trees or near the ground in coniferous forests.

SEASONALITY Adults: June through November.

Rualena cockerelli Chamberlin and Ivie, 1942

Plate 30

IDENTIFICATION This is a small to medium-sized agelenid. The cephalothorax is light tan with two brown lines extending back from the sides of the head region. The light median band is wider in the head region. In all members of this genus the anterior median eyes are distinctly smaller than the anterior lateral eyes. The abdomen has a lighter median line flanked by two dark lines with enclosed light spots. The legs have brown rings.

OCCURRENCE These spiders are found in coastal southern California, including the Channel Islands. This species builds its funnel webs in shrubs and bushes.

SEASONALITY Adults: summer through early winter.

Tegenaria agrestis (Walckenaer, 1802) • *Hobo spider*

Plate 31

IDENTIFICATION This is a large agelenid. This spider is light brown or tan in color with two darker bands extending back from the lateral eyes to the back of the carapace. There is usually a thin dark line down the center of the head region. The legs are not banded. The spinnerets are barely visible extending beyond the end of the abdomen.

OCCURRENCE This species was introduced into the Pacific Northwest. It has chiefly been found from southern British Columbia and Montana south to central Oregon, Idaho, and northern Utah. It has been expanding its range eastward and now extends as far as Colorado and Wyoming. This is usually a ground-dwelling spider with a web that may have two or more funnel-shaped entrances, in cutbanks, under rocks, or among debris. Adult males often wander into houses.

SEASONALITY Males: July through October. Females: August through November.

REMARKS The specific name *agrestis* refers to the Latin "from the field" and has no relation to aggressive behavior. Nothing in its behavior suggests that this spider is particularly aggressive. In fact, despite some early reports of medically significant bites, recent work by Greta Binford (2001) indicates that this is probably not a dangerous spider.

Tegenaria domestica (Clerck, 1757) • *Barn funnel weaver*

Plate 31

IDENTIFICATION This is a medium-sized to large agelenid. This spider is light brown or tan in color with two darker bands extending back from the lateral eyes. There is usually a thin dark line down the center of the head region. The heart area is gray. The abdomen has light chevron patterns toward the back. The legs are usually banded. The spinnerets are barely visible extending beyond the end of the abdomen.

OCCURRENCE This spider has been found in southern Canada and across the United States. It has most often been found around human habitations. It builds its funnel web among debris or extending from cracks in rocks or cement walls.

SEASONALITY Adults: all year.

REMARKS This domestic spider may survive indoors for several years.

Wadotes calcaratus (Keyserling, 1887)

Plate 32

IDENTIFICATION This is a medium-sized agelenid. The cephalothorax is dark brown to black. The chelicerae are robust, bulging, and smooth. The legs have inconspicuous banding. The abdomen is dark brown with a lighter herringbone pattern. The ends of the spinnerets are only barely visible from above.

OCCURRENCE This species occurs from Nova Scotia west to Wisconsin and south to Kentucky and North Carolina. It is most common in the northern portion of its range. This spider is found in hardwood forests under rocks or logs on the ground, or in crevices in rock walls. This species is rare in conifer forests.

SEASONALITY Adults: all year.

REMARKS The exposed portion of the funnel web of this species often forms a sagging, bowl-like shape.

FAMILY AMAUROBIIDAE • *Hackledmesh Weavers*

There are 278 species in this family worldwide. In North America north of Mexico we have 71 species in 8 genera. Many of our species are found in California. The large species of hackledmesh weavers are relatively robust, compact, and most are dark brown in color. The chelicerae are usually large and thick with a conspicuous boss. Our species are cribellate and spin hackledmesh webs with one or more entrances that are funnel-shaped. The messy cribellate strands of hackledmesh weavers' webs are often conspicuous. Most species build their webs among and hide under rocks, logs, or decomposing material on the ground.

Amaurobius borealis Emerton, 1909

Plate 32

IDENTIFICATION This is a small amaurobiid. The cephalothorax is dark brown. The chelicerae are robust, bulging, and shiny. The legs are dark brown without banding. The abdomen is brown with or without paired light spots.

OCCURRENCE This is a northern species; it has been found from Newfoundland to northern British Columbia and south to central Alberta, Wisconsin, New York, and northern Pennsylvania. This spider has been found in deciduous forests under logs, rocks, and in leaf litter.

SEASONALITY Males: April through October. Females: May through November.

REMARKS The silk of the loosely constructed funnel-shaped web is bluish.

Amaurobius ferox (Walckenaer, 1830)

Plate 32

IDENTIFICATION This is a large, dark amaurobiid. The cephalothorax is dark brown. The chelicerae are robust, bulging, and shiny. The legs are dark brown without banding. The abdomen is brown with a light heart mark in the male and light chevrons in the female.

OCCURRENCE This is a Northeastern species; it has been found from Nova Scotia to Illinois and Virginia. This spider has been found in or near buildings, often in basements or cellars.

SEASONALITY Adults: all year.

REMARKS This spider was introduced from Europe into the Northeast. It is somewhat darker but essentially looks like a scaled-up version of the native species.

Arctobius agelenoides (Emerton, 1919)

Plate 32

IDENTIFICATION This is a large amaurobiid. The cephalothorax is smooth and shiny, brown with light at the back of the head region and a light band at the edges of the thoracic region. The abdomen is dark brown, with a dark heart mark flanked by two yellowish bands. The light bands converge in the middle of the abdomen. The legs are darker toward the tips. Some individuals have thin dark stripes on the legs.

OCCURRENCE This species occurs from Alaska east to Nunavut and south to British Columbia and Manitoba. This spider has been found in rocky areas, including talus slopes. It has also been collected in coniferous forests under rocks.

SEASONALITY Males: July through November. Females: May through November.

REMARKS According to Robin Leech (1972), this spider is aggressive. It attacks almost anything approaching its retreat.

Callobius bennetti (Blackwall, 1846)

Plate 32

IDENTIFICATION This is a medium-sized to large amaurobiid. The cephalothorax is darker in the head region. The chelicerae are robust and bulging. The abdomen is dark brown with a pale dorsal patch that forms two lobes on either side of the dark heart mark. The abdomen has chevron-shaped marks at the back. The legs are plain and lighter than the cephalothorax.

OCCURRENCE This species occurs from Newfoundland to Manitoba and south to Illinois, Tennessee, and North Carolina. This spider has been found in a variety of forests. They have been collected under logs, rocks, or other debris on the ground.

SEASONALITY Males: April through June. Females: all year.

REMARKS The funnel web is constructed under rocks or debris and extends only a short distance away from cover.

Cybaeopsis tibialis (Emerton, 1888)

Plate 31

IDENTIFICATION This is a small to medium-sized amaurobiid. The cephalothorax is darker in the head region. The abdomen is lighter in color with indistinct light marks on either side of the heart mark. The femora are lighter than the carapace. The legs are darker near their ends. The spinnerets are usually not visible from above.

OCCURRENCE This species occurs from Labrador and Newfoundland west to Lake Superior and south as far as New York. Others in the genus occur across Canada and through the Appalachians. This spider has been found in rotting logs, debris, leaf litter, and under rocks.

SEASONALITY Males: June through August. Females: June through October.

Pimus fractus (Chamberlin, 1920)

Plate 31

IDENTIFICATION This is a small amaurobiid. The cephalothorax is pale brown or tan, darker in the head region. The anterior median eyes are distinctly smaller than the posterior median eyes. The abdomen is grayish brown with paired light spots on either side of the midline, appearing as chevrons at the back. In some individuals the abdomen is plain. The legs are pale brown with pale bands and are darker near the ends.

OCCURRENCE This species occurs in California. Other members of the genus are found in Oregon and California. This spider is found under rocks, logs, debris, or leaves on the ground in humid areas.

SEASONALITY Males: December through March. Females: all year.

FAMILY AMPHINECTIDAE · Amphinectids

The family includes about 175 species from Australia, New Zealand, and South America. There is only one introduced member of this southern family found in North America north of Mexico. Our introduced species, *Metaltella simoni*, is a relatively large native of South America. In North America it is associated with human-altered habitats. It builds a tangled cribellate web under debris and could be confused with a hackledmesh weaver. Adult males may be found wandering in buildings at night.

Metaltella simoni (Keyserling, 1878)

Plate 34

IDENTIFICATION This is a moderately large brown spider with conspicuously long jaws and relatively long unbanded legs. This is a cribellate spider, and because of its large size, the cribellum is sometimes visible with a magnifying glass.

OCCURRENCE The only species in North America north of Mexico was introduced from South America, native to Argentina, Brazil, and Uruguay. It has been found in the Southeast and the Southwest but may be expanding its range. These spiders have been found in or near buildings. This species builds a tangle web under logs or other debris on the ground. The males have been found wandering away from the retreat.

SEASONALITY Nothing is published.

REMARKS The first known records for North America are from the 1940s.

FAMILY ANAPIDAE • *Ground Orbweavers*

These spiders are nearly as small as the dwarf orbweavers (Symphytognathidae) and share the odd feature that the adult females lack palps. They are similarly tiny (about 1 mm) but have eight rather than four eyes. They build a horizontal orb that has a few radii attached out of the plane of the web. The webs have been found in small spaces near the ground. There are currently two species placed in this family in North America. One of these (*Gertschanapis shantzi*) is illustrated on Plate 20. The other one (*Comaroma mendocino,* Levi) is poorly known and may actually belong in the family Theridiidae (Coddington 2005).

Gertschanapis shantzi (Gertsch, 1960)

Plate 20

IDENTIFICATION When revealed by misting, the small orb web is horizontal with lines that leave the normal flat plane of the orb and extend upward (see Fig. 25, right). The adult females have no palps. The abdomen often has a series of accordion-like folds around the side. The cephalothorax has a tall clypeus and a hump just behind the eyes. The presence of eight eyes distinguishes *Gertschanapis* from *Anapistula*, which has only four eyes.

OCCURRENCE This is a species of humid forests in the coastal hills and mountains of California and Oregon. This is a tiny spider that builds its small orb web in spaces within the layers of leaves, under ferns, or in rotting wood near the ground.

SEASONALITY Adults: all year.

REMARKS This spider is rarely encountered but can be common in restricted areas when specifically searched for by using a mister or dusting the area with a fine powder such as cornstarch. Most are collected from leaf litter in the summer.

FAMILY ANTRODIAETIDAE • *Foldingdoor Trapdoor Spiders*

The members of this family are found in North America and Japan. We have 2 genera and 26 species in North America north of Mexico. Twelve of the members of the genus *Antrodiaetus* construct the foldingdoors that gave this family its common name. These spiders build a burrow with a loose silk collar that is pliable and can be pulled together, forming an invisible seam. Another group of three species, which used to be classified in a separate genus (*Atypoides*), build a burrow with a turret. There are 11 members of a third group (*Aliatypus*) that build a burrow with a thin, hinged trapdoor.

One distinctive feature of the foldingdoor trapdoor spiders is the presence of one to three hard tergites on the top of the abdomen, near the front. Among the mygalomorphs, this feature is shared only with the purseweb spiders (Atypidae) and the midget funnelweb tarantulas (Mecicobothriidae). As is the case with many burrow-inhabiting spiders, most people are likely to encounter wandering adult males.

Aliatypus californicus (Banks, 1896)

Plate 2

IDENTIFICATION This is a large burrowing mygalomorph spider. There are three pairs of spinnerets; the anterior pair has two segments. The female has one dark abdominal tergite near the front. The female has robust chelicerae with a rastellum. The fovea is a pit, not longitudinal as in *Antrodiaetus*. The chelicerae of the male are less robust than in the female. In *Aliatypus* the males' chelicerae do not have the projecting process seen in *Antrodiaetus riversi* males. The male has three abdominal tergites, the posterior two are small. Males have long palps.

OCCURRENCE This species occurs in California. There is a population in the San Francisco Bay region, one in the Sutter Buttes, and another along the western foothills of the Sierra Nevada. In the Bay Area they have been found on banks in dense forests of various types. In the Sierra foothills they have been found at elevations between 250 and 600 m.

SEASONALITY Males: wander in autumn. Females: all year.

REMARKS The burrow entrance is covered with a thin, wafer-like trapdoor at ground level.

Antrodiaetus riversi (O.P.-Cambridge, 1883) • *Turret spider*

Plate 1

IDENTIFICATION This is a large burrowing mygalomorph spider. There are three pairs of spinnerets; the anterior pair are small and difficult to see. The female has one dark abdominal tergite near the front. The female has robust chelicerae with a rastellum. The abdomen is lighter, sometimes with a purplish cast. The fovea is longitudinal. The male has a narrow, hairy process extending forward on the chelicerae that is at least twice as long as it is wide.

OCCURRENCE This species occurs in the Coast Ranges of California in shady pine or deciduous forests but is absent or rare from redwood forests. There are also populations in the Sierra Nevada in pine and fir forests above 2,000 m. Their turreted burrows have been found on slopes near shaded streams.

SEASONALITY Males: wander in fall and winter wet season, earlier at higher elevations. Females: all year.

REMARKS The silk-lined tubular burrow has a turret extending above the ground surface. The height of this turret varies from about 1.5 to 6 cm; the top may sometimes be closed with silk. This species was previously known as *Atypoides riversi*.

Antrodiaetus unicolor (Hentz, 1842)

Plate 1

IDENTIFICATION This is a large burrowing mygalomorph spider. There are only two pairs of spinnerets. The female has one dark abdominal tergite near the front. The female has robust chelicerae with a rastellum. The abdomen is lighter with a purplish cast. The fovea is longitudinal. The chelicerae of the male are narrower and have a tuft of stiff hairs at the ends. The abdomen of the male has three tergites.

OCCURRENCE This species occurs from western Pennsylvania west to southern Illinois and south to northeastern Alabama and the Appalachian region. This spider most often builds its burrow in soft sandy loam soils on slopes. These spiders have been found in shady hemlock or rhododendron stands.

SEASONALITY Males: wander in late autumn. Females: all year.

REMARKS The burrow has a flexible collar that the spider can pull closed, rendering it nearly impossible to detect. The best time to search is at night when the burrows are open and the spider is waiting near the entrance.

FAMILY ANYPHAENIDAE · *Ghost Spiders*

This family is found throughout the world with more than 500 species named, mostly from the American tropics. In North America north of Mexico, 37 species are known from 6 genera. They are generally medium-sized pale-colored spiders. These spiders are nocturnal hunters on vegetation and other surfaces. They run constantly during their active period. This continuous activity, atypical for spiders, is evidently fueled by the consumption of sugary plant nectar (Taylor and Foster 1996). This nectar is taken either from extrafloral nectaries or flowers.

Norman Platnick (1974) described an interesting behavioral difference between the anyphaenids and the members of the family Clubionidae. The courtship of the ghost spiders is active. The male moves his abdomen so rapidly during his courtship performance that it becomes a blur. In contrast, the courtship of clubionids is relatively sluggish. The tracheal spiracle is well forward of the spinnerets, near the center of the abdomen. The claw tufts are composed of two clumps of lamelliform setae. These specialized hairs provide ghost spiders with remarkable traction on smooth surfaces; they are excellent climbers.

Anyphaena maculata (Banks, 1896)

Plate 36

IDENTIFICATION This small anyphaenid is a tan spider, darker in color than *Anyphaena pectorosa*. The chelicerae are darker than the body. The cephalothorax has thin, dark lines extending back from the eyes in the cephalic region to a ring of dark around the thoracic region. The legs have a pattern of dark spots or rings. The abdomen has dark markings forming chevrons at the back.

OCCURRENCE This species occurs from coastal New York southwest to Arkansas and south to Georgia and Louisiana. It has been found in the leaf litter, foliage, and Spanish moss in both pine and hardwood forests.

SEASONALITY Males: September through February. Females: October through April.

Anyphaena pectorosa L. Koch, 1866

Plate 36

IDENTIFICATION This small anyphaenid is a tan spider, lighter in color than *Anyphaena maculata*. The chelicerae are darker than the body. The cephalothorax has a series of thin, dark lines and a pair of longitudinal stripes. The abdomen is pale with scattered spots and two dark lines at the front. The legs are banded.

OCCURRENCE This species has been found throughout most of the eastern United States except peninsular Florida and as far west as Missouri and eastern Texas. This species has been collected in low foliage in both forests and open areas as well as on the ground under rocks.

SEASONALITY Males: April through September. Females: April through August.

REMARKS A similar species, *Anyphaena pacifica*, has been found throughout the western states and southern British Columbia.

Arachosia cubana (Banks, 1909)

Plate 36

IDENTIFICATION This is a medium-sized anyphaenid. The body is tan or light orange. The chelicerae are the same color as the carapace. There is a tuning fork–shaped mark on the carapace, with the prongs pointing forward. There are three dark spots on each side of the carapace. The abdomen has a dark stripe down the center. The legs have scattered spots.

OCCURRENCE This spider has been found from Massachusetts to Michigan and south to Arizona and Florida. This species has been found in tall grass and on the ground.

SEASONALITY Males: May through August. Females: March through August.

REMARKS Previously known as *Oxysoma cubanum*.

Hibana gracilis (Hentz, 1847) • *Garden ghost spider*

Plate 36

IDENTIFICATION This medium-sized anyphaenid is a pale tan or yellowish spider. There may be two faint lines extending back from the posterior lateral eyes on the carapace. The abdomen has scattered dark marks, sometimes in lines or rows. The legs are unbanded.

OCCURRENCE This species occurs from New England west to Iowa and south to Texas and Florida. This spider has been found wandering in a variety of low foliage, including shrubs and trees.

SEASONALITY Adults: all year.

REMARKS This spider has often been found in houses. Previously known as *Aysha gracilis*.

Hibana velox (Becker, 1879)

Plate 36

IDENTIFICATION This medium-sized anyphaenid is a pale tan or yellowish spider. The cephalothorax has two faint longitudinal stripes. The chelicerae are dark orange or brown, extending forward in males. The abdomen has a series of indistinct spots that may form two lines. The legs are unbanded.

OCCURRENCE This is a southeastern species, common along the Gulf Coast and in Florida. They have been found on shrubs and trees.

SEASONALITY Adults: all year.

REMARKS This spider wanders into houses.

Lupettiana mordax (O.P.-Cambridge, 1896)

Plate 36

IDENTIFICATION This is a small anyphaenid. The male has long, dark, projecting chelicerae. The cephalothorax is brown. The banded legs are pale with orange coxae. The abdomen is pale yellow with transverse rows of dark spots or chevrons. Some individuals are brown.

OCCURRENCE This species occurs from Virginia to Florida and west to eastern Texas. It has also been found in coastal southern California. They have been found wandering in shrubs or the low branches of trees.

SEASONALITY Adults: all year.

REMARKS Previously known as *Teudis mordax*.

Pippuhana calcar (Bryant, 1931)

Plate 36

IDENTIFICATION This is a small anyphaenid. The male has long, dark, projecting chelicerae. The cephalothorax is brown. The abdomen is light gray. The femora of the legs are much darker than the other leg segments.

OCCURRENCE This species occurs in extreme southern Texas and central Florida.

SEASONALITY Males: April through July. Females: May through July.

REMARKS Previously known as *Teudis calcar*.

Wulfila saltabundus (Hentz, 1847)

Plate 36

IDENTIFICATION This small anyphaenid is pale, nearly white. Some individuals have scattered gray-green spots forming a ring on the cephalothorax. The abdomen may have small light green spots. This spider has remarkably long legs, particularly the first pair.

OCCURRENCE This species occurs from Nova Scotia west to Minnesota and south to eastern Texas and Florida. They have been found in herbaceous vegetation in fields and prairies as well as in orchards.

SEASONALITY Adults: April through August.

REMARKS This spider's species name means "jumps frequently," an apt choice. They run and jump quickly and are difficult to capture.

FAMILY ARANEIDAE • *Orbweavers*

The family Araneidae is synonymous with the orb-weaving habit. Of the seven families of spiders that spin flat orb webs, this family has the most species. This family includes 3,020 species, or nearly 7 percent of all spider species known. The family is also diverse in North America, with 31 genera and 161 species. Many orbweavers are large and colorful, including some familiar spiders. Most have roundish bodies and moderately short, thick legs. In some members of this family, the males are dwarfs, much smaller than the females.

Many species in this family have distinctive dark patches and light spots on the underside of the abdomen. These spots are particularly conspicuous at night when viewed with a flashlight. For example, the larger species of the genus *Araneus* and all of the *Neoscona* have a large rectangular black area with white or yellowish spots or comma-shaped marks at each corner. Members of the genus *Argiope* have a black patch with many small pale spots. Other genera, such as *Aculepeira* and *Metepeira,* have a black patch with white lines. Such markings are often useful in determining which genus is involved but are rarely distinctive enough to identify the species.

Orbweavers have relatively poor vision and sense prey by their vibrations. When a potential food item blunders into the web, the spider rushes to the site and quickly wraps the prey in a cocoon of silk. Only after the prey is sufficiently trussed up will the spider bite and paralyze the prey. The wrapped prey may be cut out of the web and brought back to the hub or a retreat, where the spider begins feeding. Sometimes a wrapped prey is stored for some time before it is consumed. A few species in the family have abandoned the orb web–building habit and make reduced webs or capture prey in other ways.

Acacesia hamata (Hentz, 1847) • *Difoliate orbweaver*

Plate 13

IDENTIFICATION This is a distinctive spider: the abdomen has a characteristic shape, widest about a third of the way back, tapering nearly to a point. On the dorsal surface of the abdomen are two sets of wavy, curving lines forming a double folium mark. The folium areas are darker than the rest of the abdomen, but they have a lovely gradation of color with a lighter region nearest the dark, wavy lines.

OCCURRENCE This is an eastern species with relatives in the American tropics. It occurs from Rhode Island to Iowa and south to Florida and Texas. They have been collected by sweeping in the moist understory of forests and forest edges, even bogs.

SEASONALITY Males: late summer through autumn. Females: all warm seasons.

REMARKS A fresh web is constructed every night, beginning at dusk; it is removed at dawn. The web is a vertical orb about 20 cm (8 inches) in diameter. The closely spaced spiral orb is effective at capturing moths. This spider sits at the hub at night.

Acanthepeira stellata Walckenaer, 1805 • *Starbellied orbweaver*

Plate 6

IDENTIFICATION *Acanthepeira* possess a distinctive star-shaped abdomen with 10 to 12 points around the edge. The lateral eyes are on a pointed tubercle. The body is covered with a dense clothing of appressed hairlike setae. The color is variable, usually with a dark area in the middle of the rear portion of the abdomen. When the legs are withdrawn in the resting position, the body looks a bit like a burr. The variegated pattern enhances this camouflage.

OCCURRENCE This is an Eastern species with relatives across the continent. This spider is common in grasslands, sand dune vegetation, meadows, and occasionally in trees. Other species have been found in the understory of forests as well as in fields.

SEASONALITY Adults: spring through autumn in the north, all year in the south.

REMARKS This spider occupies its web at night and usually moves to a nearby retreat during the day.

Aculepeira packardi (Thorell, 1875)

Plate 16

IDENTIFICATION The long oval abdomen with a dramatic white folium somewhat resemble *Neoscona oaxacensis*, but *Aculepeira* has denser hairiness on both the cephalothorax and abdomen. *Neoscona oaxacensis* usually has a median dark streak and lateral bands on the cephalothorax. The largest pair of light-colored lobes on the abdomen of *Aculepeira* flare out toward the sides at the back. On the underside of the abdomen, *Aculepeira* has three white lines; the middle one is widest and may have two parts.

OCCURRENCE This species has been found from the Yukon and Northern Territories south through Colorado to western Mexico. This spider has been found in mountain meadows, grasslands, chaparral, and sagebrush deserts. There have been scattered eastern records.

SEASONALITY Adults: summer in north; spring through autumn in south.

REMARKS This spider occupies its web at night. It usually moves to a nearby retreat during the day but will enter the web to capture prey at any time.

Allocyclosa bifurca (McCook, 1887)

Plate 7

IDENTIFICATION The green color and unusual abdominal shape with two points at the tip distinguish it from the somewhat similar spiders of the genus *Cyclosa*. There are two pairs of small humps at the wide portion of the abdomen.

OCCURRENCE This tropical species is restricted in the United States to southern Florida, coastal Alabama, and coastal southern Texas. Individuals have been collected in understory palms and palmettos. It is sometimes abundant in and around buildings.

SEASONALITY Males: rarely observed. Females: all year.

REMARKS This spider suspends strings of egg sacs in its web and perches below them, resembling one more lump in the series. The egg sacs match the greenish color of the spider.

Araneus andrewsi (Archer, 1951)

Plate 8

IDENTIFICATION This spider is among the largest orbweavers in western North America. The coloration is variable but typically dark. The abdomen may have a darker folium mark, not shown in the variety illustrated in Plate 8. The abdominal humps are conspicuous. The other large species with humps in the West, *Araneus gemma* and *Araneus gemmoides*, are paler in color, less hairy, and often have white marks at the front of the abdomen. The ventral abdomen lacks conspicuous black and white areas that are common on other *Araneus*.

OCCURRENCE This is a coastal species in Oregon and California. They have been found around buildings, in shrubs, and often on tree trunks. The spider has also been collected on the southern California Channel Islands.

SEASONALITY Adults: April through summer.

REMARKS The retreat is usually not close to the web, sometimes deep in a bush or in the rough bark of a tree.

Araneus bicentenarius (McCook, 1888) • *Lichenmarked orbweaver*

Plate 8

IDENTIFICATION This is one of North America's largest orbweavers. Most individuals are marked with a pattern of light green and reddish-brown blotches on the abdomen. The presence of green markings on a large *Araneus* in combination with abdominal humps is distinctive. The bases of the legs are typically orange or orange red.

OCCURRENCE This uncommon eastern species has been found in forests. It has also been collected from hedges, fences, and walls.

SEASONALITY Adults: mostly summer.

REMARKS The abdominal coloration renders the spider camouflaged against the lichen-covered bark of trees where they often rest. Individuals from the southern portion of the distribution are larger.

Araneus bispinosus (Keyserling, 1885)

Plate 11

IDENTIFICATION This relatively small species is variable in color. Some have pale abdomens with no distinctive markings; others have a dark folium and dark band at the front of the abdomen. There are two points at the widest part of the abdomen, near the front. They may resemble the plain color forms of the variable species *Araneus montereyensis*.

OCCURRENCE This species has been collected in live oak woodland, chaparral, and coastal sage habitats of central and southern California.

SEASONALITY Adults: early spring through summer.

Araneus cavaticus (Keyserling, 1882) • *Barn orbweaver*

Plate 8

IDENTIFICATION This is a large orbweaver with prominent humps on the abdomen. The humps become even more conspicuous after a molt or when the spider has not eaten recently. The nearly uniform grayish coloration is usually combined with a darker folium on the abdomen. The legs are banded. The entire spider is hairy.

OCCURRENCE This is a spider of northeastern North America and the Allegheny Plateau, west of the Appalachian Mountains and south to coastal Texas. It has often been found around barns, other wooden structures, porches, bridges, and cliffsides.

SEASONALITY Adults: summer; some females survive winter into early spring.

REMARKS This is the spider that E. B. White chose as his model for the story *Charlotte's Web*. The webs can be large, with support threads sometimes extending up to 3 m (about 10 feet). Where the species occurs in dense concentrations on cliffsides, the frames of the webs may be interconnected.

Araneus cingulatus (Walckenaer, 1841)

Plate 10

IDENTIFICATION This is a small greenish orbweaver. The abdomen is green, without the black markings present on the abdomen of *Araneus niveus*. There are two rows of red spots down the sides of the abdomen. In one color variant there is a red mask-shaped mark on the front half of the abdomen surrounding two yellow spots. The related species, *Araneus nashoba*, is similar in appearance to the typical form of *A. cingulatus*.

OCCURRENCE This spider occurs in eastern North America. The webs have been found in tall grasses, shrubs, and trees. Herbert Levi (1973) suggested that this species might build its webs high in trees, which would explain why relatively few are collected even though they frequently appear as prey of spider-hunting wasps. Jonathan Coddington (1987) provided evidence that this species does build its webs in treetops. In Florida they have been found in orchards. The spider is active at night.

SEASONALITY Males: spring, earlier than females. Females: spring through summer.

REMARKS The web of these small nocturnal spiders is sometimes built across a single leaf.

Araneus detrimentosus (O.P.-Cambridge, 1889)

Plate 10

IDENTIFICATION This species is green and brown. The cephalothorax and lower abdomen are brown. The legs are light brown with darker bands. The upper surface of the abdomen is light green with a white margin. Some individuals of *Eriophora ravilla* can have the same combination of a brownish body with a green dorsal abdomen, but they are typically much larger as adults. Immature individuals of these species could be confused.

OCCURRENCE This subtropical species is distributed from western Florida to California along the southern border of the United States and south into Central America. Most of the records are from Texas. The webs have been found in fields and a variety of shrubs, including cacti.

SEASONALITY Males: summer through autumn. Females: spring through autumn.

REMARKS The spider's orb is small, about 8 cm (about 3 inches) in diameter. The retreat is open.

Araneus diadematus Clerck, 1757 • *Cross orbweaver*

Plate 8

IDENTIFICATION This is a large and conspicuous spider, with variable coloration. Near the widest part of the abdomen, a series of white spots form a cross-shaped mark that is present in both males and females. The base color of the abdomen can be yellowish, tan, brown, or even red. There is a dark folium with a wavy margin that is most obvious on the back.

OCCURRENCE This is a European spider that was probably introduced into North America at least 100 years ago. It has been found in two regions; in northeastern North America, where it is slowly expanding its range westward, and on the West Coast, where it has been found between southern British Columbia and California. It is common around Seattle, Washington. The large conspicuous webs of this thoroughly domestic spider are frequently found in garden plantings or near buildings.

SEASONALITY Males: late summer, autumn. Females: summer, autumn; some survive winter into early spring.

REMARKS This species has been the subject of a great deal of research in Europe as well as in North America. Many fundamental aspects of orbweaver biology were first discovered studying the cross orbweaver.

Araneus gemmoides Chamberlin and Ivie, 1935 • *Plains orbweaver*

Plate 8

IDENTIFICATION This is a large and variable species that closely resembles the other humped species *Araneus gemma* and *Araneus cavaticus*. It is relatively pale with few markings on the abdomen except for a light spot at the center front. West Coast specimens can be darker, with spots and sometimes a dark folium on the abdomen.

OCCURRENCE This spider has been collected from the West Coast eastward as far as Michigan and Alabama. The similar *Araneus gemma* is mostly a West Coast species and *Araneus cavaticus* is eastern. The large webs of this species have often been found around buildings, cliffs, cave entrances, pine forests, and shrubs.

SEASONALITY Adults: autumn.

REMARKS Like most orbweavers, this species is primarily nocturnal. I have seen a few individuals in their large webs during the day. The specific name means "resembles gemma" and is appropriate. There is even some evidence of hybridization between these two species.

Araneus guttulatus (Walckenaer, 1841)

Plate 10

IDENTIFICATION This small orbweaver has a pale green cephalothorax and legs. The abdomen is greenish below but on the upper surface it has a large orange or red region with a series of white spots. Some individuals have black on the abdomen between the white spots.

OCCURRENCE The range extends from the East Coast west to Wisconsin and south into Arkansas and southern Georgia. Many of the records of this species have been from swamps, bogs, and other moist habitats.

SEASONALITY Adults: autumn.

Araneus iviei (Archer, 1951) • *Orange orbweaver*

Plate 9

IDENTIFICATION The typical form has an orange body. The plain unbanded legs distinguish this species from the marbled and shamrock orbweavers. The abdomen has a large number of small white spots, usually outlined in red or black. On the ventral abdomen of females there is a white square with orange in the center. The male is brownish with a paler abdomen.

OCCURRENCE This species occurs from eastern British Columbia to Nova Scotia and south into Montana and New England. It has been found in coniferous or mixed forests as well as open meadows and fields. It has also been reported from swamps.

SEASONALITY Adults: May through September.

REMARKS This spider builds a cone-shaped retreat in a folded leaf near the web.

Araneus juniperi (Emerton, 1884)

Plate 10

IDENTIFICATION This is a variable species. The individuals that I have collected were green with three white longitudinal stripes on the abdomen. In some specimens red spots or bands are present on the abdomen. *Araneus bivittatus* also has green and white bands on the abdomen, but the abdomen is distinctly oval, longer than wide.

OCCURRENCE This uncommon spider has been found in the East. It has been collected by sweeping junipers.

SEASONALITY Adults: early spring in south, late summer in north.

Araneus marmoreus Clerck, 1757 • *Marbled orbweaver*

Plate 9

IDENTIFICATION One of our most distinctive orbweavers. The cephalothorax is usually orange or yellow; the legs are orange at the base and have cream or white alternating with dark brown or black bands. The color of the abdomen is variable. The most common form has a pale or yellow base color. It has a characteristic dark pattern on the folium. There are a number of other color forms; the most different are the dusky and the black-spotted forms, which occur in the North. Young individuals are pale and usually rest in the web at night.

OCCURRENCE This species is found in coniferous and deciduous forests as well as in open habitats, primarily in wet or humid areas. The web is built in low bushes, tall grasses, or the low branches of trees.

SEASONALITY Adults: May through October.

REMARKS The retreat is under a folded leaf near an upper corner of the web. Younger individuals may have an exclusively silken retreat. During the day the spider waits in the retreat and holds a signal line connected to the hub. At night it may be found hanging at the hub. The appearance of large orange females, preparing to lay eggs, observed on the ground near the end of October is the source of the nickname "Halloween spider."

Araneus miniatus (Walckenaer, 1841)

Plate 11

IDENTIFICATION This small orbweaver has an abdomen that is widest at the anterior end with two lateral humps. The abdomen is reddish brown with a white transverse band near the widest point. The back of the abdomen has paired dark spots. The cephalothorax and legs are pale brown. The center of the cephalothorax is darker brown, widest at the eye region. The legs are faintly banded.

OCCURRENCE This is an eastern species with most records in southern coastal states, particularly Florida. The range extends to the Northeast as well as into eastern Texas. It has been collected in open woods, swampy woods, pines, scrub oaks, and southern hammocks.

SEASONALITY Adults: spring through summer, earlier in the South.

REMARKS Active at night, returning to the retreat during the day. Frequently captures mosquitoes.

Araneus montereyensis (Archer, 1851)

Plate 11

IDENTIFICATION This small orbweaver has several color forms. The abdomen is triangular, widest at the front. The humps at the anterior end are not as distinct as in *A. bispinosus*. At the front of the abdomen there is usually a dark band, with a rim of white or orange behind. The abdomen is densely covered with light hairs. The head region is dark brown or black and hairy. The thoracic region is brown. The legs are usually banded near the tips.

OCCURRENCE This California species has been found in the low branches of trees, shrubs in oak woodlands, chaparral, and in ornamental plantings.

SEASONALITY Adults: spring through November.

Araneus niveus (Hentz, 1847)

Plate 10

IDENTIFICATION This small greenish orbweaver resembles *Araneus guttulatus*. It has a yellow-green abdomen with an orange dorsum with white spots. Unlike *A. guttulatus*, the abdomen usually has black areas or spots near the front as well as in the center at the back. The cephalothorax and legs are pale green.

OCCURRENCE This spider has been collected in southeastern forests, orchards, as well as near the coast as far north as New Jersey.

SEASONALITY Adults: spring and summer.

Araneus nordmanni (Thorell, 1870)

Plate 8

IDENTIFICATION This is a large spider with dramatic color forms and well-defined humps on the abdomen. The abdomens of some are mottled brown with a light spot at the front centerline of the abdomen. Others have a dramatic white-and-black abdomen with a black folium, a black band between the humps, and a black-bordered white spot or spots at the front centerline. The dark folium on the abdomen is bordered by a light wavy margin usually with three constrictions. All varieties have banded legs. The similar-looking *Araneus saevus* has a more dramatic white centerline at the front of the abdomen as well as a folium that usually has four constrictions.

OCCURRENCE This species is widely distributed in the West but is not found in the Great Plains. It also occurs in the Northeast and throughout the Appalachian range. Individuals are usually found near large boulders and trees in dark mature forests. In the West it has been found high in mountain coniferous forests as well as at lower elevations in shrubs. The species was first described from Europe.

SEASONALITY Males: summer. Females: summer through autumn.

REMARKS There is an interesting relationship between this species and the closely related *Araneus saevus*. Where the two species overlap, *A. nordmanni* is relatively small. Where *A. nordmanni* lives outside of the range of *A. saevus*, it is larger in size, suggesting an ecological interaction between these similar spiders (Levi, 1971). After capturing prey, this spider returns to the retreat to feed.

Araneus partitus (Walckenaer, 1841)

Plate 11

IDENTIFICATION This is a small orbweaver. The cephalothorax and legs are greenish yellow. The abdomen is distinctive in shape, being wider than long, triangular with lateral humps. The dorsal surface of the abdomen is red with yellow spots and a transverse band. At the front corners of the abdomen there are a series of dark spots, surrounded by yellow. There are also small dark spots along the sides of the abdomen, each surrounded by yellow.

OCCURRENCE This is an eastern species. Most records are from coastal states, but the range extends throughout the eastern deciduous forests, west to Arkansas.

SEASONALITY Adults: spring through summer, earlier in the South.

Araneus pegnia (Walckenaer, 1841)

Plate 11

IDENTIFICATION This medium-sized orbweaver has a distinctive abdomen with four oval areas that some describe as looking like a set of butterfly wings. These oval areas are usually pale, sometimes even pink in color. Behind these light spots on the abdomen are paired dark bands. The cephalothorax and legs are dull colored, as illustrated in Plate 11, or yellow. The legs are either banded or plain.

OCCURRENCE The species is widely distributed in the East but found only in southern areas of the West. It builds its distinctive web in shrubs or the low branches of trees. It has often been found in wet areas, including bogs. In the South it is common in garden shrubs.

SEASONALITY Adults: early spring through summer in south, summer in north.

REMARKS The web of this species is somewhat similar to that of the labyrinth orbweaver, with both an orb and a tangle web. The spider usually waits in its retreat, holding a taut line that is attached to the hub of the orb. The silk retreat is a relatively large tent. This species commonly falls victim to mud dauber wasps.

Araneus pratensis (Emerton, 1884) • *Openfield orbweaver*

Plate 14

IDENTIFICATION This is a small orbweaver with a long oval abdomen. Most individuals are yellowish, but some are darker with brownish longitudinal bands. The lack of black in the head region and brown rather than black banding on the abdomen distinguishes this species from other small orbweavers with oval abdomens.

OCCURRENCE This is an eastern species found in open field habitats. They are often plentiful in sweep collections from fields of goldenrod and moist meadows.

SEASONALITY Males: late spring. Females: summer and autumn.

REMARKS The spider occupies the small orb (about 20 cm) even during the day. When disturbed, it drops to the ground.

Araneus saevus (L. Koch, 1872)

Plate 8

IDENTIFICATION This is a large dark spider with conspicuous humps on the abdomen. Some individuals are black. Other individuals closely resemble those of *Araneus nordmanni*. *Araneus saevus* typically has more lobes in the folium on the abdomen. There is a distinct white median line at the front of the abdomen.

OCCURRENCE This species occurs in forests across both Eurasia and North America. In North America it is primarily a northern spider, but the range extends as far south as New Mexico at the southern end of the Rocky Mountains. It is usually found on large trees.

SEASONALITY Adults: late summer through autumn.

REMARKS Previously known as *Aranea solitaria*.

Araneus thaddeus (Hentz, 1847) • *Lattice orbweaver*

Plate 11

IDENTIFICATION This is a medium-sized orbweaver. It has a pale dorsal abdomen with a dark band around the sides. Some individuals have small, dark oval-shaped spots on the abdomen. The cephalothorax and legs are orange or pale brown. The legs, particularly the hind legs, may be faintly banded.

OCCURRENCE This species is primarily eastern, but there are scattered specimen records from the West. It is not known from Florida. Webs are built in shrubs and tall herbaceous growth at the edge of woods. This species is sometimes fairly common in agricultural shrubs such as blueberry and in ornamental plantings around houses.

SEASONALITY Males: late summer. Females: late summer through early winter.

REMARKS The conspicuous silk retreat of this spider is the source of its common name. The silken tube has a pattern of holes that resemble latticework.

Araneus trifolium (Hentz, 1847) • *Shamrock orbweaver*

Plate 9

IDENTIFICATION The name of this spider is derived from the distinct pattern of light spots on the abdomen, said to resemble a shamrock. The base color of the abdomen is often cream-colored but may be green, yellow, or red. The cephalothorax is distinctly marked with three broad dark lines—one down the center expanding to cover the head region, the other two being on the sides of the carapace. The legs are boldly banded.

OCCURRENCE They have been found in a variety of open habitats, particularly in moist areas, marshes, or swamps. This spider builds its web in low bushes or grasses.

SEASONALITY Adults: July through October.

REMARKS This spider can change the color of the body to match the background. The red form is often found among red- or orange-colored leaves in the autumn. The spider waits in a retreat among the vegetation near the web, holding a signal line connected to the hub. The large egg sac with a great many eggs is laid in the autumn. The young usually emerge in spring but sometimes in late fall.

Araniella displicata (Hentz, 1847) • *Sixspotted orbweaver*

Plate 12

IDENTIFICATION The abdomen of this medium-sized orbweaver resembles that of *Araneus thaddeus* in that it is usually pale above with darker sides. Unlike that species, it has a series of six black spots at the back of the abdomen, larger nearest the end. The spots are ringed with pale orange or yellow. In one color form the abdomen is red or pink. The legs are usually unbanded.

OCCURRENCE This species is widespread in the North and the West but absent from the southern Great Plains and the Southeast. There is a second species that shares the northern portion of its range. The sixspotted orbweaver has been found in tall grasses, shrubs, and the lower branches of trees.

SEASONALITY Adults: early spring through summer.

REMARKS The web is small, often suspended near the tip of a conifer branch, or just spanning one curled leaf. The spider has been found resting near the center of the small web during the day.

Argiope argentata (Fabricius, 1775) • *Silver garden spider*

Plate 17

IDENTIFICATION The name is a hint that this species is densely covered with reflective silvery hairs. The back portion of the abdomen is variable but is usually brown with a number of silver or orange spots. In some individuals the entire abdomen is light. There are three humps along each side of the abdomen. The abdomen is less elongated than in the Florida garden spider. The femora of the legs are usually gray, sometimes tan with faint bands. The male is tiny.

OCCURRENCE This is a tropical species; it is common in Central and South America. The range extends into southern Florida including the Keys, southern Texas, southern coastal California, and extreme southern Arizona. This species is common in gardens and humid areas with suitable shrubs. The web is large.

SEASONALITY Males: spring. Females: spring through autumn.

REMARKS Adults usually have four white zigzag decorations, or stabilimenta, in the web forming a cross shape. The spider often sits with the legs held in pairs, sometimes aligned with the bands of the stabilimentum. Immatures may build a circular or spiral stabilimentum. This spider is diurnal.

Argiope aurantia (Lucas, 1833) • *Black-and-yellow garden spider*

Plate 17

IDENTIFICATION Adults of this large species are unmistakable. The abdomen has a series of yellow or pale spots on a black background. The cephalothorax is covered with reflective white hairs. The legs have femora that are pale or orange at the base and darker, often black, at the tips. There are two humps at the front of the abdomen that may be inconspicuous in well-fed females. Some growing females with thin abdomens are paler. On the underside of the abdomen there are two light lines with a dark area in the center that has four to six small white spots. Juvenile females are silver. The males are tiny and usually pale or orange.

OCCURRENCE This species occurs across the United States but is less common in the western Great Plains, the Rocky Mountains, and the western deserts. In Canada there are records from southern Ontario, Quebec, and Nova Scotia. It is most common in humid fields, bogs, and the edges of lakes and ponds.

SEASONALITY Males: late summer. Females: late summer through autumn, through December in south.

REMARKS The legs are often held in pairs. The web of adult females usually has a conspicuous vertical zigzag stabilimentum extending above and below the center where the spider rests (see Fig. 26, left). Young females build a wider platform of zigzag stabilimentum. This spider is diurnal. The egg case is as large as a grape, tan in color, and pointed at the top.

Argiope florida Chamberlin and Ivie, 1944 • *Florida garden spider*

Plate 17

IDENTIFICATION This species resembles the silver garden spider except that it has a longer, thinner abdomen. The abdomen has only two humps on each side, and the dark regions on the back part of the abdomen usually form two parallel dark lines. These dark lines are often marked with yellow. The femora of the legs are orange. The front of the abdomen and the cephalothorax are silvery. The male is tiny.

OCCURRENCE This is a southeastern species that has been found from North Carolina to Alabama. This species has been found in drier sandy areas more than the other species of garden spiders.

SEASONALITY Males: summer through autumn. Females: late summer through November.

REMARKS The legs are often held in pairs. The stabilimentum is composed of four white zigzags extending from the points of the paired legs. This spider is diurnal.

Argiope trifasciata (Forskal, 1775) • *Banded garden spider*

Plate 17

IDENTIFICATION As the name suggests, the abdomen is crossed by a series of narrow black lines. Between the black lines are silver or yellow bands. Some females are so densely hairy that the bands are indistinct. The cephalothorax is silvery. On the underside of the abdomen there are two light lines with a dark area in the center that has six or eight small white spots. The legs are pale orange banded with black. The male is tiny.

OCCURRENCE This species is cosmopolitan. In North America it is the most widespread *Argiope* found throughout the conterminous United States and southern Canada. They have been found in fields, shrubby areas, and gardens. This species seems to tolerate drier conditions than *Argiope aurantia*.

SEASONALITY Males: summer through autumn. Females: late summer through December.

REMARKS The legs are often held in pairs. There is sometimes a vertical white zigzag stabilimentum in the web above and below where the spider is perched. This spider is diurnal. The cream-colored egg case is flat on the top.

Colphepeira catawba (Banks, 1911)

Plate 6

IDENTIFICATION This small orbweaver has a series of points, in two groups at the back of the tall abdomen. The abdomen varies from pale to dark, nearly black. The dark brown cephalothorax has light spots. The legs are variegated light and dark, or banded. This species could be confused with *Dolichognatha* (Plate 17), but that species has elongate chelicerae. In *Dolichognatha* the abdominal tubercles are not grouped.

OCCURRENCE This spider has been rarely observed, but it occurs throughout the southeastern United States. The horizontal web is built near the ground among the roots of trees in second-growth woodlands, pinewoods, and grassy areas. The spider may hide under loose bark near the web.

SEASONALITY Adults: late spring through October.

REMARKS This spider preys on small ants. The egg case is hidden in debris hanging near the web.

Cyclosa conica (Pallas, 1772) • *Trashline orbweaver*

Plate 7

IDENTIFICATION This small orbweaver is often identified by the appearance of its web that usually has a line of accumulated debris, a trashline, placed vertically through the center on a band of silk (see Fig. 21, lower left). The debris includes the indigestible remains of prey and fragments of dry leaves. Sometimes only the silk stabilimentum is present, or the web is not decorated. The spider has a variegated brown-and-white abdomen and a brown cephalothorax. The end of the abdomen is extended into a variable lobe. She usually sits in the center of the web but occasionally rests in the trashline. This is an effective camouflage. There are four species in this genus in North America; *conica* is the most common one. *Cyclosa conica* can be distinguished from *Cyclosa turbinata* by the absence of two humps near the wide part of the abdomen.

OCCURRENCE This spider is found across North America and Eurasia. The other species in this genus, including *Cyclosa turbinata*, largely replace *conica* in the Southeast.

SEASONALITY Males: spring. Females: spring through autumn.

REMARKS Adult females attach their egg cases to nearby twigs or leaves, not the web. Small individuals and sometimes adults vibrate the web violently when disturbed. This is probably a defense against wasp or hummingbird predators.

Cyclosa turbinata (Walckenaer, 1841) • *Humped trashline orbweaver*

Plate 7

IDENTIFICATION This small orbweaver is often identified by the appearance of its web that usually has a line of accumulated debris, a trashline, placed vertically through the center. The spider has a variegated brown-and-white abdomen, a dark brown cephalothorax, and is similar to *Cyclosa conica*. *Cyclosa turbinata* can be distinguished from *Cyclosa conica* by the presence of two forward-facing humps near the wide part of the abdomen. The humped trashline orbweaver incorporates her well-camouflaged egg cases into the trashline (see Fig. 26, right).

OCCURRENCE This spider is found across central and southern North America but is less common in the center of the continent. The species has been collected in many habitats. It may be more common in open areas than *Cyclosa conica*.

SEASONALITY Males: summer. Females: spring through summer in north, all year in south.

REMARKS This spider may vibrate the web violently when disturbed. This is probably a defense against wasp or hummingbird predators.

Eriophora ravilla (C. L. Koch, 1844) • *Tropical orbweaver*

Plate 15

IDENTIFICATION The genus *Eriophora* includes some of the largest orbweavers of the southern states. The abdomen may have humps near the front. The species *ravilla* is chosen to represent the group. The color is exceedingly variable. Some individuals are nearly black, others pale. Most individuals are bristly hairy and brownish in color with or without a large light spot in the center of the abdomen. The light spot, when present, can be broad, lobed, or even reduced to a thin white line. Adult females have an unusually long pointed scape on the epigynum, extending nearly to the end of the abdomen. There is also a black triangle on the underside of the abdomen. Adult males possess peculiarly curved tibiae on their second legs. Some immature individuals, as well as occasionally males, have a pair of shiny bumps on the end of the abdomen.

OCCURRENCE This is chiefly a species of Florida and the Gulf Coast, particularly common in south coastal Texas. A similar species occurs in southern Arizona and southern California. The web is often attached to large oak trees. The spider may have a retreat in the bark of such trees or in folded leaves. The species has also been found in shrubs, palmettos, and other low vegetation at the edge of oak hammocks. This spider is occasionally found near buildings.

SEASONALITY Adults: all year.

REMARKS The spider is strictly nocturnal, building its web after dark and removing the web before dawn. The spider waits in the web at night with its legs spread apart.

Eustala anastera (Walckenaer, 1841) • *Humpbacked orbweaver*

Plate 13

IDENTIFICATION The common name of this species highlights one unusual feature: the abdomen is tallest in the back. This species was chosen to represent five similar ones. The abdomen varies in color and pattern, but it is usually dark brown or reddish-brown. Some individuals have a dark folium or spots on the abdomen. The cephalothorax is completely covered with light hairs.

OCCURRENCE This species has been found across the continent, but it is most common in the eastern and southern states. Several other *Eustala* species are more common in the West. The webs are usually built among dead branches in the understory of conifer and broadleaf forests, or in open woodlands. Occasionally they have been collected by sweep netting in fields and marshes.

SEASONALITY Adults: spring through autumn. The young emerge from the egg case, grow during the summer, and then overwinter as subadults.

REMARKS Typically the web is rebuilt each evening and removed before dawn. The spider waits in the web at night with her legs slightly spread apart. The spider does not build a retreat but rests curled up on branches, looking like a piece of bark. If disturbed, the spider may run a short distance then resume this cryptic pose.

Gasteracantha cancriformis (Linnaeus, 1758) • *Spinybacked spider*

Plate 5

IDENTIFICATION This is perhaps the most distinctive American spider. There are many color forms. The abdomen has a hard exoskeleton, with six spines around the edge. In western populations the spines are reduced to points. The spinnerets are located on a raised tubercle under the abdomen, instead of the more normal position at the end of the abdomen. There is also a small projection from the underside of the abdomen that the male grasps while mating. The male is tiny and rarely seen.

OCCURRENCE This is a southern species with populations in the southeastern states as well as in California. The species is particularly common in Florida. Webs are built in shrubs and the low branches of trees as well as on structures. The spider constructs its web in the morning and can be found at the center of the web in the daytime.

SEASONALITY Males: autumn through winter in Florida, spring and summer in Texas and California. Females: all year.

REMARKS The hard, spiny abdomen may be an adaptation to avoid predation by small birds. According to Willis Gertsch (1979), the spiny body is no deterrent to wasps. The egg sacs are usually attached to the underside of leaves, typically near the web. According to Herbert Levi (1978), the egg case is constructed with various colors of silk and includes a longitudinal dark green stripe. The web is sometimes constructed at an angle to the vertical. The web is occasionally decorated with small white tufts of silk that may attract prey. Males have been found hanging from single silk lines near the webs of females.

Gea heptagon (Hentz, 1850)

Plate 17

IDENTIFICATION This species resembles a miniature version of a garden spider (*Argiope*), and like those spiders it holds its legs together in pairs. The pattern on the abdomen is extremely variable. One common form has a dark central folium on the back half of the abdomen. The abdomen has a series of humps or points that give it the shape of a seven-sided polygon when viewed from above, hence the name. There are four white spots on the dark underside of the abdomen. The cephalothorax is yellowish to light brown. The male has a small abdomen.

OCCURRENCE This species has been found in the eastern United States as well as in southern California and Arizona. The web is usually built in low vegetation, including agricultural fields.

SEASONALITY Nothing is published.

REMARKS This spider is diurnal and shy. When the web is disturbed, she will drop quickly to the ground. Such frightened individuals may darken in color. There is sometimes an open sector of the web below the spider. This species may have been introduced to the Americas from the South Pacific.

Hypsosinga pygmaea (Sundevall, 1831)

Plate 14

IDENTIFICATION This spider has an oval-shaped abdomen and orange-and-black coloration. Some individuals of this species can have two thick longitudinal black bands on the abdomen; others resemble the form illustrated in Plate 14. There is a black area around the eye region. The legs are orange and unmarked. The males of this genus often have a black abdomen.

OCCURRENCE This species has been found from Alaska to Cuba. They have been found in grassy areas and wet meadows. They are usually detected by sweep netting.

SEASONALITY Males: spring. Females: late spring through summer, as late as winter in Florida. Overwinters as half-grown immatures.

REMARKS The orb web is built close to the ground.

Hypsosinga rubens (Hentz, 1847)

Plate 14

IDENTIFICATION This spider has an oval-shaped abdomen. The abdomen is either orange or orange with black at the rear and two black spots at the sides. There is a black area around the eye region. The legs are orange, darker near the ends. The males of this genus often have a black abdomen.

OCCURRENCE This is primarily an eastern spider but has also been found in the prairie provinces of Canada. This species has been collected from shrubs or other understory vegetation in forests. It may hide under bark or in the leaf litter on the forest floor.

SEASONALITY Adults: spring and summer.

Kaira alba (Hentz, 1850) • Frilled orbweaver

Plate 5

IDENTIFICATION The most distinctive feature is a frill of pointy spikes on either side at the front of the abdomen. Most individuals have a pale abdomen, but some have a dark folium. The leg segments are usually dark at the ends.

OCCURRENCE This species is southeastern as are two of the other species. The fourth species has been found in Arizona. Individuals of *Kaira alba* have been collected from tall grasses, shrubs, and the low branches of mangroves.

SEASONALITY Males: early summer. Females: all year.

REMARKS This species does not construct an orb, but it builds a minimal web consisting of a few threads. At night it hangs from these threads by its hind legs and opens the other legs wide. It captures moths by grasping them out of the air. The moths are evidently attracted by an odor lure.

Larinia directa (Hentz, 1847)

Plate 16

IDENTIFICATION The abdomen is tan in color, long and thin, with a series of longitudinal stripes. There are usually two rows of five or six small dark spots. On the underside of the abdomen there is a light median stripe with black stripes on either side. The somewhat similar marsh orbweaver (*Neoscona pratensis*) has shorter legs and an abdomen that is not as thin. *Larinia* has an abdomen that is about three times as long as wide. Longjawed orbweavers (*Tetragnatha*) also have long abdomens, but they have distinctive large, protruding jaws.

OCCURRENCE This species is distributed along the coasts and the southern boundary of the United States. This spider has been found in open country, grasslands, as well as fields. It is replaced by a similar species, *Larinia borealis*, in the North.

SEASONALITY Males: spring through early autumn. Females: all year.

REMARKS The pale tan or yellow coloration perfectly matches the dried grasses where the spider rests during the day.

Larinioides cornutus (Clerck, 1757) • *Furrow orbweaver*

Plate 9

IDENTIFICATION The folium mark on the abdomen is darkest at the margins, fading gradually toward the center. There are a regular series of lobes at the sigilla, places where the heart musculature is attached inside. The upper abdomen, lateral to the folium mark, is pale and unmarked unlike the other members of this genus. Another distinguishing feature is that the abdomen and carapace both appear shiny with few hairs.

OCCURRENCE This is a widely distributed species. It is most common near bodies of water, particularly lakeshores. It often builds on buildings and other structures, sometimes at extraordinary densities.

SEASONALITY Adults: all year.

REMARKS The spider often sits in the center of its web at night and occasionally during the day. The retreat is near an upper corner of the web in a crevice or similar hiding place. The spider does not usually employ a signal line to the hub but may hold any radial strand.

Larinioides patagiatus (Clerck, 1757)

Plate 9

IDENTIFICATION This spider usually has a reddish-brown coloration, particularly on the abdomen. The thin white lines formed by conspicuous white hairs at the margins of the folium on the abdomen are usually continuous. The head region often appears to be clothed in light-colored hairs.

OCCURRENCE This species is common in over much of the northern portion of the continent south to California and Virginia, less common in the southern portion of its range. Unlike the other members of this genus, this species often builds in shrubs and bushes. Also, this species seems less associated with open-water habitats.

SEASONALITY Adults: all year; most common June through September.

REMARKS Like the other members of this genus, the spider may go to the web using any radial thread but on occasion does construct a signal thread to the hub. During the daytime this spider hides in a silk retreat.

Larinioides sclopetarius (Clerck, 1757) • *Bridge orbweaver*

Plate 9

IDENTIFICATION This is the darkest member of this genus. There are thin white lines formed by conspicuous white hairs at the edge of the carapace and margins of the head region. Similar light-colored lines on the abdomen are visible at the boundary of the folium as well as surrounding the central dark mark at the front of the abdomen.

OCCURRENCE This species was probably introduced from Europe. Its distribution is concentrated in the Northeast and the Northwest. The web is most often found on bridges, fences, or buildings.

SEASONALITY Adults: all year.

REMARKS The web is usually reconstructed each evening. The spider often sits in the center of its web at night. When not in the web, the spider rests near an upper corner of the web. The spider does not usually employ a signal line to the hub but may hold any radial strand.

Mangora gibberosa (Hentz, 1847) • *Lined orbweaver*

Plate 12

IDENTIFICATION This species has a yellowish-green body with thin parallel lines on the abdomen, down the center of the cephalothorax, and on the ventral surfaces of the first two legs. The color of this spider's body can be yellow or green. The green form is more common among younger individuals.

OCCURRENCE This is a species of the eastern part of North America. It is fairly common in fields, forest edges, roadsides, and gardens. The web is built in low vegetation.

SEASONALITY Adults: summer and autumn. The hatched young remain in the egg case through the winter and emerge in spring.

REMARKS The web of this spider is a fine-meshed orb built horizontally or at an angle to the vertical. The spider may remain in the web during the day.

Mangora maculata (Keyserling, 1865) • *Greenlegged orbweaver*

Plate 12

IDENTIFICATION This spider is usually grass green to yellow green, but some individuals are yellowish. The abdomen is pale, often almost completely unmarked except for two rows of small black spots at the back. This is the only *Mangora* in North America north of Mexico that lacks a black median line on the carapace.

OCCURRENCE This is primarily a species of eastern deciduous forests. It is usually found in understory vegetation in moist deciduous forests, particularly in floodplains. It has also been collected from the edges of marshes, swamps, dune grasses, and bogs.

SEASONALITY Adults: early summer through autumn.

REMARKS The orb web has closely spaced strands and is usually oriented in a horizontal plane. The egg case is constructed among dry leaves.

Mangora placida (Hentz, 1847) • *Tuftlegged orbweaver*

Plate 12

IDENTIFICATION This species has a pale brown cephalothorax with a dark brown central band and some dark near the outer edge of the carapace. The abdomen is pale with a dark area in the center, wider at the back. There are two rows of three black or dark brown spots along the edge of the dark folium on the abdomen. There is usually a pair of white spots near the widest part of the folium. In some individuals the folium is pale.

OCCURRENCE This is a species of the eastern part of North America. This is one of the most common spiders collected in low understory vegetation of deciduous forests as well as in fields and tall grasses.

SEASONALITY Adults: spring and summer. Young are common in the autumn and overwinter to become mature the following spring.

REMARKS The web of this spider is a fine-meshed orb.

Mangora spiculata (Hentz, 1847)

Plate 12

IDENTIFICATION This species has a green body with a black line running down the center of the cephalothorax and abdomen. Some individuals have parallel lines of black spots rather than a central black band on the abdomen. The abdomen usually has faint yellow crosswise bands.

OCCURRENCE This is a southeastern species most common along the Gulf Coast and in Florida. This spider has been found in low vegetation in open fields as well as in the understory of pine oak woodlands.

SEASONALITY Adults: spring and summer.

Mastophora cornigera (Hentz, 1850) • *Southern bolas spider*

Plate 7

IDENTIFICATION In this species the rough warty bumps on the cephalothorax are relatively high. This species has an abdomen that is blotched with brown and has two distinct humps. The egg case is nearly spherical and is hung in the web by a stout stalk.

OCCURRENCE This is mostly a spider of the southern and coastal states. It is the most common species of *Mastophora* in California.

SEASONALITY Adults: warm months in east; all year in California.

REMARKS At night they hunt using a bolas (see "The Remarkable Bolas Spiders"). During the day the spider rests on the surface of leaves with the legs folded tightly around the body. The lumpy brown body with a white lower abdomen gives the spider the appearance of a bird dropping. The tiny males emerge from the egg case as mature adults.

The Remarkable Bolas Spiders

Spiders in the genus *Mastophora* (bolas spiders) are known for their unusual hunting method and their rarity. Thirteen species have been described from North America north of Mexico, where they are found in all but the Northwestern states. Their abdomen is wider than long, often with humps. The cephalothorax has a warty crown.

The bolas spiders' unique snare has been reduced to two lines. When preparing to hunt, the female constructs a few silk lines, at least one horizontal. The spider attaches a silk line to the horizontal line and moves away, extending the new silk line a few centimeters long. Then she combs additional glue and silk by drawing her back legs along the strand from the spinnerets to a growing globule or ball. This ball is composed of coiled silk as well as glue. When a sufficient amount of material has accumulated, the spider cuts the line between her spinnerets and the sticky ball. Now the line swings free, suspended from the original attachment on the horizontal frame line. The spider turns around and picks up the line with its ball with one of her front legs and assumes the posture shown for *Mastophora hutchinsoni* on Plate 7.

Bolas spiders' minimal web is explained by the fact that they have evolved the ability to mimic moth sex pheromones (Stowe, Tumlinson, and Heath 1987). Male moths approach from downwind. When the moth is close enough, the spider flings the line of silk with its gluey ball at the moth. Recent slow-motion movies made by Mark Stowe and Tonia Hsieh have revealed that the ball actually unwinds, producing an entangling strand that wraps around the moth. The moth is instantly caught, beating its wings to no effect, spinning at the end of the line. The spider pulls in the line, grasps and bites the prey. The name bolas spider is derived from the fancied similarity of the spider's line with a ball at the end to the bolas used by the Argentinean gauchos. Both the tiny males and immature females hunt without a bolas. They use a different odor lure to attract male psychodid flies by hunting at the edges of leaves. Another remarkable feature of these spiders is that they give off an odor when disturbed that is detectable at close range.

Mastophora hutchinsoni Gertsch, 1955 • *Cornfield bolas spider*

Plate 7

IDENTIFICATION In this species the rough warty bumps on the cephalothorax are relatively low. This species has an abdomen that is blotched with brown and has discernable humps. The egg case is vase-shaped and attached at a joint on a small stem of a tree or shrub. It closely resembles a bud or short broken stem and is the same color as the stem, making it hard to see.

OCCURRENCE This is a northeastern species. They have been found in small trees in orchards and rural yards. They may be common near cornfields, perhaps because one of their moth prey is a corn pest.

SEASONALITY Males: summer. Females: late summer through fall; may survive until the first frost.

REMARKS At night they hunt using a bolas (see "The Remarkable Bolas Spiders"). During the day the spider rests on the surface of leaves with the legs folded tightly around the body. The lumpy brown body with a white lower abdomen gives the spider the appearance of a bird dropping. The tiny males emerge from the egg case as subadults.

Mastophora phrynosoma Gertsch, 1955 • *Toadlike bolas spider*

Plate 7

IDENTIFICATION This species has an abdomen that is almost triangular in shape and wider than long when viewed from above; it is swollen laterally at the front and lacks dorsal humps. The legs are long and covered with white velvety hairs. The egg case can be hidden in a folded leaf and has a long stalk as well as irregular tabs attached to the round part.

OCCURRENCE This is an eastern species with a range that extends from New England to Texas. They have been found in the low branches of shrubs and cane.

SEASONALITY Males: spring. Females: May through November.

REMARKS At night they hunt using a bolas (see "The Remarkable Bolas Spiders"). During the day the spider rests on a silk disk that she has attached to the surface of a leaf. With her legs folded tightly around her green and ocher body, she has the appearance of a bird dropping. The tiny males emerge from the egg case as subadults.

Mecynogea lemniscata (Walckenaer, 1841) • *Basilica orbweaver*

Plate 16

IDENTIFICATION This spider is usually identified by the appearance of its unusual web. It begins as an orb with radial and spiral strands spaced so closely that it resembles a fine mesh screen. The holes in the mesh are less than 1 mm and almost square. This orb is held in a dome shape by fibers forming a tangle both above and below. The spider hangs within the dome. The cephalothorax is brown with a black median stripe and margin. The legs are green. The abdomen is relatively long and shiny white with dark green, brown, and yellow lines, and patches of red orange. The coloration is reminiscent of the two *Leucauge* species, but the dome-shaped web eliminates confusion.

OCCURRENCE This is a spider of the southeastern states. It builds its distinctive web in shady areas within or under shrubs or in the understory of deciduous forests.

SEASONALITY Males: summer. Females: summer through autumn.

REMARKS Both males and females build a domed web but the males' web is smaller and less curved. Adult males build webs close to those of females. The egg sacs are attached to each other in a string that is hung within the web.

Metazygia calix (Walckenaer, 1841)

Plate 13

IDENTIFICATION The most distinctive feature of this species is the pair of large black spots at the back of the abdomen, absent in other similar species. The smooth oval abdomen with a wide folium is typical. In some individuals the abdomen is orange with white spots.

OCCURRENCE This species occurs from the Chesapeake Bay region east to Illinois and south to Alabama and Florida. This spider builds it webs in low vegetation or shrubs in open habitats or low branches in forests.

SEASONALITY Adults: spring through September.

REMARKS Previously known as *Alpaida calix*. A related species, *Metazygia wittfeldae*, builds its webs on buildings, railings, bridges, and other structures. It is similar to *Metazygia calix* but has a darker head and lacks the black spots on the abdomen.

Metepeira labyrinthea (Hentz, 1847) • *Labyrinth orbweaver*

Plate 13

IDENTIFICATION *Metepeira* are easily identified by their web, which consists of an orb web among a tangle of lines, the labyrinth. The spider has a distinctive lobed folium on the hairy abdomen with the lobes in the front marked with white. There is also a white line on the underside of the abdomen behind the epigynum and another down the middle of the sternum.

OCCURRENCE This species is widespread in the in the East, extending as far west as Texas. Other species are found further west as well as into southern Canada. The distinctive web is usually constructed in shrubs in forests or forest edges.

SEASONALITY Females: spring through autumn, in Florida all year. Males: spring and summer.

REMARKS The several egg cases are incorporated into the detritus above a tubular retreat in the tangle. After emerging from an egg case, the spiderlings may reside in the labyrinth and even feed on insects captured in the web.

Micrathena gracilis (Walckenaer, 1805) • *Spined micrathena*

Plate 5

IDENTIFICATION The body somewhat resembles a burr or crumpled leaf. There are a ring of eight dark points around the top of the abdomen, and another pair at the back. The underside of the abdomen is a cone ending in a tubercle with the spinnerets. The abdomen color is variable, usually black or black and white but some are orange or yellow. The carapace is dark, either black or orange with a light rim. The male is tiny. The web is built at an angle to the vertical, has finely spaced sticky lines and a circular opening at the hub.

OCCURRENCE This is an eastern species, but there is a somewhat similar species *Micrathena funebris* that occurs in southern Arizona and California. *Micrathena gracilis* builds its webs in the understory of forests and open woodlands.

SEASONALITY Males: late summer. Females: summer through autumn.

REMARKS The webs are often built across trails at about face height; the spiders either drop or cling to hikers who accidentally walk through the webs, but the spiders do not bite. The spider hangs under the hub during the day. It builds no retreat, but may move to a line near the web, away from the hub at night. They do not wrap their prey.

Micrathena mitrata (Hentz, 1850) • *White micrathena*

Plate 5

IDENTIFICATION This species usually has a white abdomen with several dark markings, particularly near the center and at the rear. There are four points at the back of the abdomen. The abdomen is not as deeply cone-shaped as the other species in the genus. The carapace is brown and the legs are lighter brown, darkest at the ends each segment. The male is tiny with a flat body. The web is built at an angle to the vertical, has finely spaced sticky lines and a circular opening at the hub.

OCCURRENCE This is an eastern species, whose range overlaps with the other two species in the genus. It is usually less common than *Micrathena gracilis*. The web is built low in the shady understory of deciduous forests and woodlands, particulary in humid areas. It does not occur in the Florida peninsula.

SEASONALITY Adults: late summer through autumn.

REMARKS This is both the smallest and least common member of the three eastern species of this genus. The spider is diurnal, resting at the hub of the web.

Micrathena sagittata (Walckenaer, 1841) • *Arrowshaped micrathena*

Plate 5

IDENTIFICATION This is a beautiful spider with a brightly colored body. The abdomen has two prominent points at the back and is narrow in front, suggesting an arrowhead shape. There are four other small points on the abdomen. Most of the abdomen is yellow with dark maroon-colored points, often with pink or purple colors near their bases. The web is built at an angle to the vertical, has finely spaced sticky lines and a circular opening at the hub. There may be a white stabilimentum above the hub.

OCCURRENCE This is an eastern species. It is usually less common than *Micrathena gracilis*. The web is built in the humid understory of deciduous forests and open woodlands.

SEASONALITY Adults: summer and autumn.

REMARKS The spider is diurnal, resting at the hub of the web. It builds no retreat, but may move to a line near the web, away from the hub at night. They do not wrap their prey.

Neoscona arabesca (Walckenaer, 1841) • *Arabesque orbweaver*

Plate 15

IDENTIFICATION This is a medium-sized orbweaver with a series of dark streaks on the abdomen in two rows. These streaks are narrow, oval-shaped, and at an oblique angle to the long axis of the abdomen. In the other species of this genus, the dark spots are typically perpendicular to the long axis. There is variation in color; some individuals are orange with few markings.

OCCURRENCE This species is widespread in the conterminous United States and occurs commonly up into the Prairie Provinces of Canada. It is the most common member of this genus. This spider builds it webs in low vegetation, shrubs, and small trees in a wide variety of habitats. They are most abundant in open areas.

SEASONALITY Males: late spring through summer. Females: spring through autumn.

REMARKS The name of this spider refers to the web-building movements that suggest the precise movements of a ballet dancer. They may be found in the web at any time, but typically they return to a retreat nearby during the day.

Neoscona crucifera (Lucas, 1839) • *Arboreal orbweaver*

Plate 15

IDENTIFICATION This species is one of the most polymorphic of all our common spiders. They are usually shades of brown or orange, but some individuals are nearly without any pattern while others, often in the same area, are almost as dramatically marked as *Neoscona domiciliorum*. Where they occur together, the bold-patterned variants are difficult to distinguish from *Neoscona domiciliorum*. The most common forms have a light mark down the center of the abdomen, with light and dark lobes extending toward the sides. These lobes form a crosslike shape and are the source of the species name, *crucifera* or cross bearing. This species is much larger than *Neoscona arabesca*.

OCCURRENCE This is mostly an eastern species, but the southern part of the range extends west to Arizona and California. They are common in deciduous forests, open woodlands, and shrubs. They can be abundant in suburban habitats, frequently building their large webs on fences, patios, buildings, or near lights.

SEASONALITY Adults: late summer through autumn; females survive until the first hard frosts in the North.

REMARKS This species is nocturnal and typically rebuilds its web each evening. Previously known as *Neoscona hentzii*.

Neoscona domiciliorum (Hentz, 1847)

Plate 15

IDENTIFICATION This species has a boldly marked abdomen with a broad white or yellowish central band and several lateral extensions. The light band is usually surrounded by black. Some show pairs of dark spots parallel to the light central band, but unlike *Neoscona arabesca*, these are usually perpendicular to the long axis. The femora of the legs are usually a dull red and the rest of the leg segments are banded with dark near the ends of the segments. This species is intermediate in size between the small *Neoscona arabesca* and the larger *Neoscona crucifera*.

OCCURRENCE This is an eastern species, more common in the southern part of its range. This spider has been collected in open woods and a variety of shrubby habitats. It often builds its webs on fences and in yards near houses.

SEASONALITY Adults: midsummer through autumn.

REMARKS This is a nocturnal species, usually found in a folded-leaf retreat near the web during the day. This species is also known by the name *Neoscona benjamina*.

Neoscona oaxacensis (Keyserling, 1864) • *Western spotted orbweaver*

Plate 15

IDENTIFICATION This is a large orbweaver with an oval-shaped abdomen. The abdomen is usually dark, often black, above with a broad lobed white- or cream-colored band down the middle. Within this light band is another dark band, often lobed at the front. The shape and pattern on the abdomen resembles *Aculepeira packardi* (Plate 16), but that species is much hairier and lighter in color with different lobes. The cephalothorax in *Neoscona oaxacensis* has a dark centerline and dark lateral bands. The legs are banded.

OCCURRENCE This is a southwestern species whose range extends down into Mexico and Central America. The species builds its webs in a variety of habitats including fields, orchards, shrubs, and suburban yards.

SEASONALITY Adults: late summer and autumn.

REMARKS On occasion, this harmless spider's webs are so common around buildings that it is considered a nuisance.

Neoscona pratensis (Hentz, 1847) • *Marsh orbweaver*

Plate 15

IDENTIFICATION This is a medium-sized orbweaver has an oval abdomen with alternating light and dark stripes. The abdomen is wider than the somewhat similar but smaller members of the genus *Larinia* (Plate 16). In *Neoscona pratensis* the central stripe on the abdomen is the darkest, usually with two narrow light stripes on either side. There are dark spots visible along the light bands. In *Larinia* the central abdominal stripe is lighter, with two narrow dark stripes on either side

OCCURRENCE This is primarily an eastern species found in marshes, swamps, and wet grasslands. There is an isolated record from British Columbia.

SEASONALITY Adults: spring and summer.

Ocrepeira ectypa (Walckenaer, 1841) • *Asterisk spider*

Plate 6

IDENTIFICATION The body is brown and covered with a delicate pattern that resembles bark. There are two prominent humps at the front of the abdomen. The distinctive web is rarely observed. It is nearly horizontal and has a hub in the middle connected to a few radial threads that extend to surrounding branches. It has been called an asterisk web because of this shape.

OCCURRENCE This is a species of moist eastern deciduous forests. This spider can be common but is rarely noticed.

SEASONALITY Males: May and June. Females: spring through summer.

REMARKS There is no sticky silk in the web. The radial fibers of the web transmit vibrations to the spider. When a potential prey walks along one of the thin branches to which the radial lines are attached, the spider runs up and rapidly spins a silk band around both the prey and the twig, pinning it there. During the daytime the spider perches nearby on a branch with legs withdrawn, and is nearly invisible. Even so, this species falls victim to wasps and many have been found paralyzed in wasps nests. Previously known as *Wixia ectypa*.

Scoloderus nigriceps (O. P.-Campbridge, 1895) • *Ladderweb spider*

Plate 6

IDENTIFICATION This small orbweaver has an unusually shaped cephalothorax. The four median eyes form a square near the front of the carapace. The lateral eyes are small and touching at the lateral edge of the carapace. There is a large hump in the center of the cephalothorax. The back of the cephalothorax slants steeply to the back edge. The abdomen is held in a nearly vertical position. It has a series of lumps—one at the center in the front and two lateral humps slightly behind.

OCCURRENCE *Scoloderus* has been collected in Florida and extreme southern Texas. The range extends into South America. It is a spider of forests and woodlands.

SEASONALITY Males: spring and late fall. Females: all year.

REMARKS The web of this small spider is a very long ladderlike extension of several sections of the spiral, up to 70 cm long, with a complete orb and hub near the bottom. This web is particularly effective at capturing moths. As the moth tumbles down the web, it is stripped of its scales and eventually the body becomes entangled.

Singa eugeni Levi, 1972

Plate 14

IDENTIFICATION This small orbweaver has an oval-shaped abdomen with longitudinal black and white bands. The cephalothorax is dark in the head region with a band down the center and orange on the sides. The male is similar to the female but may have the black bands on the abdomen reduced to four spots.

OCCURRENCE This species has been collected from the Great Lakes area. There are isolated records from Washington, D.C., and coastal Georgia. Has been found low in wet forests, riverbanks, grasses, and marsh vegetation.

SEASONALITY Males: autumn. Females: spring through autumn.

REMARKS This is a nocturnal spider that hides in hollow stems of grass during the day.

Singa keyserlingi McCook, 1894

Plate 14

IDENTIFICATION This small orbweaver has an oval-shaped abdomen with longitudinal black and white bands. The cephalothorax is orange with a black area around the eyes extending back as a central line. The black in the head region is not as extensive as in *Singa eugeni*. The abdomen of the male is black.

OCCURRENCE This species has been found across southern Canada and south into the Midwest. This spider has been found in low vegetation in forests and around lakes.

SEASONALITY Adults: spring and early autumn.

Verrucosa arenata (Walckenaer, 1841) • *Triangulate orbweaver*

Plate 6

IDENTIFICATION This spider has a number of distinct color forms. The dark parts of the abdomen and the cephalothorax are usually black, brown, or rusty red. There is a large triangular light-colored spot on the top of the abdomen; this spot can be either white or yellow. Different color varieties are sometimes found together. At the back of the triangular abdomen there are six to eight small humps. The spider hangs in its orb web head upward, an unusual position for an orbweaver.

OCCURRENCE This is the only species of the genus that occurs north of Mexico. It is common in eastern deciduous forests. It has also been found in suburban plantings.

SEASONALITY Males: summer. Females: summer through autumn.

REMARKS The web is rebuilt each evening and usually removed near dawn, but I have seen spiders still in their webs in midmorning in Ohio.

Wagneriana tauricornis (O.P.-Cambridge, 1889)

Plate 6

IDENTIFICATION The abdomen has more than a dozen humps or spines distributed on the dorsal surface. This species is somewhat similar to *Acanthepeira*, but *Wagneriana* is smaller and does not have a point in the center at the front of the abdomen.

OCCURRENCE Our single species of this chiefly tropical group is found in the South from Florida to Texas and into the tropics. The webs have been found within 3 m (10 feet) of the ground in shrubs or trees along streams, ravines, in hammocks, or swamp forests.

SEASONALITY Adults: March through December.

REMARKS With its legs withdrawn the spider resembles a bit of debris.

Zygiella x-notata (Clerck, 1757) • *Opensector orbweaver*

Plate 13

IDENTIFICATION The abdomen has a regular oval shape and is typically gray in color and often shiny. There is a wide dark folium on the upper surface of the abdomen, darkest at the edges in a series of lobes. The cephalothorax has a pale thoracic area and a darker head. The dark color of the head extends back as a central line in the thoracic part. The legs are banded.

OCCURRENCE This species is nearly cosmopolitan, having been accidentally introduced to North America on both coasts. The ranges of the other four species include much of the rest of the continent. The webs of *Zygiella x-notata* are often constructed on fences, railings, and other structures.

SEASONALITY Adults: summer through autumn. The eggs are laid in the fall and typically hatch the following spring. In California they may emerge in the autumn of the same year that the egg case was laid.

REMARKS The members of this genus are notable for the fact that they usually build orb webs with a sector that lacks the spiral sticky lines. It looks as though a pie slice has been removed from the web. Thus there is a one radius that is not connected to spiral lines. This nocturnal spider often hides in her retreat touching that single radial thread.

FAMILY ATYPIDAE • *Purseweb Spiders*

The family Atypidae is represented by 33 species worldwide. In North America north of Mexico there are eight species. The purseweb spiders are the probably the most peculiar of our mygalomorphs because of their unique habits. The purseweb spiders build a silken tube. The tube includes a subterranean burrow, but it also extends up above the ground. In most species the exposed part of the tube is attached to a tree trunk, rock, or other vertical surface (Plate 1). In one species that is not illustrated in this guide, *Sphodros niger*, the tube lies along the ground, often hidden within leaf litter. All of the purseweb spiders cover the surface of the tube with bits of debris and soil, so that the tube matches the surroundings well. The hunting spider waits within the tube. When prey walk over the tube, the spider strikes through the tube wall. After capture, a small temporary slit is cut in the tube and the prey is drawn inside. This slit is quickly repaired. Thus the spider lives its entire life hidden from view. Some females are known to live at least seven years. Only dispersing juveniles or adult males searching for mates are found out in the open.

Another distinctive feature of these spiders are their large, forward-projecting chelicerae. These are often about two-thirds as long as the carapace and have long heavy fangs. Viewed from underneath, the endites extend forward to a point. Despite their fearsome appearance, particularly when they rear up in a defensive posture, these are harmless spiders and do not bite humans unless provoked.

Sphodros abboti Walckenaer 1835

Plate 1

IDENTIFICATION This is a large mygalomorph spider. The chelicerae are large, often longer than half of the carapace. The female has a dark brown cephalothorax and legs. The abdomen is dark reddish or purplish brown. The male has a nearly black cephalothorax and legs, and a purple or even iridescent blue abdomen. The posterior spinnerets have four segments.

OCCURRENCE This species is known from southern Georgia and northern Florida, as far south as Marion County. This spider lives in moist forests, swamps, and hammocks.

SEASONALITY Males: wander in June. Females: all year.

REMARKS The purseweb tube of this species may extend up the trunk of a tree as much as 35 cm. It is usually camouflaged with sand, bits of debris, or even moss.

Sphodros rufipes (Latreille, 1829) • *Redlegged purseweb spider*

Plate 1

IDENTIFICATION This is a large mygalomorph spider. The chelicerae are large, often longer than half of the carapace. This is our largest purseweb spider. The female has a dark brown cephalothorax and legs as well as a dark brown abdomen. The male has a dark reddish brown to black cephalothorax. The legs of adult males beyond the trochanters are bright orange. The posterior spinnerets have three segments.

OCCURRENCE This species occurs from New York west to southern Illinois and south to eastern Texas and northern Florida.

SEASONALITY Males: wander in June even during daytime. Females: all year.

REMARKS The purseweb tube of this species may extend up the trunk of a tree as much as 35 cm and is often wider than in *abboti*, reflecting the larger occupant. It is usually camouflaged with sand, bits of debris, or even moss. Previously known as *Sphodros bicolor*.

FAMILY CAPONIIDAE · *Bright Lungless Spiders*

Worldwide, this family includes 70 species. North of Mexico there is one species with eight eyes and eight others with only two eyes, represented in this guide by the eight-eyed *Calponia harrisonfordi* and the two-eyed *Orthonops gertschi* (Plate 80). They usually have orange or orange-brown bodies and legs with paler abdomens. These spiders lack book lungs. They are ground wandering hunters that are known to eat other spiders. When not active, they live in a silk retreat, often under a rock.

Calponia harrisonfordi Platnick, 1993

Plate 80

IDENTIFICATION This is perhaps the most distinctive member of the family because it retains the typical spider complement of eight eyes. Others in the family have only two. The eight eyes are in a tight group at the front of the head region. The tarsi have conspicuous claws. All parts of the spider are orange.

OCCURRENCE This species is restricted to the coast ranges of California, typically in oak woodland on the ground under rocks, leaves, or other debris.

SEASONALITY Adults: all year.

REMARKS Other spiders are frequent prey of these small wandering hunters. Yes, this spider was named in honor of the actor, recognizing his support of the American Museum of Natural History.

Orthonops gertschi Chamberlin, 1928

Plate 80

IDENTIFICATION This small orange spider has only two eyes on a dark spot above a high clypeus. This species was chosen to represent eight similar spiders. The abdomen is hairy and gray. The legs lack spines.

OCCURRENCE This species occurs in the intermountain West including parts of eastern California, Nevada, Utah, and northwestern Arizona. Other similar species are found throughout the Southwest. These are spiders of arid scrub habitats, often in rocky areas. They have been found hunting on the ground at night or under rocks during the day.

SEASONALITY Adults: December through September; mostly collected in spring.

REMARKS They are nocturnal wanderers known to eat other spiders.

FAMILY CLUBIONIDAE · *Sac Spiders*

A great many spiders have been placed in this family at one time or another. Recently many of these have been reclassified into other related families (for example, Corinnidae, Liocranidae, Miturgidae, Tengellidae, and others). As currently defined, there are about 530 species worldwide. North of Mexico there are 58 species, in 2 genera. These spiders are mostly medium-sized, light-colored nocturnal spiders that forage on the ground or in the foliage. The sac spider name is derived from the fact that they build a compact silk retreat each morning before they become inactive for the daytime.

They can be distinguished from ground spiders (Gnaphosidae) by their conical spinnerets. These are arranged in a compact group that rarely extends beyond the end of the abdomen. In contrast, the ground spiders have conspicuous cylindrical spinnerets, often visible from above (see Figs. 12G,H). The sac spiders have dense claw tufts that provide them with excellent climbing ability.

Clubiona abboti L. Koch 1866

Plate 37

IDENTIFICATION This is a small pale sac spider. The eye area is dark. The chelicerae are light brown or orange. There is usually a dusky heart mark.

OCCURRENCE This species is widespread in distribution; most records are from the Northeast. This spider has been collected from a wide variety of open and wooded habitats, often near the ground. It has been found among the leaf litter, roots, rocks, herbaceous vegetation, and low shrubs.

SEASONALITY Males: March through September. Females: March through October.

REMARKS Like a number of foliage hunting spiders, this species is known to eat insect eggs when it encounters them.

Clubiona canadensis Emerton, 1890

Plate 37

IDENTIFICATION This is a medium-sized sac spider. The cephalothorax is tan, darker in the eye region. The chelicerae are brown. The legs are pale and unmarked. The abdomen is pale reddish brown with a darker heart mark.

OCCURRENCE This is a northern species that occurs across the continent. It is most common in northern forests. It has been found on a variety of trees and shrubs. It hides under bark, leaf litter, moss, or rocks during the day.

SEASONALITY Adults: April through October in south; June through October in north.

Clubiona kastoni Gertsch, 1941

Plate 37

IDENTIFICATION This is a small sac spider. The cephalothorax is tan, darker in the eye region. The chelicerae are a shade darker but not dark brown. The abdomen is tan with a dark heart mark. The legs are pale without banding.

OCCURRENCE Most of the records for this spider are in the Northeast, but it occurs across Canada and in the Northwest. They have most often been found on the ground in leaf litter or pitfall traps in deciduous forests. This species has also been found in open habitats such as bogs, sand dunes, grasslands, and parks.

SEASONALITY Adults: April through September.

Clubiona riparia L. Koch, 1866

Plate 37

IDENTIFICATION This is a medium-sized sac spider. The cephalothorax is olive, darker in the eye region. The chelicerae are dark brown. The abdomen is light red brown with a dark central stripe flanked by two light stripes. The legs are olive and unbanded.

OCCURRENCE This is primarily a northern species, occurring from Newfoundland to British Columbia and Washington. There are many records from the Great Lakes region. It has also been found in the Rocky Mountains. It is a spider of tall grasses in wet meadows and along streams.

SEASONALITY Adults: April through September. Subadults overwinter under bark or stones on the ground.

REMARKS The adult female builds a distinctive nest by folding a blade of grass twice to create an angular chamber (see Fig. 3, left). Within this nest she spins a silk cocoon where she lays her eggs. She stays with the eggs and usually dies in the nest.

Elaver excepta (L. Koch, 1866)

Plate 37

IDENTIFICATION This is a medium-sized sac spider. The body is tan or light brown. There are eight similar species in the genus. The chelicerae are dark brown. The abdomen has a distinctive pattern of dark spots and chevrons.

OCCURRENCE This is primarily an eastern species but similar relatives are found in the West. This spider has been found under bark or rocks during the day. It occurs in deciduous forest and open habitats nearby. It forages on the ground among the leaf litter or lawns.

SEASONALITY Males: April through August. Females: May through October. Some adults may overwinter.

REMARKS Previously known as *Clubionoides excepta*.

FAMILY CORINNIDAE · *Antmimic Spiders, Antlike Runners*

This is a large worldwide family with more than 900 species. A total of 127 species have been named in this family in North America north of Mexico. Most of these are fast-running ground spiders. They are either nocturnal or diurnal. There are usually several sets of paired spines under the tibiae of the front legs, but these are difficult to see in a living spider. The convergent cone-shaped spinnerets distinguish them from the ground spiders (Gnaphosidae).

Most species are shades of brown, but some species are colorful. A number are thought to be ant or wasp mimics. The coloration, body shape, and behavior all contribute to a resemblance to ants. Presumably this resemblance provides some protection because many predators avoid ants. When inactive, the antmimic spiders live in a tubular silk retreat often under rocks or other debris on the ground.

Castianeira amoena (C. L. Koch, 1842) • *Orange antmimic*

Plate 40

IDENTIFICATION This is a striking orange-colored spider.

OCCURRENCE This is a southeastern species. It has been found in rocky outcrops, under stones, and on the ground in woodlands and forests. This spider is active during the day and has often been seen crossing trails.

SEASONALITY Nothing is published.

REMARKS The bright color suggests that of velvet ants. They are orange- or red-stinging wasps of the family Mutilidae. It is possible that the colors of this spider provide protection from predators because of their resemblance to these vicious wasps.

Castianeira cingulata (C. L. Koch 1841) • *Twobanded antmimic*

Plate 40

IDENTIFICATION This species is dark maroon with a pair of white bands on the front half of the abdomen. The cephalothorax is plain. The legs are pale yellow with dark lateral and dorsal longitudinal stripes. The lighter areas are emphasized with white hairs. There is another color form with a dark brown cephalic region that contrasts with a light-colored thoracic region.

OCCURRENCE This is an eastern species. Other similar species occur in the West. This species has been collected in a variety of habitats, including prairies and sand dunes as well as leaf litter of forests and woodlands. They are sometimes found under rocks, logs, or other debris.

SEASONALITY Males: summer. Females: all year.

REMARKS These active runners have been found in the daytime or at night and are often seen in areas with similar-looking ants. When these spiders pause, they wave their front legs in the air. This behavior enhances their antlike appearance.

Castianeira descripta (Hentz, 1847) • *Redspotted antmimic*

Plate 40

IDENTIFICATION This spider is nearly black except for a bright red-orange spot on the abdomen. The spot may cover the entire central region or be restricted to the back of the abdomen.

OCCURRENCE This is an eastern species. Other similar species are found farther to the south and west. This species has been found primarily on the ground in deciduous forests, pastures, prairies, and along the shoreline. It has been collected among leaves as well as under logs or rocks.

SEASONALITY Adults: summer.

REMARKS According to Jon Reiskind (1969), the redspotted antmimic has more often been found in open areas than the twobanded or manybanded antmimics.

Castianeira longipalpa (Hentz, 1847) • *Manybanded antmimic*

Plate 40

IDENTIFICATION This is a dark gray, brown, or black spider with numerous light transverse bands on the abdomen. White hairs enhance the abdominal light bands. The femora are dark with a band of white hairs at their ends. The cephalothorax is often covered with scattered white hairs.

OCCURRENCE This is the most widespread member of the genus. They have been found in prairies, deciduous forests, shrubby areas, and wooded sand dunes. They live on the ground among the leaves or under rocks, logs, or other debris.

SEASONALITY Adults: early summer in the South, late summer in the North until autumn.

REMARKS This is a fast-running spider, often active during the day. They have often been observed running on trails.

Castianeira occidens Reiskind, 1969

Plate 40

IDENTIFICATION This is a gray spider with a broad orange top surface of the abdomen. The cephalothorax has a light median band as well as a light rim. There are numerous white hairs on the femora of the legs, sometimes forming bands.

OCCURRENCE This is a western species from California and Nevada south to Arizona and southwestern New Mexico. This species runs rapidly on the ground. It has been collected in grasses, mesquite woodlands, and in mountains as high as 2,400 m.

SEASONALITY Nothing is published.

REMARKS Jon Reiskind (1969) suggested that this spider might be a velvet antmimic.

Castianeira variata Gertsch, 1942

Plate 40

IDENTIFICATION This species is similar to the manybanded antmimic. There are usually fewer light bands on the abdomen and these are narrower.

OCCURRENCE This is a primarily a southeastern species. They have been collected in dry woods and open areas. They are often captured in pitfall traps.

SEASONALITY Adults: summer.

REMARKS This is a fast-running species. According to Jon Reiskind (1969), they are known for a characteristic waving of the abdomen while pausing between sprints.

Drassinella gertschi Platnick and Ubick, 1989

Plate 40

IDENTIFICATION This is a small plain spider. The chelicerae are fairly robust. The abdomen of the male has a shiny scutum over most of the dorsal surface.

OCCURRENCE This species has been found only in California and Baja California. The related species have been collected in a number of adjacent western states. Most individuals have been collected under rocks in canyons or hilly areas. Others have been found in leaves on the ground. This species is also captured in pitfall traps.

SEASONALITY Adults: December through September.

Meriola decepta Banks, 1895

Plate 39

IDENTIFICATION This species has a dark maroon or red cephalothorax and pale abdomen. They typically rest with the long first two pairs of legs extended parallel in the front. In the male the front legs are nearly as dark as the cephalothorax. The color pattern is somewhat similar to the bullheaded sac spider. *Meriola* is smaller with a thinner, proportionately longer, abdomen.

OCCURRENCE This species occurs from New England to California and south into the tropics. These spiders have been captured on the ground among debris or in leaf litter samples. I have found them in discarded cardboard boxes on the ground. They have also been collected in meadows and soybean fields

SEASONALITY Adults: summer and autumn.

REMARKS Previously known as *Trachelas deceptus*.

Phrurotimpus alarius (Hentz, 1847)

Plate 80

IDENTIFICATION The tibiae of the front legs are conspicuously bicolored, black at the base. There are two common species. *Phrurotimpus alarius* is best distinguished from the similar *P. borealis* by a series of distinct black spots or rings on the hind legs, absent in *borealis*. Both share the habit of folding their front legs up over the cephalothorax when at rest. The abdomen coloration is variable, some individuals have the distinct markings as shown in Plate 80, and others are plain.

OCCURRENCE This species is found across the eastern half of the continent. Other similar species are present in the West. These spiders are found on the ground, frequently under rocks or logs. They run very fast when disturbed. They have often been collected in pitfall traps or leaf-litter samples.

SEASONALITY Adults: early spring and summer. Subadults overwinter under logs or rocks.

REMARKS They are sometimes active in the daytime. If they pause while running, they sometimes wave their front legs in the air. The compact, circular, and shiny red egg sacs are attached to the underside of rocks or logs.

Scotinella fratrella (Gertsch, 1935)

Plate 80

IDENTIFICATION This species was chosen to represent 35 similar species. They are united by their dark coloration and antlike behavior. The cephalothorax is uniform in color, brown and shiny. Some individuals are yellowish (Penniman, 1985). They have thin, unmarked legs.

OCCURRENCE This species is eastern, but other species are found throughout southern Canada and the United States. They have usually been found on the ground. They have been collected in a variety of habitats, including deciduous forests and open grassy habitats.

SEASONALITY Adults: all year.

REMARKS Species in this group run in short bursts, pause, wave their front legs, then dash off again. This behavior resembles the ants that they are often found with. Some species actually live in ant nests. Previously known as *Phrurolithus fratrella*.

Trachelas tranquillus (Hentz, 1847) • *Bullheaded sac spider*

Plate 39

IDENTIFICATION The bullheaded sac spider has a maroon-colored cephalothorax and a gray or tan abdomen. It closely resembles *Meriola* in color but has a much heavier head region, swollen and squared off in the front, and with a thick exoskeleton. The color pattern is similar to the unrelated woodlouse spider (*Dysdera*). This spider has eight eyes—unlike *Dysdera*, which has only six. The jaws of the bullheaded sac spider are not protruding or divergent as they are in the woodlouse spider. Like the woodlouse spider and *Meriola*, these spiders lack spines on the legs.

OCCURRENCE The bullheaded sac spider occurs in eastern North America. Individuals have occasionally been found in the West, where humans may have transported them. These spiders have been found wandering on the ground among debris but also on fence posts or the trunks of trees. This spider has often been found in sheds and buildings. During the day it is found in a silken sac, typically in a rolled leaf.

SEASONALITY Males: June through November. Females: all months except February and March.

REMARKS Both *Trachelas* and *Dysdera* occur around buildings and have been implicated in bites. The bites are not considered medically significant. The egg sac is a distinctive smooth white disk raised in the center. I have often found them under the loose bark of standing dead trees. Previously known as *Trachelas ruber*.

FAMILY CTENIDAE • *Wandering Spiders*

This family is primarily tropical in distribution with more than 450 species. Only eight species have been found north of Mexico, two of which have been introduced. As their name implies, these spiders are known for their wide-ranging movements during the night. The tropical species are excellent climbers, but our native species are found on the ground. Occasionally tropical species have been accidentally imported with commercial fruit shipments and are subsequently found among the produce in grocery stores. The most distinctive feature of these spiders is their peculiar eye arrangement. The eyes appear to be in three rows, the anterior median eyes in front, followed by a row of four including the anterior lateral eyes and posterior median eyes, and finally the posterior lateral eyes behind. The median eyes form a distinctive trapezoid at the front of the face (see Fig. 19B). Some South American species of wandering spiders are dangerously venomous.

One species in this family (*Cupiennius salei*, Keyserling) has been the subject of extensive research by Friedrich Barth (2002). This work includes some of the most detailed examination of the sensory structures and behavior of any spider. This is an excellent source of information on all aspects of sensory physiology of spiders, and the influence of the senses on behavior.

Anahita punctulata (Hentz, 1844)

Plate 39

IDENTIFICATION This yellow spider has conspicuous spots on the femora, chiefly at the bases of the spines. There is a light median band with a dark margin on the cephalothorax. The legs are spiny. The front legs are nearly twice as long as the body. The tips of the legs are sparsely scopulate.

OCCURRENCE This species is primarily southeastern, extending north into the Western Allegheny Plateau. It has been found primarily in moist forests but also in woodlands and hammocks in Florida. They hunt on the ground at night. They have been found under rocks and debris during the day. They have frequently been captured in pitfall traps.

SEASONALITY Males: spring and summer. Females: all year.

REMARKS Previously known as *Anahita animosa*.

Ctenus hibernalis (Hentz, 1844)

Plate 39

IDENTIFICATION These relatively large wandering spiders have been confused with wolf spiders, but the pattern of the eyes is completely different. There is a broad pale line extending down the center of the cephalothorax and abdomen. The legs are conspicuously spiny. The legs have dense scopulae.

OCCURRENCE This is a southeastern species. There are three similar species that have been found further south and west. It wanders, primarily on the ground or low foliage. There are records from caves. Other spiders in this group are excellent climbers.

SEASONALITY Adults: early spring through late autumn.

FAMILY CTENIZIDAE · *Trapdoor Spiders*

This worldwide mygalomorph family includes 118 species. Most of the species are tropical in distribution. In North America north of Mexico we have 14 species in 4 genera. Ctenizids are the most famous of the trapdoor-building spiders. Most species build a thick, hinged door that they can hold tightly closed with their fangs by virtue of two tiny holes near the lip of the lid. When hunting, they wait under the slightly cracked opening, ready to pounce on any prey that wander within reach. After a lightning-fast capture, they return to the burrow and close the lid quickly. The snug-fitting lids of these spiders resemble a cork stopper and are sometimes referred to as cork doors. They are well camouflaged. The two species of the genus *Cyclocosmia* build thin doors.

Bothriocyrtum californicum (O.P.-Cambridge, 1874) · *California trapdoor spider*

Plate 2

IDENTIFICATION This large mygalomorph is a stocky brown spider with a dark shiny cephalothorax. The chelicerae have a rastellum. The fovea is U-shaped with the open end toward the rear; in *Ummidia* the U is open toward the front. The abdomen is lighter in color, gray, orange, or yellow. The legs are stocky.

OCCURRENCE This spider has been found in California south of Fresno and Santa Barbara Counties. The burrows have most often been found on dry sunny slopes.

SEASONALITY Males: wander from autumn through spring. Females: all year.

REMARKS This spider builds a hinged, thick, corklike trapdoor to its large burrow. The inside of the lid has two small holes corresponding to the place where the spider grasps the lid with her fangs to hold it shut.

Cyclocosmia truncata (Hentz, 1841)

Plate 2

IDENTIFICATION This is a large mygalomorph. The abdomen is beautifully sculptured with a corduroy pattern of ridges. The back of the abdomen is a black, nearly flat disk, the source of this spiders' specific name.

OCCURRENCE This species is found in the Southeast from central Tennessee, northern Alabama, and northwestern Georgia. A similar species occurs as far south as northern Florida. They build their burrows in the slopes of wooded ravines. This spider has rarely been collected, probably due to its populations being restricted to small areas of suitable habitat.

SEASONALITY Males: wander in late summer and autumn. Females: all year. Females may live to be at least 12 years old.

REMARKS The burrows of this species often have a side branch. The spider hides in the burrow with the plate facing the entrance. It is unclear whether this deters predatory wasps or reduces the chance of flooding.

Ummidia audouini (Lucas, 1835)

Plate 2

IDENTIFICATION This is a large mygalomorph. The chelicerae have a rastellum. The members of this genus have a U-shaped fovea with the open end forward. The female of this species is a dark brown or black spider with a round body and short legs. The male is black with longer legs and sometimes with a lighter-colored abdomen.

OCCURRENCE This species is widespread from Virginia west to Illinois and south to Oklahoma and the Gulf States. They have been found along banks and hillsides in wooded habitats.

SEASONALITY Males: wander in summer. Females: all year.

REMARKS The burrows of this spider have a hinged, thick cork-type trapdoor. When closed, the burrow and its door are well camouflaged and difficult to detect.

FAMILY CYBAEIDAE · Soft Spiders, Water Spiders

The 80 species in 8 genera representing this group in North America north of Mexico have sometimes been placed in the family Argyronetidae (Jocqué and Dippenaar-Schoeman 2006). This family is Holarctic in distribution, including about 150 species. In the past our species were classified with the Agelenidae. They possess conical spinnerets, the anterior pair (AS) being somewhat longer than the posterior pair (PS). They have spiny legs. The carapace is usually hairless and shiny.

Our species of soft spiders are ground living, often preferring humid environments. The majority of North American species are western. They can be common on the ground in western forest habitats. One species is known to forage on the trunks of trees (Bennett 2005). These spiders build thin, loose, funnel-shaped webs. Typically the webbing is hidden under a rock or log with only a small amount, or even just a few signal lines, extending into the open.

Cybaeus giganteus Banks, 1892

Plate 33

IDENTIFICATION This is a relatively large dark brown spider with robust chelicerae. The cephalothorax is dark brown. The chelicerae are tall and robust. The legs are brown without rings. The abdomen is dark brown or black, with or without a pattern of light areas.

OCCURRENCE This is an eastern species occurring from New York west to Ohio and south to Georgia and Alabama. These spiders have been found in the understory of moist forests, usually near the ground. They build a loose web with somewhat funnel-like openings at each end. The web is placed under rocks or among debris on the forest floor and along streambeds.

SEASONALITY Adults: May through October.

REMARKS This is the largest member of the genus in the eastern states. They resemble the species of *Wadotes* that are common in the same areas.

Cybaeus reticulatus Simon, 1886

Plate 33

IDENTIFICATION This is a relatively large brown spider with robust chelicerae. The cephalothorax is dark orange brown with black markings. The chelicerae are tall and robust. The legs are orange brown with black rings. The abdomen is brown with a pattern of pale spots shaped like chevrons toward the back.

OCCURRENCE This is a western species occurring from Alaska to California along the West Coast. They have been found in the understory of moist forests, usually near the ground. The loose web has somewhat funnel-like openings at each end. It is often placed under debris or in the forest floor litter. They have also been collected under rocks in alpine meadows.

SEASONALITY Adults: June through October.

REMARKS This is probably the most frequently encountered member of a large group of similar species. The females guard their eggs in protected places under bark or rocks.

FAMILY CYRTAUCHENIIDAE (NOW EUCTENIZIDAE) • *Waferlid Trapdoor Spiders*

This family contains 126 species, but only 17 are described from North America north of Mexico. There are many undescribed species in this group; most of these are from California. Members of the family Cyrtaucheniidae are known for their wafer-thin and cryptic trapdoors. Exceptionally, the members of the genus *Apomastus* build burrows with a turret but without a hinged trapdoor. Some species in this family are nearly as large as tarantulas (Theraphosidae) and may be confused with them. The legs of waferlid trapdoor spiders are generally thinner, with longer segments than those of tarantulas. The fovea is typically oriented as a transverse groove, either straight or curved.

Aptostichus simus Chamberlin, 1917

Plate 3

IDENTIFICATION This is a medium-sized mygalomorph. In the female the head region of the carapace is elevated above the thoracic region. The fovea is transverse or curved into a U-shape opening forward. The cephalothorax is brown or reddish brown. The abdomen of the female is light colored with a subtle pattern of dark chevrons. The male is lighter tan in color with longer legs and a flatter cephalothorax.

OCCURRENCE This spider occurs in southern California. This is a sand-loving species found in dune areas along the coast.

SEASONALITY Males: wander after autumn or winter rains. Females: all year.

REMARKS This species builds a burrow with a wafer-thin trapdoor.

Myrmekiaphila comstocki Bishop and Crosby, 1926

Plate 3

IDENTIFICATION This large mygalomorph is a variable reddish-brown spider. The cephalothorax is usually darker than the abdomen and the femora of the legs are darker than the distal segments. In the female the head region is raised well above the thoracic portion of the carapace. The chelicerae are massive. The fovea is a transverse slot. The abdomen is dusky, often with a hint of a banded pattern.

OCCURRENCE This species occurs in Oklahoma, eastern Texas, Arkansas, and northern Louisiana. Others in the genus have been found in the southeastern states. The burrows have been found in areas of thick leaf-litter in woods.

SEASONALITY Males: spring. Females: all year, long lived.

REMARKS The burrow has a wafer-thin trapdoor. There may be a branch within the burrow that has a second trapdoor.

FAMILY DEINOPIDAE • *Ogrefaced Spiders, Netcasting Spiders*

There are 57 species in this family; most species are tropical. There is only one species of this family north of Mexico, and it is restricted to subtropical habitats in the southeastern states. These spiders have an odd face with huge forward-facing posterior median eyes (see Fig. 19A). They also have long bodies and rest in positions that resemble small sticks, hence the alternate common name "stick spider."

The hunting behavior of these spiders is distinctive; it is often referred to as net casting. They hang from a small frame of threads and hold a tiny flexible orb web between their front four legs (Plate 47). When potential prey walk or fly nearby, the spider rapidly extends its legs and thrusts the web onto the prey. The sticky hackleband threads of the orb entangle the prey. As the spider retracts its legs, the net collapses around the victim.

Deinopis spinosa Marx, 1889 • *Netcasting spider*

Plate 47

IDENTIFICATION There is no difficulty identifying this species, at least once it is recognized as a spider! The posterior median eyes are extraordinarily large and placed on the front of the cephalothorax forming a unique face (see Fig. 19A). Some consider it ugly, so an alternate name for this spider is the ogrefaced spider. The body is long and thin. When resting during the day, these spiders cling to a twig and are difficult to find.

OCCURRENCE This species is found in Florida, the Caribbean, and South America. According to J. H. Comstock (1912), the species has also been recorded from Alabama.

SEASONALITY Males: July to following February. Females: April to following February.

REMARKS Bert Theuer (1954) studied this spider in detail. The spider spins a miniature cribellate orb web, which it holds in the front four legs. When prey approach, the spider extends its legs and throws the web onto the prey. As the legs are being extended, the web stretches and on retraction it collapses around the prey. A victim is usually captured as it walks beneath a spider hanging in a nearly vertical posture over a likely substrate. Even more remarkably, the spider occasionally casts its net at flying prey approaching from the side or behind. It accomplishes this by swinging the body around as the legs are extended. It often misses.

FAMILY DESIDAE • *Desids, Saltwater Spiders*

Most of the 180 species in this family have a Southern Hemisphere distribution. We have only one native species and one introduced species north of Mexico. The native species (*Paratheuma insulana*) is an intertidal resident found in southern Florida and the Caribbean. Related species occur in western Mexico and across the Pacific region. Our introduced species (*Badumna longinqua*) is found in and around buildings in coastal California.

The two species in our area could hardly be more different. *Badumna* is a large, dark brown, robust cribellate spider that spins conspicuous messy webs in and around buildings. *Paratheuma* is a small, inconspicuous spider that spends the daytime hidden in silk retreats among the coral rubble or in dead barnacle shells. It has forward-projecting and divergent chelicerae.

Badumna longinqua (Koch, 1867)

Plate 33

IDENTIFICATION This is a large brown spider. The chelicerae are tall and robust, dark brown, and shiny. The abdomen is dark brown densely covered with white or gray hair. There is usually a darker heart mark on the abdomen with paired dark spots toward the back. The femora are darker than the other leg segments. The legs have indistinct banding.

OCCURRENCE This species has been introduced to California and Oregon from Australia or New Zealand. It has adapted well to human habitations, and its messy cribellate webs are found in and around buildings (see Fig. 27, upper right). The spider hides deep in the web during the day, but it may be observed waiting near the funnel opening at night.

SEASONALITY Adults: probably all year.

REMARKS This spider may bite. The bite is painful but is not considered medically serious. Previously known as *Ixeuticus martius*.

Paratheuma insulana (Banks, 1902)

Plate 47

IDENTIFICATION This small spider has relatively large chelicerae that are separated at the ends. The cephalothorax is light brown and shiny. The legs are pale and unmarked. The abdomen is greenish gray. The spider has few distinctive features except that it lives among the coral rubble near the tide line.

OCCURRENCE This species is known from the Florida Keys and the Caribbean islands. It lives among the coral rubble, building a silken retreat in a gap in the broken coral or among vegetation debris. The spider remains in its retreat even when covered by high tide. They emerge from their retreat at night during low tide and hunt.

SEASONALITY Adults: March, May, and December.

REMARKS These spiders are considered rare, but in the correct habitat they have been found in moderate numbers (Beatty and Berry 1988). A related species in the intertidal zone of the Pacific coast of northwestern Mexico builds its retreat in empty barnacles.

FAMILY DICTYNIDAE · *Meshweavers*

This is a large family, occurring worldwide, and throughout North America with 290 species found north of Mexico. Worldwide there are more than 550 species. The meshweavers are cribellate spiders, so their webs contain the distinctive hackled meshwork. This gives the webs an unkempt look. Often there are sections of the web with parallel lines separated by gaps, which are filled with a zigzag of the reflective hackled mesh (see Fig. 22, left). Most of the meshweavers are small and their compact webs are found contained within one folded leaf, spread across the surface of a leaf, or covering the end of a small branch or seed head. Nearly half of our species are members of the closely related genera *Dictyna, Emblyna, Mallos,* and *Mexitlia.* The males of these genera typically have odd bow-shaped cheliceral bases (see Fig. 15). During mating these males hold the females' chelicerae with their bowed chelicerae. With practice, it is fairly easy to recognize the web of a meshweaver as distinct from that of a cobweb weaver (Theridiidae) or hackledmesh weaver (Amaurobiidae). The webs of cobweb weavers have more open structure with unadorned silk lines, without the cribellate zigzags. Unlike the space-filling webs of most cobweb weavers (Theridiidae), the meshweavers often build their webs in the open, without overhanging protection. The hackledmesh weavers are larger spiders that have a spacious web often with a funnel-like depression leading to a tubular retreat.

Some meshweavers do not build obvious webs. They are either free-ranging spiders found near the ground, or they build inconspicuous webs under logs or other debris. Examples of these ground-living spiders are the species of the genera *Blabomma, Cicurina,* and *Yorima.*

Argenna obesa Emerton, 1911

Plate 81

IDENTIFICATION This is a small plain spider. The cephalothorax is brown and the legs are lighter and plain. Unlike many members of the family Dictynidae, the cephalothorax has few hairs and appears shiny. The abdomen is dusky with light gray markings.

OCCURRENCE This species has been found under debris and in leaf litter in wet habitats, including humid forests. The range is primarily in the Northeast, mostly from the New England region. There are also records from Oregon, Idaho, Utah, and Colorado.

SEASONALITY Adults: March through late November.

REMARKS The ground-living habits of these spiders distinguish them from most members of the family Dictynidae. Previously known as *Argenna akita* or *Lathys hesperus*.

Blabomma californicum (Simon, 1895)

Plate 35

IDENTIFICATION This species has six eyes. The anterior median eyes, when present in other members of this genus, are tiny. The body is pale with only an indistinct pattern on the abdomen. These spiders are similar to members of the genera *Cicurina* and *Yorima* (Plate 35).

OCCURRENCE This species occurs along the Pacific Coast between southern California and British Columbia. Other related species have been collected further inland, including the Sierra Nevada. This species has been collected under rocks, logs, and in deep leaf litter.

SEASONALITY Adults: early spring through December.

REMARKS Individuals collected in Washington have usually been found within 500 m of beaches. Many related species have not yet been formally described.

Cicurina arcuata Keyserling, 1887

Plate 35

IDENTIFICATION This species has eight eyes in two regular rows. Some species in this large genus, particularly cave-inhabiting forms, possess only six eyes. The members of the genera *Yorima* and *Blabomma* also have six eyes. The cephalothorax is usually unmarked, but the abdomen sometimes has a pattern of dark spots or chevron-shaped marks.

OCCURRENCE This species is eastern, but there are a large number of western species. All occur on the ground. They have usually been found under rocks or debris or in hollow logs. Some species are known from caves. Many specimens have been obtained from leaf litter using a Berlese funnel.

SEASONALITY Adults: all year.

REMARKS They build inconspicuous thin sheets of silk that are often destroyed when the debris are disturbed.

Dictyna bostoniensis Emerton, 1888

Plate 21

IDENTIFICATION This is a small compact spider. The cephalothorax is plain. The abdomen is covered with white hairs, often with black spots. The legs are yellow or pale brown.

OCCURRENCE This species occurs from Massachusetts to Washington and as far south as southern Oklahoma. It builds a small tangle web, often in fences.

SEASONALITY Adults: May through September.

REMARKS James Emerton (1888) mentioned that this spider looks like a bird dropping.

Dictyna coloradensis Chamberlin, 1919

Plate 21

IDENTIFICATION This is a small, dark compact spider. The cephalothorax is grayish brown with five bands of light hairs on the head region, extending down the midline to the thoracic region as one band. The abdomen is gray with a brown band down the center flanked by brown spots in the back. The legs are yellowish with dark bands near the joints.

OCCURRENCE This spider occurs from Quebec and Maine west to Washington and south to Arizona and Texas. This species builds a small mesh web at the tips of vegetation.

SEASONALITY Adults: summer.

REMARKS This species is similar to *Dictyna volucripes*. It is typically darker in color and replaces *volucripes* in the North.

Dictyna foliacea (Hentz, 1850)

Plate 21

IDENTIFICATION This is a small compact spider. The female has a dark brown cephalothorax with a line of white hairs down the center of the head region. Her abdomen is reddish brown with a yellow median band of variable width. The male has an unmarked dark brown cephalothorax. The male's abdomen is reddish brown with faint lighter median blotches. The legs of both sexes are yellow.

OCCURRENCE This is an eastern species. It occurs from New England to North Dakota and south to Texas and Florida. This is one of the most common meshweaver species. It builds a small web at the ends of vegetation. It has been collected in large numbers by sweep netting in fields.

SEASONALITY Adults: May through late August.

REMARKS The male and female are sometimes found together in the web

Dictyna volucripes Keyserling, 1881

Plate 21

IDENTIFICATION This is a small compact spider. The cephalothorax is grayish brown with five bands of light hairs on the head region, extending down the midline to the thoracic region as one band. The abdomen is gray with a dark spot in the midline at the front and a variegated pattern of brown spots behind. The pattern on the abdomen is variable. The legs are yellowish with dark bands near the joints.

OCCURRENCE This spider occurs in the East from Maine to Wisconsin and Colorado and south to Texas and Florida. This species builds a small mesh web on fences, walls, or at the tips of grasses and weeds.

SEASONALITY Adults: April through December.

Emblyna annulipes (Blackwall, 1846)

Plate 21

IDENTIFICATION This is a small compact spider. The cephalothorax is brown with three broad bands of white hairs on the head region. The abdominal pattern is extremely variable. Some individuals are gray with brown markings as shown; others have few dark marks. The legs are gray.

OCCURRENCE This northern species occurs from Labrador and Newfoundland west to Alaska and south to Oregon, Colorado, Missouri, and Virginia. This species builds a small mesh web on fences, walls, or at the tips of grasses and weeds. This species has frequently been collected by sweep netting in fields.

SEASONALITY Adults: spring through late September.

REMARKS Previously known as *Dictyna annulipes* or *Dictyna muraria*.

Emblyna sublata (Hentz, 1850)

Plate 21

IDENTIFICATION This is a small compact spider. The cephalothorax is dark brown with three lines of white hairs on the head portion, rendering it much lighter in color. The abdomen is reddish brown with a broad yellow band down the center. The legs are yellow. The male has an orange cephalothorax and orange-brown abdomen with or without a central light band. The chelicerae of the male are large and bowed.

OCCURRENCE This species is eastern, it occurs from Maine and southern Canada west to North Dakota and south to Texas and Florida. This spider builds a thin mesh web on the upper surface of a broad leaf, usually in low herbaceous vegetation.

SEASONALITY Males: early spring through June. Females: early spring through September. The young emerge from the egg case after two weeks. These immature spiderlings overwinter under bark or among leaves on the ground.

REMARKS The male and female are sometimes found together in the web. Previously known as *Dictyna sublata*.

Lathys maculina Gertsch, 194

Plate 81

IDENTIFICATION This is a small dictynid. It has a gray or pale yellowish abdomen with darker brown or black markings. The cephalothorax varies in color from pale to yellowish, orange, or light brown. There is usually a black area surrounding the eyes.

OCCURRENCE This spider has been recorded from Long Island, New York, to southeastern Texas. Most records are from coastal states. It has been found in leaf litter and debris on the ground in hardwood forests, where it is often common.

SEASONALITY Adults: all year.

REMARKS Previously known as *Scotolathys maculatus*.

Mallos pallidus (Banks, 1904)

Plate 22

IDENTIFICATION This is a small compact spider. The cephalothorax of this species has a distinct white band around the edges of the thoracic part. On the head region there are three light hair bands. The abdomen has a dark spot in the center and at least two pairs of dark curved spots at the back. The legs are sometimes banded.

OCCURRENCE This is a western species found from California to western Montana and south to Arizona and eastern New Mexico. It occurs in foothills as well as mountains. In southern California this species has been collected from grasslands and chaparral. In more arid regions it has been found close to streambeds and arroyos. The webs are built in the tips of vegetation or on the upper surface of single leaves.

SEASONALITY Adults: all months except February.

REMARKS This spider usually remains still when disturbed and is hard to see because it closely resembles the crumpled prey remains in its web.

Mexitlia trivittata (Banks, 1901)

Plate 22

IDENTIFICATION This is a small to medium-sized compact spider, the largest species in this group. The cephalothorax is dark, lacking a marginal white band. The head region has five white hair bands. The abdomen is gray with a variable width brown spot at the front and a series of paired curved spots in the back. The legs are either plain or faintly banded.

OCCURRENCE This spider has been found from the mountains of central California east to Idaho and Colorado and south to New Mexico and Arizona. It has been found to be common in mountains at elevations up to at least 3,100 m.

SEASONALITY Adults: July through August, scattered reports of females into October.

REMARKS This species has been found in aggregations of up to 50 with contiguous webs.

Nigma linsdalei (Chamberlin and Gertsch, 1958)

Plate 22

IDENTIFICATION This is a small compact spider. The female is green with a mottled pattern on the abdomen. There are white hair bands on the head region and a light marginal band in the thoracic region. In some individuals there is a red central spot on the abdomen. The legs are pale green without banding. The male has a dark brown cephalothorax with the same white marginal band as the female but lacks the white hair bands on the head. The abdomen of the male is green with darker spots and lines.

OCCURRENCE This spider occurs in California. It has been collected in the coastal mountains and foothills in oaks.

SEASONALITY Adults: April through July.

REMARKS This is the only species of *Nigma* discovered in North America. There are 11 other species known from Eurasia and Africa.

Phantyna bicornis (Emerton, 1915)

Plate 22

IDENTIFICATION This is a small compact spider. The cephalothorax of this species is dark, without white hair bands. The abdomen has a single spot in the front and a series of paired spots in the back. The legs are pale.

OCCURRENCE This is an eastern species, with a range that extends as far as western North Dakota. It has been found building its web in vegetation near the ground, or even among rocks on the ground.

SEASONALITY Adults: spring through early autumn.

Saltonia incerta (Banks, 1898)

Plate 22

IDENTIFICATION This is a small plain-colored spider. The chelicerae are relatively robust. The cephalothorax is light brown. The abdomen is gray with scattered spots. The legs are not banded. On close inspection there is a large colulus between the anterior spinnerets. Also visible with a lens, the tracheal opening is about three-quarters of the way between the epigastric furrow and the spinnerets.

OCCURRENCE This spider is rarely encountered. Individuals have been found near dried lakebeds in the Mojave Desert. Some have been located in thin silk webs under sticks, rocks, or other debris.

SEASONALITY Adults: spring and summer.

REMARKS This spider was long thought to be extinct. Until its rediscovery, it was only known from a few specimens found in salt crusts near the Salton Sea in southern California.

Yorima angelica Roth, 1956

Plate 35

IDENTIFICATION This species has six eyes. The body is pale with an indistinct pattern of dark chevrons on the abdomen. These spiders look like members of the genera *Blabomma* and *Cicurina*.

OCCURRENCE The range extends from coastal central California south to Baja California. They have been collected in humid shady areas in deep leaf litter, rotting logs, or under debris on the ground, particularly in chaparral.

SEASONALITY Adults: all year.

FAMILY DIGUETIDAE • *Desertshrub Spiders*

This small family contains only 15 desert-inhabiting species. There are seven species north of Mexico. Their space-filling webs are three-dimensional tangles suspended between the branches of desert shrub vegetation. There is often a sheetlike layer surrounded by a tangle web. They often build in the open central region of spreading cacti. The spider constructs a cone-shaped structure in the center of the web and usually rests within this retreat. The bodies of these spiders are densely white or tan and hairy. These light-colored setae are often reflective, and they may be an adaptation to the intense desert sunshine. The desertshrub spiders possess only six eyes, arranged in three pairs; evidently the anterior median eyes have been lost.

Diguetia canities (McCook, 1889) • *Desertshrub spider*

Plate 22

IDENTIFICATION This is a medium-sized tan spider with six eyes in three groups of two. The body has a dense wooly covering of scale-like setae. The joints of the legs are dark brown.

OCCURRENCE This species occurs from southern California to southwestern Oklahoma and south into Mexico. This spider builds a relatively large space-filling tangle web in low shrubs, particularly cacti, in arid deserts.

SEASONALITY Males: June through September. Females: May through December; egg sacs are produced in August and September.

REMARKS This spider builds a tube-shaped retreat suspended in the center of the tangle web. This retreat is constructed of silk, incorporating a variety of other materials including dead leaves, bits of sand, and prey remains. Females lay their egg sacs within the tube and if disturbed, she will remain to defend them.

FAMILY DIPLURIDAE • *Funnelweb Spiders*

The funnelweb spider family has been made famous by the dangerously venomous Sydney funnelweb (*Atrax robustus*) of Australia. The funnelweb spiders in North America north of Mexico are not considered dangerous. Worldwide, the family includes 175 species. There are two genera and five species in our region. One small species, the spruce-fir moss spider (*Microhexura montivaga*), is a federally listed endangered species occurring in the southern Appalachians. The funnelweb species native to North America build extensive silken retreats usually hidden under rocks, logs, or in moss clumps. The retreat may have a branched funnel-shaped entrance with a small sheet. Typically most of the webbing is hidden, but the sheet may extend out into the surrounding debris or leaf litter.

These spiders have only four spinnerets; the anterior pair has been lost. The median spinnerets have one short segment. The posterior spinnerets of the funnelweb spiders are long, widely spaced, and conspicuous. The species in North America have the smallest chelicerae among our mygalomorphs. The members of the genus *Euagrus*, which includes three species in our region, are among the most common mygalomorphs in Mexico and Central America.

Euagrus comstocki Gertsch, 1935

Plate 3

IDENTIFICATION This medium-sized mygalomorph is a brown or black spider with forward-projecting chelicerae. The abdomen is covered with a velvety coat. The conspicuous long and flexible spinnerets are often held in an upwardly curved position.

OCCURRENCE This species is found in the southern Rio Grande Valley of Texas. It inhabits low elevation acacia-mesquite grasslands. They have most often been found in their webs under a rock or other debris on the ground. A similar species, *Euagrus chisoseus*, occurs from central Texas west to Arizona in oaks and pinion juniper woodlands.

SEASONALITY Males: all year; wander in spring and fall. Females: all year.

REMARKS The web consists of fine but tough sheets with several funnel-shaped openings mostly hidden under a rock, log, or other debris. Some of the sheets and funnel webbing may extend well out into the open. The spider usually remains in the retreat but may rest near the mouth of a funnel at night.

Microhexura idahoana Chamberlin and Ivie, 1945

Plate 3

IDENTIFICATION This tiny mygalomorph has a shiny brown sculptured-looking cephalothorax with a black area around the eyes. The velvety brown abdomen is either unmarked or has a series of paired light spots. The legs and spinnerets are lighter in color, appearing somewhat greenish.

OCCURRENCE This species has been found from western Montana to Washington and south to Oregon and Idaho. This species has been found in moist conifer forests at elevations from 600 to 2,200 m. The webs have been found in the duff of the forest floor, under wood or rocks.

SEASONALITY Males: winter and spring, wander in spring. Females: all year.

REMARKS The web is a thin silken tube extending into the retreat, a depression in the duff. The tube usually forks into two long funnel-like extensions. A second species, *Microhexura montivaga*, is restricted to spruce-fir forests at the tops of the Appalachian Blue Ridge.

FAMILY DYSDERIDAE · *Woodlouse Spiders*

This family has more than 500 species in Eurasia but only one is found in North America north of Mexico. Our species, *Dysdera crocata*, was accidentally introduced by humans. These spiders have only six eyes, arranged in a row of two in front and four behind. They sometimes appear to be in an arc-shape with the gap at the front (see Fig. 29). There are two pairs of conspicuous spiracles under the abdomen.

They have huge chelicerae and long fangs, making them look frightening. In truth, they will usually retreat if given a chance. There are scattered reports of bites, but these have not been medically serious. The bites sometimes result from a spider building its retreat in the fingers of gloves. The large jaws and fangs are adaptations to biting through the hard shells of their principal prey; pill bugs (also known as woodlice, slaters, or roly pollies). Pill bugs are isopod crustaceans with a particularly hard calcareous exoskeleton, necessitating the robust chelicerae of their woodlouse spider predators.

Dysdera crocata C. L. Koch, 1838 · *Woodlouse spider*

Plate 76

IDENTIFICATION This spider has an unusually bright reddish-colored, shiny cephalothorax and legs with a pale abdomen. The large jaws extend forward and have long fangs. The six eyes are arranged in an arched group at the front of the head. There is a somewhat similar-looking spider, *Trachelas tranquillus* (Plate 39), but it has more typical jaws and eight eyes in two rows.

OCCURRENCE This is a spider usually associated with human activities and was probably introduced from its native home in the Mediterranean region. It is found around buildings, gardens, farms, and fields—wherever pill bugs can be found. They seem to prefer moist areas, often under rocks or other debris on the ground. They do not build a capture web but do spin a silk retreat.

SEASONALITY Adults: all year.

REMARKS The name of this spider refers to its preferred prey: pill bugs. These prey have hard bodies, but the large and powerful jaws of the woodlouse spider can pierce their defenses.

FAMILY FILISTATIDAE • *Crevice Weavers*

There are 113 species known worldwide, primarily in the tropics. Only seven occur in North America north of Mexico. Five of these are large spiders in the genus *Kukulcania*. The other two are relatively small spiders, less than 5 mm long. The body is densely covered with setae of relatively uniform length, giving them a velvety appearance. The eight eyes are tightly grouped on a central mound.

The name crevice weaver refers to the peculiar retreat and web of these spiders. They have a retreat in a crack or other space usually either among rocks or dead wood. The spider builds a silken tube in this space then extends a circular band of hackle-banded silk in a circular pattern around the entrance. From this circular area the spider spins a series of long trip lines. The resulting web has a characteristic appearance (see Fig. 28, left). They have adapted well to human structures, and their webs can be found in and around buildings.

Kukulcania hibernalis (Hentz, 1847) • *Southern house spider*

Plate 4

IDENTIFICATION This is a large crevice weaver. The females' cephalothorax is dusky gray or black with a velvety covering of hairs. The legs have velvety hairs and many short spines. The abdomen is dark gray or dark brown, usually lighter in color than the cephalothorax. The eyes are in a tight group on a prominence. The high clypeus is slanted forward rather than vertical. The male is tan with long legs and palps. His palps are often held in a folded position, extending directly in front of the chelicerae. This spider can be distinguished from the mygalomorph spiders by the compact chelicerae and shorter spinnerets that are tightly grouped.

OCCURRENCE This is a southeastern species occurring from Florida to eastern Texas. Other similar species are found in the Southwest. This spider builds its tubular retreat in cavities in dead wood, under rocks, among debris, or in old structures. It has frequently been found inside buildings.

SEASONALITY Males: wander in summer. Females: all year.

REMARKS The messy cribellate web spreads away from the tubular retreat (see Fig. 28, left). The strands of the web serve as trip lines. When potential prey brush against them, the spider is stimulated to attack.

FAMILY GNAPHOSIDAE · *Ground Spiders, Stealthy Ground Spiders*

The family Gnaphosidae is one of the most diverse spider groups. There are more than 2,000 species worldwide. We have 255 species in North America north of Mexico. They are generally dull-colored spiders that are rarely seen in the open. Most observers encounter them by looking under rocks, logs, or other debris on the ground. They hunt on the ground and do not build a capture web. When they are seen in the open, they are usually dashing from one shelter to another. Most of the species are nocturnal or active at dawn and dusk. The antlike members of the genus *Micaria* are an exception. They are often found in the same sunny areas where ants are common. A few species of ground spiders have adapted well to living in buildings. These urban dwellers include the parson spider (*Herpyllus ecclesiasticus*) and mouse spider (*Scotophaeus blackwalli*). The anterior spinnerets of ground spiders are cylindrical and widely separated (see Fig. 12G,H). They are often easy to see extending beyond the end of the abdomen.

Many species of ground spiders have unusually bright posterior median eyes. These eyes have an unusual flattened surface and are oval in shape. The long axes of the ovals are perpendicular to each other. In some species of ground spiders it has been shown that such eyes perceive the plane of polarized skylight and that these spiders use this ability to assist them in navigation back to their retreat.

Callilepis imbecilla (Keyserling, 1887)

Plate 41

IDENTIFICATION This is a small, dark ground spider. The highly reflective posterior median eyes are unusually thin, oval, and oriented transverse to the body. The posterior lateral eyes are larger and nearly circular. The leg tips have claw tufts that are not conspicuous.

OCCURRENCE This is a southeastern species, but there are other similar-looking northern and western species. This spider has been found under logs, boards, and leaf litter in a variety of habitats from woodlands and forests to sand dunes and beaches.

SEASONALITY Males: Winter through June. Females: November through August.

REMARKS Members of this genus eat ants.

Cesonia bilineata (Hentz, 1847)

Plate 42

IDENTIFICATION This is a small but distinctive ground spider. This species has a pair of parallel dark lines on the cephalothorax and the abdomen. The legs are tan.

OCCURRENCE This species has been recorded between New England and Manitoba, south to New Mexico and Florida. This spider has been found among the leaf litter in both pine and deciduous forests. It has also been recorded in prairies, grasslands, sand dunes, mesquite woodlands, as well as in buildings and greenhouses.

SEASONALITY Males: all year except February and March. Females: all year; egg sacs have been recorded in summer and early autumn.

REMARKS These are swift runners, sometimes seen during the day. They attack other spiders from behind.

Cesonia josephus (Chamberlin and Gertsch, 1940)

Plate 42

IDENTIFICATION This is a small but distinctive ground spider. This species has a pair of parallel dark lines on the cephalothorax. On the abdomen the dark marks have a wavy margin and are connected at the back. The legs are tan.

OCCURRENCE This species is known from California. It has been recorded from a variety of habitats including canyons, coniferous forests, oak woodlands, and woodrat nests. It has been in collected in lowlands as well as in the foothills of mountains up to 1,200 m.

SEASONALITY Adults: December through July.

REMARKS This spider is a swift runner. They attack spiders from behind.

Drassodes auriculoides Barrows, 1919

Plate 43

IDENTIFICATION This is a medium-sized red ground spider. The cephalothorax and legs are reddish brown without markings. The abdomen is tan. The posterior median eyes are bright, irregular oval, or nearly rectangular and converging at the back. The spinnerets are all about the same length. The tarsi have scopulae and tufts.

OCCURRENCE This is a northeastern species but other similar species are found throughout North America. This species has been collected between New England and Wisconsin and south to Arkansas and Tennessee. It has been found under boards or rocks in fields and pastures. They have also been found among the leaf litter of woods. They have often been captured in pitfall traps.

SEASONALITY Males: May and June. Females: May through October; females have been found guarding egg cases with emerging young in August and September.

REMARKS Spiders in this genus are known to attack other spiders, sometimes larger than themselves. They use a trailing silk band to entangle the legs of the other spider.

Drassyllus depressus (Emerton, 1890)

Plate 41

IDENTIFICATION This is a small ground spider. The female usually has an orange cephalothorax and legs and a dark abdomen. Some females are darker. The males usually have a dark brown, nearly black cephalothorax. The front legs of males are brown, lighter toward the ends. The femora of legs III and IV of males are lighter. The abdomen is black with a hard scutum at the front. In both sexes the posterior median eyes are relatively large, somewhat rectangular, shiny, and close together.

OCCURRENCE This species is widespread from southern Canada south to Oregon and Arizona in the West and Arkansas to Virginia in the East. There are other similar species throughout North America. This species has been captured in a wide variety of habitats, including grasslands, marshes, bogs, prairies, oak and coniferous forests, agricultural fields, and buildings. They have been recorded in mountains up to 2,700 m. They hide under rocks or logs during the day.

SEASONALITY Males: late April through late September. Females: January through September.

REMARKS This is a fast-running spider that has been found in houses.

Drassyllus insularis (Banks, 1900)

Plate 41

IDENTIFICATION This is a small ground spider. The cephalothorax is dark brown. The front legs are brown, lighter toward the tips. The patellae and tibiae are dark on the rear legs. The abdomen is dark gray or black. The male has a hard brown scutum at the front of the abdomen. The posterior median eyes are relatively large, somewhat rectangular, shiny, and close together.

OCCURRENCE This is a western species. It has been found between southern British Columbia and western Colorado south to Mexico. They have been recorded from a variety of habitats, including agricultural fields, orchards, chaparral, riparian woodlands, deserts, and at elevations up to 2,100 m. They hide under rocks or logs during the day.

SEASONALITY Adults: all year.

Gnaphosa fontinalis Keyserling, 1887

Plate 41

IDENTIFICATION This is a small to medium-sized ground spider. There are two color forms. One has a bright-orange cephalothorax and legs, with a black abdomen. The other form is somewhat larger, has a dark brown cephalothorax and legs, and a dark gray abdomen. The posterior median eyes are similar to the posterior lateral eyes in size, slightly separated and oval in shape.

OCCURRENCE The species occurs from New England west to Wisconsin and Kansas, south to Texas and northern Georgia. The larger brown form is more common in the northeastern and southwestern parts of the range. This species has been found in deciduous and coniferous forests as well as in agricultural fields. Individuals have often been collected in pitfall traps set in leaf litter.

SEASONALITY Males: spring and summer. Females: April through October.

REMARKS This is one of the few polymorphic species in the Gnaphosidae.

Gnaphosa muscorum (L. Koch, 1866)

Plate 43

IDENTIFICATION This is a medium-sized ground spider. The cephalothorax and legs are rusty brown. The abdomen is usually gray, but sometimes a mottled brown and gray. The posterior median eyes are oval and closer together at the back.

OCCURRENCE This species is widespread in Canada and has been found as far south as West Virginia in the East and southern New Mexico in the West. It has been found in a wide variety of habitats, including coniferous forests, beaches, grasslands, sandy areas, and pitcher plants. This species has been recorded at elevations up to 3,900 m. They hide under rocks or logs during the day.

SEASONALITY Males: May through September. Females: April through November.

REMARKS The female has sometimes been found with her large white cocoon in a silk-lined depression under a rock.

Haplodrassus signifer (C.L. Koch, 1839)

Plate 41

IDENTIFICATION This is a relatively large ground spider. The body is brown. The posterior median eyes are shiny and nearly touching. They have an oval shape with a somewhat transverse orientation, occasionally converging at the back. There is some variation in the coloration of the abdomen; a few individuals show a pattern of dark chevrons.

OCCURRENCE This species occurs in Eurasia as well as North America. This species has a wide distribution in North America. This species has been found under rocks or other debris in open, relatively dry, areas. It has also been recorded from a variety of woodlands, particularly western conifers. Records extend up to elevations of 4,500 m.

SEASONALITY Adults: all year.

REMARKS This is a widespread and common species, occasionally found in houses. They have been captured in molasses-baited traps.

Herpyllus ecclesiasticus Hentz, 1832 • *Parson spider*

Plate 44

IDENTIFICATION This is a dark-colored ground spider with a white band down the center of the abdomen. The body is densely covered with hairs, often laying flat. This species has a dark reddish-brown cephalothorax and legs that appear black in low light. The abdomen is dark gray with a lobed white median band and a white spot above the spinnerets. Adult males have a dark brown scutum at the front of the abdomen.

OCCURRENCE This species occurs in the East; the similar looking *Herpyllus propinquus* replaces it in the West. This species hides under rocks and other debris on the ground. It has been found in low vegetation. It has been recorded at elevations up to 2,450 m. It overwinters under bark. It has often been found in buildings, including houses.

SEASONALITY Adults: all year.

REMARKS The white markings on the black abdomen are the source of the common name. Some think that these marks resemble the necktie worn by certain ministers. This spider is a fast runner and difficult to catch.

Herpyllus hesperolus Chamberlin, 1928

Plate 44

IDENTIFICATION This is a medium-sized ground spider with an orange-brown cephalothorax and legs and a gray abdomen. The body is densely hairy. The spinnerets are the same color as the legs.

OCCURRENCE This is a western species. It has been found from southern British Columbia east to Saskatchewan and south to southern California and western Texas. This species has been collected under rocks and in desert shrub habitats as well as in houses. It has been recorded at elevations up to 2,800 m.

SEASONALITY Adults: all year.

REMARKS Previously known as *Herpyllus validus* or *Prosthesima valida*.

Litopyllus temporarius Chamberlin, 1922

Plate 43

IDENTIFICATION This is a small to medium-sized ground spider. It has an orange cephalothorax and legs with a gray abdomen. The posterior median eyes are large and bright. In the male there is a scutum covering the front third of the abdomen.

OCCURRENCE This is an eastern species occurring from New England west to Missouri and south to Mississippi and Florida. It has been recorded from both deciduous and pine forests. It has been found under rocks, in leaf litter, and in pitfall traps.

SEASONALITY Males: April through August. Females: April through October.

Micaria longipes Emerton, 1890

Plate 42

IDENTIFICATION This is a small, antlike ground spider with 40 similar relatives. The body is reddish brown, darker at the back. The abdomen also has two thin white cross bands, one near a constriction at the midpoint. The body looks iridescent in sunlight.

OCCURRENCE This species is widespread in North America but is replaced by a similar species in the far west. It has been found in a variety of open habitats, including grasslands, prairies, pastures, sandy areas, and agricultural crops. It has also been recorded from oak woodlands, oak-pine barrens, and spruce plantations. This spider occasionally wanders into buildings. It has been collected at elevations up to 2,100 m.

SEASONALITY Males: June through October. Females: April through November.

REMARKS These spiders are diurnal and have often been seen running in open sunny areas where ants are found.

Micaria pulicaria (Sundevall, 1832)

Plate 42

IDENTIFICATION This is a small, antlike ground spider with 40 similar relatives. This species has a nearly black body. There are four thin white lines at the bases of the legs on the cephalothorax as well as a wavy white line across the middle of the abdomen. The legs are reddish brown with darker femora. The body looks iridescent in sunlight.

OCCURRENCE This species occurs in Eurasia as well as North America. It is widespread in the North and occurs throughout the West as far south as southern California and Texas. It has been found in a wide variety of dry open and forested habitats, including salt marshes. They have been recorded in the mountains at elevations up to 3,700 m.

SEASONALITY Adults: all year.

REMARKS This small, dark-colored spider runs rapidly over the ground, often in sunny areas. Their darting motions resemble those of ants.

Nodocion voluntarius (Chamberlin, 1920)

Plate 44

IDENTIFICATION This is a small ground spider with a reddish-brown cephalothorax and legs and a gray abdomen. The spinnerets are gray. It is one of six species in this genus. The posterior median eyes are shiny and are the largest eyes.

OCCURRENCE This is a western species. It has been found between Washington and Montana south to Mexico. Most records are from foothills and mountains at elevations up to 2,600 m. It has been recorded from under rocks and logs in both arid and humid areas.

SEASONALITY Males: May through August. Females: April through September.

Orodrassus coloradensis (Emerton, 1877)

Plate 43

IDENTIFICATION This is a medium-sized ground spider. The cephalothorax and legs are orange brown and the abdomen is gray. The posterior median eyes are oval and separated by more than one diameter. The sigilla on the upper abdomen are prominent.

OCCURRENCE This is a western species with most records from the Rocky Mountain states. It has been found in both aspen and conifer forests under rocks, logs, or bark. Sometimes it wanders into buildings. This spider has been collected at elevations between 1,250 and 3,800 m.

SEASONALITY Males: May through September. Females: March through November.

Scopoides catharius (Chamberlin, 1922)

Plate 43

IDENTIFICATION This is a small ground spider with an orange-brown cephalothorax and legs and a gray abdomen. The posterior median eyes are similar in size to the other eyes and set farther back so that the posterior eye row is procurved. These posterior median eyes are relatively smaller than those of the similar-looking species in the genus *Litopyllus*.

OCCURRENCE This species has been found in southern California. Individuals have been captured under rocks, in woodpiles, in pitfall traps, and in buildings. They have been found at elevations up to 1,500 m.

SEASONALITY Males: May through September. Females: April through September.

REMARKS This species has been observed feeding on a variety of other spider species.

Scotophaeus blackwalli (Thorell, 1871) • *Mouse spider*

Plate 41

IDENTIFICATION This is a large, dark-colored ground spider. It is covered with a dense clothing of flat hairs. The cephalothorax and legs are dark red brown and the abdomen is dark gray. The spinnerets are widely spread and prominent. The posterior median eyes are bright and oval. The male is similar to the female but with a small scutum at the front of the abdomen.

OCCURRENCE This is a European species that was long ago accidentally introduced into North America. There are records from the Southwest as well as the Southeast. They have been found under bark and in shrubs. It has frequently been found in and around buildings including on walls.

SEASONALITY Adults: all year.

REMARKS This spider has often been found wandering on walls and even the ceiling at night. Previously known as *Herpyllus blackwalli*.

Sergiolus capulatus (Walckenaer, 1837)

Plate 42

IDENTIFICATION This is a medium-sized and dramatically colored ground spider. It has a bright-orange cephalothorax and femora. The abdomen is black with a distinctive pattern of white bands.

OCCURRENCE This is an eastern species. It has been found in open areas such as meadows and lawns. It has also been recorded from leaf litter in deciduous forests. It has been found inside buildings.

SEASONALITY Males: February through November. Females: May through November.

REMARKS This spider is often seen wandering out in the open during the day where its bright colors make it conspicuous. It may be a velvet ant mimic. The velvet ants are actually wasps with a venomous sting and many animals wisely avoid them; perhaps its resemblance protects *Sergiolus* from predators.

Sergiolus montanus (Emerton, 1890)

Plate 42

IDENTIFICATION This is a medium-sized ground spider. This spider has a reddish-brown cephalothorax with a covering of flat light hairs. The front femora are dark but most of the legs are lighter. The abdomen is gray, usually with light bands; the one at the center is often broken into two lateral spots. There is also a white band at the front of the abdomen and sometimes one near the spinnerets.

OCCURRENCE This species is widespread but much more common in the West. It is found in open and forested habitats up to 3,400 m in elevation. It has been collected under rocks, logs, and other debris on the ground.

SEASONALITY Males: Februrary through October. Females: all months except December.

Sosticus insularis (Banks, 1895)

Plate 42

IDENTIFICATION This is a small, dark-colored ground spider. The cephalothorax is dark brown as are the femora of the legs. The tips of the legs are lighter in color. The abdomen is dark gray, nearly black. The male has a scutum over more than half of the abdomen. The posterior median eyes are widely spaced and about the same size as the other eyes.

OCCURRENCE This is primarily an eastern species, occurring from New England west to the Great Lakes region and south to northern Texas and eastern Georgia. It has been captured primarily in pitfall traps. It has also been found in buildings and under bark.

SEASONALITY Males: May through August. Females: April through September.

Talanites echinus (Chamberlin, 1922)

Plate 43

IDENTIFICATION This is a small ground spider with a spotted abdomen. The cephalothorax and legs are pale orange brown. The chelicerae project forward. The abdomen is gray with a distinctive pattern of dark spots.

OCCURRENCE This is a southeastern species. It has been found from Virginia west to Missouri and south to Texas and Florida. Has been collected in a variety of habitats, mostly deciduous forests. Most records have been from under rocks or captured in pitfall traps in leaf litter.

SEASONALITY Males: November through the following June. Females: all year.

REMARKS Previously known as *Rachodrassus echinus*.

Trachyzelotes lyonneti (Audouin, 1826)

Plate 44

IDENTIFICATION This is a medium-sized ground spider. The cephalothorax and legs are purplish brown and not very hairy. The abdomen is gray. The posterior median eyes are oval and shiny, oriented sideways. There is a tuft of stiff hairs on the front of the chelicerae.

OCCURRENCE This species is Mediterranean in origin but has been accidentally introduced into North America. There are records scattered across the United States from California to Illinois. It has been recorded from a wide variety of habitats, including suburban yards and inside buildings. This species hides under rocks in the daytime.

SEASONALITY Adults: all year.

Urozelotes rusticus (L. Koch, 1872)

Plate 44

IDENTIFICATION This is a medium-sized ground spider. The cephalothorax and legs are orange red. The abdomen is cream-colored or light gray. The eyes are closely spaced. The posterior median eyes are oval, converging at the back. The male has a small scutum at the front of the abdomen.

OCCURRENCE This is a Eurasian species that has been accidentally introduced around the world. It hides under rocks or other debris near buildings. It has also commonly been found inside buildings.

SEASONALITY Adults: all year.

REMARKS This species is often found in sinks or bathtubs.

Zelotes fratris Chamberlin, 1920

Plate 44

IDENTIFICATION This is a medium-sized black ground spider, one of 47 in this genus. The cephalothorax is shiny and dark ebony or black. The abdomen is also black. The legs are ebony near the body but the tips may be lighter.

OCCURRENCE This species has been found from Nova Scotia to Alaska and south to California, New Mexico, and South Carolina. There are other similar relatives throughout North America.

These spiders have been found in many kinds of forests, woodlands, and open habitats. In the mountains they have been recorded as high as 3,600 m.

SEASONALITY Males: April through September. Females: all year.

FAMILY HAHNIIDAE • *Comb-tailed Spiders*

The common name for this family refers to the fact that many of the species share the unusual feature that all six of their spinnerets arranged in one line. Three genera (*Antistea, Hahnia, Neoantistea*) with a total of 19 species fit this description. This distinctive feature is usually visible on these small spiders with a magnifying glass. These are spiders of the ground environment, usually encountered when searching through leaf litter or examining pitfall trap samples. The members of the genus *Neoantistea* build a thin, inconspicuous sheet web across a small depression in the soil.

The second group in the family (subfamily Cryphoecinae) includes spiders with the more typical arrangement of their spinnerets. They may be small wandering species, such as *Cryphoeca montana,* or web builders, such as *Calymmaria*. Many members of the genus *Calymmaria* are cave dwellers. Nearly 40 species are known from North America north of Mexico, placed in 6 genera. Of these, only *Cryphoeca* and *Calymmaria* are commonly encountered.

Calymmaria persica (Hentz, 1847)

Plate 33

IDENTIFICATION This spider is pale yellow with dark brown or black markings. The cephalothorax has a thin black rim as well as black spots at the bases of each leg. The femora have four black bands separated by pale yellow. The abdomen is variegated pale yellow or gray with dark markings. The spider hangs upside down above its distinctive cone-shaped web (see Fig. 27C).

OCCURRENCE This species occurs primarily in the Appalachian Mountains, the Western Allegheny Plateau, and a few locations further west. Most of the other 30 species have been found along the West Coast an in the Sierra Nevada. They have been collected from dark humid areas, often under overhanging rocks, or in caves or cave entrances. They have often been collected from rocks behind waterfalls.

SEASONALITY Adults: all year.

REMARKS The web is has a cone- or basket-shaped part and a flat sheet. The flat sheet is within the cone near the rock face to which the web is attached. The spider hangs under this flat sheet suspended above the cone.

Cryphoeca montana Emerton, 1909

Plate 81

IDENTIFICATION This is a small spider that somewhat resembles *Hahnia cinerea* in coloration but the spinnerets are not in one line; the anterior ones are well separated and the posterior lateral ones are longest. The coloration is variable and sometimes much paler than illustrated in Plate 81. The cephalothorax has a dark rim and dark marks radiating from the center. The abdomen usually has a series of paired light spots. The femora have a dark band most of the way toward their ends.

OCCURRENCE Occurs in the northeastern region. Has been found in damp forests beneath the bark of dead trees as well as under leaf litter, rocks, or moss. A similar species is common in Washington.

SEASONALITY Adults: srping through late autumn.

REMARKS This species is nocturnal.

Hahnia cinerea Emerton, 1889

Plate 81

IDENTIFICATION This is a small hahniid with an unusual arrangement of all six spinnerets in one line, the outermost being the longest. These can usually be seen, even on these small spiders. The carapace lacks hairs and is shiny. *Hahnia cinerea* differs from *Neoantistea* by having a pattern of dark markings on the cephalothorax. *Neoantistea* usually have plain dark brown carapaces (Plate 35).

OCCURRENCE This species is found throughout most of North America except the western states from Oregon to Texas. Other similar species in this genus have been recorded from these areas. These spiders are usually found in leaf litter.

SEASONALITY Adults: autumn.

REMARKS According to Opell and Beatty (1976), members of this genus in North America are not known to build webs. European species build webs that are similar to those of *Neoantistea*.

Neoantistea agilis (Keyserling, 1887)

Plate 35

IDENTIFICATION This is a small hahniid with an unusual arrangement of all six spinnerets in one line, the outermost being the longest. These can usually be seen, even on these small spiders. The carapace is brown, lacks hairs, and is shiny. The femora have two dark rings. Species in the genus *Hahnia* (Plate 81) differ from *Neoantistea* by having a more distinct pattern of dark markings on the cephalothorax, dark marks at the ends of each femur, and being smaller.

OCCURRENCE This species is found throughout most of North America except the far north. These spiders build a thin horizontal sheet, typically across a small depression in the soil, between small rocks, or mosses. This species has been recorded from the ground in a wide variety of habitats.

SEASONALITY Adults: all year.

REMARKS The delicate sheet web is only noticeable when it is wet with dew. When disturbed, the spider runs to the edge of the web or off into the surrounding litter.

FAMILY HERSILIIDAE · *Longspinneret Spiders*

This family has 140 species from the tropics and subtropics. The members of this family have very long posterior spinnerets. They are often as long or longer than the rest of the body. We have two species in different genera north of Mexico. The more common one of the two is *Neotama mexi-*

cana, which has been found in trees in southern Texas. Another similar spider, *Yabisi habanensis*, has been collected in the Florida Keys, where it is thought to live among rocks on the ground. The long spinnerets are used to capture prey by producing a circular net of silk as the spider runs around the prey.

Neotama mexicana (O.P.-Cambridge, 1893)

Plate 47

IDENTIFICATION This spider has extraordinarily long posterior spinnerets that are sufficient to identify the spider. They variegated gray in color and cling to bark, often on lichen. This is effective camouflage.

OCCURRENCE This species has been found in the vicinity of Brownsville, Texas. It has an extensive range in Mexico as well as in Central and South America. They have been found mostly on the trunks and branches of trees, holding the body flat to the surface. Their camouflage makes them easy to miss.

SEASONALITY Adults: May through November.

REMARKS The remarkable spinnerets are used to spin a dome-shaped web that pins the prey on the surface of the substrate. The spider rapidly circles around the prey with the spinnerets held over the prey, creating the capture web.

FAMILY HOMALONYCHIDAE • *Dusty Desert Spiders*

This is a very small family with only three species. Two of these species have been found in North America north of Mexico (*Homalonychus selenopoides* and *H. theologus*), both from the southwestern United States and northwestern Mexico. They are ground-living spiders that do not spin a capture web. According to S. C. Crews (2005), they do not even use a dragline. Silk is used to construct the egg sac, and the spider usually incorporates sand in its covering. These spiders possess unusual setae over the body that trap sand and dirt particles. They often sit motionless, partly buried in the sand, where they are well camouflaged and difficult to detect.

Homalonychus selenopoides Marx, 1891

Plate 46

IDENTIFICATION This species has a pentagonal-shaped abdomen. The posterior median eyes are the largest. The clypeus is high. The spiders' bodies are often covered with particles of dust or sand. The relatively long legs lack heavy spines and are typically held widespread.

OCCURRENCE These spiders have been collected in southern Nevada, Arizona, southern California, and northwestern Mexico. They have been found on the slopes of hills and gullies in deserts and shrublands on the ground under rocks or on fallen vegetation. Adult males and juveniles wander at night.

SEASONALITY Adults: all year.

REMARKS This spider has been observed eating other homalonychids.

FAMILY HYPOCHILIDAE • *Lampshade Weavers*

The most unusual of all the funnel web builders in this book are the enigmatic lampshade weavers in the family Hypochilidae. Ten of the eleven species in this family are from North America; one has been found in China. These remarkable spiders are evidently an ancient group, taxonomically placed near the origin of the Araneomorphae. They possess a number of primitive features. The ten members of this family in North America have localized distributions in mountain regions. Half are found in the Appalachians and other species are in the western United States. These spiders are found on rocky cliff faces, usually in undercut areas. Sometimes the webs are near waterfalls, the entrance to caves or mines, or in human-made culverts. The web is unique—a broad circular structure attached at its base and shaped something like a lampshade surrounded by an open series of lines (see Fig. 27, lower right). The spider rests against the rock face at the center of the web. The legs are long and held spread apart in a distinctive posture.

Hypochilus thorelli Marx, 1888 • *Lampshade weaver*

Plate 57

IDENTIFICATION Members of this genus can be identified by their distinctive lampshade-shaped web. The spider itself has long thin legs that are often held bent at a right angle between the femur and patella. This permits the spider to cling flat to the rock surfaces where it lives. On the under-side of the abdomen four book lungs are visible.

OCCURRENCE This species occurs in the Appalachian region. The spider often builds its web on a rock face above a stream or in a humid location.

SEASONALITY Adults: June through November.

REMARKS The web of these spiders is circular, widening as it spreads from the rock face where it is attached. The overall appearance is as if a lampshade has been attached to the rock. There is an extended tangle of threads spreading away from the lampshade. The spider rests in the center, usually flat against the rock.

FAMILY LEPTONETIDAE • *Midget Cave Spiders*

This family includes 190 described species, but many more are known in collections and remain to be formally described. There are 40 known in North America north of Mexico. The features that are key to distinguishing the species require dissection or specialized microscopy to reveal their structure. These are tiny spiders of moist, dark environments near the ground or in caves. Some species are found wandering; others build flimsy tangle or sheet webs. Even when they do build a web, they are easy to overlook because the spider drops if disturbed. Midget cave spiders usually have six eyes; either tightly grouped or with one pair well separated behind the other four. Some cave species lack eyes entirely. The males of many species have relatively long palps, and the females' reproductive structures are not visible externally. One unusual feature of this family is that if a spider looses one of its legs, it usually separates at the joint between the patella and tibia. Most spiders lose legs by separation at the coxa/trochanter joint.

Archoleptoneta schusteri Gertsch, 1974

Plate 82

IDENTIFICATION This tiny spider is a member of a group of similar spiders whose six eyes are grouped closely together. They are pale tan in color. In strong light the legs of living spiders often show an iridescence of blue or green.

OCCURRENCE This spider lives in California, restricted to relatively moist shaded or protected areas near the ground, under rocks, leaves, debris, or in caves. They are rarely found in a web but sometimes build a flimsy tangle or sheet web. Other species have been found in Texas and Mexico.

SEASONALITY Adults: all year.

REMARKS The egg sacs are flat disks attached under rocks. If disturbed, they drop and curl up, making them difficult to find.

Neoleptoneta myopica (Gertsch, 1974) • *Tooth Cave spider*

Plate 82

IDENTIFICATION This tiny spider is a member of a group of similar species whose posterior median eyes are located well behind the other four. They are uniformly pale in color. In strong light the legs of living spiders often show an iridescence of blue or green.

OCCURRENCE Known from Tooth Cave in Texas. There are many other similar species found in Arizona, California, Georgia, Oregon, Texas, and the Appalachian region. In addition to the cave forms, some species are found near the ground, under rocks, or other debris.

SEASONALITY Adults: probably all year.

REMARKS The egg sacs are small, circular, and covered with sand. If disturbed, these spiders drop and curl up, making them difficult to find. This is one of very few spiders that are protected by the Endangered Species Act.

FAMILY LINYPHIIDAE • *Sheetweb Weavers*

The Linyphiidae is a large worldwide family with more than 4,400 species, second only to the jumping spiders (Salticidae). This family is particularly diverse in the North and includes 157 genera and over 950 named species in North America, making it the family with the most representatives here. Many of the larger species in this family are members of the subfamily Linyphiinae. These spiders include the familiar hammock spider (*Pityohyphantes costatus*), the bowl and doily spider (*Frontinella pyramitela*), and the filmy dome spider (*Neriene radiata*). This subfamily includes approximately a third of the members of the family.

The dwarf sheetweavers, including about two-thirds of the species in the family, are classified in the subfamily Erigoninae. These spiders were formerly known as the "micryphantids" and you may find information about them under this older name. These interesting spiders are among the smallest in the world, some are less than 1 mm (about 1/64 inch) long when fully grown. They either build inconspicuous webs near the ground or no web at all. They are most often noticed because of their ballooning habit. On occasion, when the weather is just right, by sheer numbers, thousands of tiny ballooning spiders sometimes drift down together and coat a fence or field with silk. This coating of silk is composed of thousands of thin strands and is called gossamer. Such events are among the most beautiful displays in nature. This behavior is most frequent in autumn but can occur whenever the sunshine creates a warm rising current of air. Close examination of the gossamer sheet often reveals hundreds of busy spiders, running about trailing silk. This behavior is often an attempt to become airborne again. If rising air currents reoccur, many of the spiders will rise and drift away.

Some of the dwarf sheetweaver males have peculiar heads with horns, lumps, even pits. This ornamentation is related to courtship or mating. Females have been observed grasping the head ornamentations of the males with their fangs during mating. In a few species it has been demonstrated that the pits contain numerous glands, which may secrete substances that attract females.

The hunting webs of sheetweb weavers form either curved or flat sheets, usually accompanied by a tangle of knockdown threads either above or below the sheet, often both. The knockdown threads intercept flying insects that blunder into them and are slowed or tumble down onto the sheet. Similar threads under the sheet may act to warn the spider of approaching danger or prey emerging from below. The spider is most often found hanging upside down under the sheet or near the edge of the sheet. When a prey item strikes the web, the spider rushes out to make the capture. Because the spider is hanging under the sheet, it bites through the web to capture and subdue the prey. After capture, the prey is pulled through the web, wrapped with more silk, then consumed. The silk of the sheet webs is generally not sticky. There are few viscid glue droplets. The prey is merely slowed by the difficulty of walking over the netlike sheet interspersed with trip lines. The spiders are naturally quicker and more agile than many of their prey in the web. Tiny spiders, such as sheetweb weavers, are also adept at walking on the surface of water.

When seen well, usually through a magnifying glass, the sheetweb weavers often reveal large chelicerae with numerous teeth. The small eyes are usually well up on the face, above the tall clypeus characteristic of this group. This combination of tall chelicerae and high clypeus give sheetweb weavers a characteristic face (see Fig. 15). For example, look at the face of *Drapetisca alteranda* on Plate 78. If they weren't so tiny, they would look intimidating.

For more information about spiders in the family Linyphiidae, begin by consulting Michael Draney and Donald Buckle's 2005 work (Draney and Buckle 2005). This valuable reference provides the citation to the most recent monograph for each genus. Identification of sheetweb weavers in most genera to the level of species is beyond the scope of this book. There are just too many similar species for confident identification without examination of the tiny reproductive parts using a good dissecting microscope. Because the information about this group is scattered through the professional literature, often in languages other than English, this work can be challenging. A useful website for information on the genera of the Linyphiidae has been created (Hormiga et al. 2008).

Bathyphantes alboventris (Banks, 1892)

Plate 78

IDENTIFICATION This species has a light abdomen with dark bands.

OCCURRENCE This species occurs in forests of northeastern North America. These spiders live at ground level among the fallen leaves, and their small webs are rarely observed.

SEASONALITY Adults: March through September.

Bathyphantes pallidus (Banks, 1892)

Plate 78

IDENTIFICATION This species has a dark abdomen with light bands. Some individuals have unbroken light bands. The color of the abdomen and bands varies, and some individuals' abdomens are nearly black and white.

OCCURRENCE This species is widespread in North America and has been found in moist areas of woods as well as fields, where it can be remarkably common. These spiders live at ground level among fallen leaves, and their small webs are rarely observed.

SEASONALITY Adults: all year.

REMARKS This spider has often been observed running over the ground surface during the daytime.

Centromerus cornupalpis (O.P.-Cambridge, 1875)

Plate 79

IDENTIFICATION This species is dark in color, the abdomen being dark gray to black. The cephalothorax and legs are dark brown. It has few spines and often has a shiny appearance.

OCCURRENCE This spider has been found among leaves at ground level in swampy areas and wet forests. The males are active wanderers. Females are most often detected by pitfall trapping.

SEASONALITY Adults: October through June.

Ceratinopsidis formosa (Banks, 1892)

Plate 79

IDENTIFICATION This small spider is bright orange, with only a bit of dark around the eyes and the spinnerets. The name *formosa* refers to the beauty of this tiny animal. For this guide it represents a large group of at least 90 tiny dwarf sheetweavers that are either red or bright orange in color.

OCCURRENCE This spider is among the many species that are rarely seen because they live primarily in low vegetation near the ground or in the leaf litter. It has also been found among rocks at the shoreline. They turn up in a variety of samples, either by sweeping low vegetation, sifting litter, or in pitfall traps.

SEASONALITY Adults: late summer and autumn.

REMARKS One obvious mystery is why would a spider be so brightly colored if it lives in dark recesses near the ground and is rarely found out in the open during the day?

Diplostyla concolor (Wider, 1834)

Plate 79

IDENTIFICATION This species is relatively uniform in color and lacks the light bands present in many species of the closely related genus *Bathyphantes*. The male's palps are distinctive, with an unusually long tarsal segment. There are two long spines extending from the tibia of the palp that lie parallel to the long tarsus.

OCCURRENCE This species is found at ground level in moss, grass, and leaf litter. They have been collected from a wide variety of habitats, including both open and wooded areas. This species is tolerant of disturbance and has often been found in degraded woodlands.

SEASONALITY Adults: all year.

REMARKS Previously known as *Bathyphantes concolor*.

Drapetisca alteranda Chamberlin, 1909

Plate 78

IDENTIFICATION This species is light in color with distinctive dark markings on the carapace. The dark bands on the legs are conspicuous. The abdomen is whitish with a dark horizontal band at the front and a series of paired black spots spreading from the center along the back. The chelicerae have three or four pairs of dramatic crossing spines on the front. Adult females have a wide protruding epigynal scape.

OCCURRENCE This spider is found on the trunk of trees in forests or near the base of the same trees among the leaves and moss. The inconspicuous web is made flat against the bark of a tree trunk and is rarely observed.

SEASONALITY Adults: late summer through winter.

REMARKS These spiders can sometimes be collected by brushing the trunks of trees with a long stiff brush of the kind used in wallpapering.

Estrandia grandaeva (Keyserling, 1886) • *Conifer sheetweaver*

Plate 78

IDENTIFICATION This is a relatively small species. The cephalothorax and legs are evenly colored light brown, tan, or orange brown. The abdomen has a median dark band of variable width.

OCCURRENCE This spider appears to be a conifer specialist, reported from elevations above 610 m in the West and lower elevations across the eastern part of its range. The flat webs have been found among the lower branches of evergreen trees. It is widespread across North American and Eurasian conifer forests in the North.

SEASONALITY Adults: summer.

Florinda coccinea (Hentz, 1850) • *Scarlet sheetweaver*

Plate 79

IDENTIFICATION This species is illustrated in its typical foraging position, hanging upside down in its web. The typical individual has bright-red coloration, but some are orange and even yellow-orange. At the end of the abdomen there is a black tubercle above the spinnerets. There is variation in the size of the tubercle. It is typically smaller, sometimes absent, in males.

OCCURRENCE This spider builds its web in damp grassy areas, even lawns. This species is a common and conspicuous spider in appropriate habitat throughout the Southeast.

SEASONALITY Adults: summer.

Frontinella pyramitela (Walckenaer, 1841) • *Bowl and doily spider*

Plate 77

IDENTIFICATION This species is usually identified by its web (see Fig. 23, left). The spider hangs below the distinctive bowl-shaped sheet. Below this is a flat sheet or "doily." There is usually a loose tangle of threads above the bowl. The spider is dark brown, nearly black, with a characteristic pattern of white markings on the abdomen.

OCCURRENCE This is one of the most widespread species in North America, but it is far more common in the humid eastern part of the continent. There is a similar species in Arizona. The web is built in a variety of habitats, usually from low vegetation but occasionally in the low branches of trees. This spider is common in peat bogs. Sometimes the male and female may cohabit the same web. The young balloon long distances.

SEASONALITY Adults: spring through late summer.

REMARKS Also known by the name *Frontinella pyramitela*.

Helophora insignis (Blackwall, 1841)

Plate 78

IDENTIFICATION This small spider is variable in color. The abdomen is paler than the cephalothorax, cream or tan and marked with a series of irregular dark spots on the sides and ventral surface. The legs are plain. Both adult and subadult females have a conspicuous epigynal scape.

OCCURRENCE This spider builds a flat sheet web in low vegetation, particularly grasses. This species has been reported from woods, cedar swamps, and open areas. It has been found all across the northern part of the continent as well as in Europe and Asia.

SEASONALITY Adults: late summer through early winter.

REMARKS They have been found near the entrances to small mammal burrows.

Hypselistes florens (O.P.-Cambridge, 1875) • *Splendid dwarf spider*

Plate 79

IDENTIFICATION This spider has conspicuous coloration of red-orange and black.

OCCURRENCE This spider is common in a band across the northern part of the continent. It is lives in a tiny sheet web among the leaves and rocks on the ground but is occasionally seen higher in the vegetation, particularly at night. This spider has sometimes been found on the surface of snowfields in winter. Like many dwarf sheetweavers, it disperses by ballooning.

SEASONALITY Adults: spring through late summer; subadults in winter.

REMARKS The palps of the immature males are conspicuous, bright orange, smooth, and shiny.

Megalepthyphantes nebulosus (Sundevall, 1829)

Plate 77

IDENTIFICATION The abdomen of this species varies; they are pale or have extensive dark spotting. The individual illustrated in Plate 77 is an intermediate form. The legs have dark rings. There is a dark tuning fork–shaped mark on the carapace.

OCCURRENCE This species occurs across North America and may have been introduced from Europe. This spider builds a large, flat sheet web often around human habitations, even damp cellars. It seems to prefer shady, humid localities.

SEASONALITY Adults: all year.

Meioneta fabra (Keyserling, 1866)

Plate 79

IDENTIFICATION This small spider usually has a relatively narrow abdomen with a gray band around the middle or near the front. This species was chosen to represent more than 50 species in our region. In the males of this group the cymbium of the palp forms a conspicuous angular structure.

OCCURRENCE This spider has been found among the leaf litter in wooded areas.

SEASONALITY Adults all year.

REMARKS Species in this genus are sometimes called *Agyneta*.

Microlinyphia mandibulata (Emerton, 1882) • *Platform spider*

Plate 77

IDENTIFICATION This spider bears a superficial resemblance to the bowl and doily spider but is lighter brown in color with variable white markings. There are usually two broad white spots at the upper front end of the abdomen. The web is a flat sheet close to the ground in low grass or vegetation. As the spider grows, the web can become slightly domed in the middle, perhaps with a raised portion near the center where the spider typically hangs. In a close-up view the face is dominated by the large chelicerae.

OCCURRENCE The platform-shaped web is found near the ground often in moist situations in open habitats or grassy areas. Webs are usually built within 10 cm of the ground. They are obvious on dewy mornings but can be nearly invisible on dry afternoons.

SEASONALITY Adults: spring through September.

REMARKS The spider rapidly drops to the ground when disturbed.

Microneta viaria (Blackwall, 1841)

Plate 79

IDENTIFICATION This small spider is easily confused with a large number of other small sheetweavers. The abdomen is dark gray to black, sometimes appearing iridescent. The cephalothorax and legs are brownish. The legs are unmarked. The chelicerae are robust.

OCCURRENCE This spider is among the most widespread sheetweavers in northern latitudes of both North America and Eurasia. It is usually found among fallen leaves at ground level in woods, ravines, and in other moist habitats, even gardens.

SEASONALITY Adults: all year.

REMARKS Sometimes it has been found in association with ant nests.

Neriene clathrata (Sundevall, 1830)

Plate 77

IDENTIFICATION This is a dark brown spider. The abdomen may have a large number of small white spots. The legs are lighter in color. The male is darker and slimmer; his cephalothorax is the same size as that of the female. Each chelicera of the male has two projections called mastidia.

OCCURRENCE This spider is usually found near the ground. The loose sheet web has been found in grassy areas, marshes, and in low vegetation in forests, often near the base of trees. They are sometimes common in disturbed habitats, including residential yards. This species builds a small sheet web and hangs underneath, usually near the edge.

SEASONALITY Adults: spring through late summer; subadults in winter.

REMARKS The male and female have sometimes been found together in the same web.

Neriene litigiosa (Keyserling, 1886) • *Sierra dome spider*

Plate 77

IDENTIFICATION The carapace of is light yellowish or greenish with a darker longitudinal central stripe and dark near the sides. The long legs are dark green. The abdomen is oblong and mostly white with a dark central line. The light areas are white or creamy, not yellow as in the filmy dome spider. This species has a large dome-shaped web that suggests that of the filmy dome spider but is larger and more variable. The web may have a raised area near the center above where the spider hangs.

OCCURRENCE This spider is more common in the West than the filmy dome spider. Sierra dome spiders can often be found in forested slopes of mountains. They are also seen in wooded areas near the coast in the northern part of their range. Their webs are usually built in low vegetation.

SEASONALITY Adults: spring through late summer.

REMARKS Mature individuals of this species are probably the largest sheetweavers in North America.

Neriene radiata (Walckenaer, 1841) • *Filmy dome spider*

Plate 77

IDENTIFICATION This species is usually identified by its dome-shaped web (see Fig. 23, upper right). The web can be difficult to see in the shadows, where this species frequently resides. There is a loose tangle of threads above the dome. The spider is relatively well marked by a light band around the dark carapace. The abdomen has a distinctive pattern of light white-and-yellow bands on a dark background. Western individuals have less yellow. The related western species, the Sierra dome spider, is larger and paler with dark green legs.

OCCURRENCE This spider occurs throughout much of North America, but it is common in forests of the East. The dome web is usually found in low vegetation or among rocks.

SEASONALITY Adults: spring through late summer.

REMARKS The spider hangs below the dome, usually near the apex. Male and female have been found in the same web during the spring. Previously known as *Linyphia marginata*.

Neriene variabilis (Banks, 1892) • *Variable sheetweaver*

Plate 77

IDENTIFICATION This is a light brown spider. The abdomen is variable in color, often dark below and paler above with a series of paired dark and light spots on the dorsal surface. The shape of the abdomen is distinctive, taller at the back. The legs are pale. The male has one projection or mastidion on each chelicera.

OCCURRENCE The web of this species has been found in low vegetation near the ground, often among the roots of trees. It has often been found in shady woods. The web is a relatively flat sheet but may be slightly raised in the area where the spider rests.

SEASONALITY Adults: spring through August.

REMARKS The fact that this spider is darker below and lighter above may be counter shading. These spiders hang upside down in their webs so the dark underside is in the light and the lighter upperside is in the shadows; the reversed coloration makes the spider hard to see in the dim light of the forest understory.

Pityohyphantes costatus (Hentz, 1850) • *Hammock spider*

Plate 78

IDENTIFICATION This is a large sheetweaver. The abdomen is long and oval, light in color with a dark dorsal band of arrowhead-shaped. The cephalothorax and legs are pale. There is a tuning fork shaped dark mark on the cephalothorax, the two tines ending at the posterior median eyes. The legs have dark bands near the joints and are spotted at the bases of the larger spines.

OCCURRENCE This species is the most common member of this genus in the East. The relatively large hammock-shaped sheet is constructed in both wooded and open areas. The web is built in shrubs and the low branches of trees. The spider waits near the edge of the web or under a leaf or other cover nearby.

SEASONALITY Adults: spring through late summer; subadults in winter.

REMARKS There are many undescribed species in this genus.

Pityohyphantes rubrofasciatus (Keyserling, 1886) • *Red hammock spider*

Plate 78

IDENTIFICATION This is a large sheetweaver. The abdomen is long and oval-shaped, light in color with a dark red dorsal band that has a ragged dark edge. The cephalothorax and legs are dark reddish. The tuning fork–shaped dark mark on the cephalothorax is more diffuse than in the hammock spider. The legs are dark reddish with darker bands near the ends of the segments.

OCCURRENCE This species is found in the Northwest. The relatively large hammock-shaped sheet is constructed in both wooded and open areas. The spider waits near the edge of the web or under a leaf or other cover nearby.

SEASONALITY Adults: spring through late summer; subadults in winter.

REMARKS There are many undescribed species in this genus.

Stemonyphantes blauveltae Gertsch, 1951

Plate 78

IDENTIFICATION This is a large sheetweaver. The orange or tan cephalothorax and legs are distinctively marked with dark brown or black. The legs are banded. The abdomen is whitish with a discernable pink cast in some individuals. There is a line of dark spots along the top and sides of the abdomen.

OCCURRENCE This spider is usually found away from its rarely observed web. It lives near the ground among the leaves, mosses, logs, or under rocks. This species has been reported near the entrance to mammal burrows. They prefer moist habitats.

SEASONALITY Adults: all year.

REMARKS Individuals move up in the vegetation at night.

Tapinopa bilineata Banks, 1893

Plate 78

IDENTIFICATION The cephalothorax is longer than wide, and the anterior median eyes are unusually large and project forward. It has a light mark down the center of the dark cephalothorax. The legs are banded.

OCCURRENCE This spider builds its web in low leaves or grass. This species is most common in wooded areas or grassy areas near the edge of woods.

SEASONALITY Adults: late summer through winter.

REMARKS The web glistens and may resemble the slime trail left by a slug or snail.

Tennesseellum formica (Emerton, 1882) • *Antlike sheetweaver*

Plate 79

IDENTIFICATION This small spider is easily confused with many other small sheetweavers. The abdomen is relatively long and thin, usually with a light brownish band across the middle. Some individuals are pale. The legs are paler at their ends and without banding. It moves about with an antlike gait.

OCCURRENCE This spider is among the most widespread sheetweavers. It has been reported from the leaf litter of forests as well as open habitats including lawns. It is among the most common spiders of agricultural fields.

SEASONALITY Adults: spring and summer.

REMARKS This species is known for the fact that the opening to the respiratory system is in a forward position, forming a crease in the abdomen, enhancing the antlike appearance.

Tenuiphantes tenuis (Blackwell, 1852)

Plate 77

IDENTIFICATION In females the abdomen is marked with large dark brown spots on a light background of white speckles. The cephalothorax is dark. The legs are not banded. The male has a dark abdomen with two elongated white spots near the anterior end.

OCCURRENCE This species has been introduced from Europe around the world and is a recent immigrant, expanding its range, in North America. It can be common in a variety of habitats. The webs are built in moss, leaf litter, or low vegetation as well as in crop fields. It has also been recorded from greenhouses.

SEASONALITY Adults: all year.

Walckenaeria directa (O.P.-Cambridge, 1874) • *Money spider*

Plate 79

IDENTIFICATION This tiny black spider is one of hundreds of species in several genera of dwarf sheetweavers that are occasionally given the name money spider. The name is apparently derived from English country dialect and may refer to the good luck that releasing these spiders unharmed is supposed to confer.

OCCURRENCE This spider spends most of its life in the leaf litter but on warm days has been found higher in vegetation and on fences.

SEASONALITY Adults: all year.

REMARKS Many males of dwarf sheetweaver species, including those in the genus *Walckenaeria*, possess dramatically modified head regions. Some have protruding horns, such as this species. Others are even more fantastic, with multiple raised lumps. Some species have eyes perched on these stalks.

FAMILY LIOCRANIDAE • *Spinylegged Sac Spiders*

Many species that were once classified as members of this family have been transferred to the antmimic spiders (family Corinnidae). This family, as currently constituted, has a worldwide distribution with 178 species. There are 11 species remaining in the family that are found in North America north of Mexico, including some tiny species. These spiders are ground-living nocturnal hunters. During the day they are found in silken retreats in the leaf litter, under rocks, or other debris. Some have been found in the burrows of other animals. They usually have a series of paired spines under their tibiae and metatarsi of their front two pairs of legs.

Agroeca pratensis Emerton, 1890

Plate 45

IDENTIFICATION This species has a brown cephalothorax with a pattern of radiating dark markings on the thoracic part. The abdomen that is pale with a complex pattern of dark spots, chevrons, or bands. Their anterior spinnerets are separated, so they can be confused with ground spiders. The posterior median eyes are unmodified, unlike the oval posterior median eyes of ground spiders.

OCCURRENCE This species is found in the eastern half of the continent but other similar species are found in the West. These have often been found in leaf litter or under debris. This species has been collected in a wide variety of open habitats, both wet and dry as well as deciduous forests.

SEASONALITY Males: summer and autumn. Females: March through November.

REMARKS Spiders in this genus cover their egg sacs with dirt and hang them in low vegetation.

Hesperocranum rothi Ubick and Platnick, 1991

Plate 80

IDENTIFICATION This is a small, plain-colored spider. The cephalothorax and legs are reddish-brown and the abdomen is dark gray. There is a row of distinctive bristles or short spines on the underside of the tibiae, metatarsi, and tarsi of the first three pairs of legs. The spider has a flat appearance.

OCCURRENCE These spiders have been found at moderate elevations in the Cascade Mountains of Oregon and the Sierra Nevada in California. They have been collected under logs and bark litter.

SEASONALITY Males: August and September. Females: May through September.

Neoanagraphis chamberlini Gertsch and Mulaik, 1936

Plate 45

IDENTIFICATION This is a small, plain-colored spider. The cephalothorax is darker than the abdomen. The tips of the spinnerets can sometimes be seen extending beyond the end of the abdomen. The tarsal claws of legs III and IV are conspicuously long and look smooth.

OCCURRENCE This spider has been collected from California east to western Texas and south into Mexico. They are ground living and have been found in the burrows of larger animals, including tarantulas. Many records are from pitfall traps.

SEASONALITY Adults: April through November.

FAMILY LYCOSIDAE · *Wolf Spiders*

This is one of the most common and widespread families of spiders. There are at least 238 species of wolf spiders north of Mexico. They can be found in all habitats, including the intertidal zone at the coast. Wolf spiders are easy to recognize—they all have an unusual arrangement of their eight eyes (see Figs. 11A, 20B). Four of the eyes, the posterior ones, are large, forming a trapezoid at the top of the high carapace. The posterior median eyes are usually the largest and face forward. The posterior lateral eyes are also large and are well behind the PME on the head region, usually facing to the side or even backward. This gives the wolf spiders the ability to see well in all directions. In front of the PME there is a row of four smaller anterior eyes. Wolf spiders have excellent vision. They employ this ability in their courtship. Typically they combine visual signals, such as waving the legs or palps, with sounds created by vibrating body parts against other body parts or the substrate (the ground or leaves).

Two genera have many of our species. For example, there are 80 species in the genus *Pardosa* north of Mexico that can be recognized by the nearly perpendicular spines on the hind legs. In other ways the members of this genus are variable. The other common genus is *Pirata*. The 30 species of little spiders in this genus usually have a dark tuning fork–shaped mark on the cephalothorax.

Many wolf spiders are nocturnal or active at dawn or dusk. Others forage during the day. Most wolf spiders do not build capture webs, but they do employ a dragline. This silk line is laid down wherever they wander and provides a secondary means of communication among individuals. When a male encounters the dragline of a female, he may follow it to find her. Female wolf spiders exhibit unusual parental care. After constructing the egg sac, the female attaches it to her spinnerets and carries it with her as she continues to wander in search of prey. She may shuttle between the sun and shade to regulate the temperature for the developing young within the sac. When the young emerge from the egg sac, they climb up onto the body of their mother, typically on the abdomen, but sometimes there are so many that they also cover much of the cephalothorax. They ride here for a week or two while they complete their development and then disperse.

Most wolf spiders are ground hunters. They don't actually chase down their prey by endurance running like their namesake wolves, but rather these spiders sit and wait for prey. If no suitable victim is detected, the spider moves on to a new location. Thus they patrol a large area, seeking prey. If an insect walks by, the wolf spider will make a rapid dash to capture it. If the prey is small, the spider may pounce on it and bite immediately. If the prey is larger, the spider may grapple with the prey, grasp with its legs and briefly tumble upside down, holding the prey in a basket formed by the eight legs. Then the spider bites its prey. Within moments the spider is upright again, holding the prey below its body, caged by its legs. In addition to insects, wolf spiders often capture and eat other smaller wolf spiders.

A number of species of wolf spiders build burrows. The burrow may be a temporary, shallow hole, only a few centimeters deep, or a silk-lined tube that extends well into the ground. The entrance to the burrow may be unmarked, or it may incorporate a silken sleeve, extending up from the ground forming a sort of turret. This turret is composed of silk with bits of grains of soil, dead grass, dry leaves, bark, or other debris. Some wolf spiders close their burrows with a thin silk sheet or hinged doors, similar to those of the trapdoor spiders. The members of the genus *Sosippus* build a funnel-shaped web leading from the burrow out over the substrate.

Allocosa funerea (Hentz, 1844)

Plate 49

IDENTIFICATION These small wolf spiders have a shiny ebony or black cephalothorax and femora. The other leg segments are lighter in color. The abdomen is yellowish with dusky purplish markings and a dark heart mark. Occasionally in the North and frequently in the South, the abdomen is plain black.

OCCURRENCE This species is common in the eastern half of the United States. It is mostly a spider of open country—grasslands, lawns, fields, and meadows. They have also been collected in pine forests.

SEASONALITY Adults: April through October.

REMARKS This is one of the most common wolf spiders found in lawns. I have seen subadults running across the snow in early spring in Ohio.

Alopecosa aculeata (Clerck, 1757)

Plate 52

IDENTIFICATION This is a medium-sized, dark wolf spider with a wide light median band on the cephalothorax. The light median band is constricted just behind the head. The legs are unmarked. The abdomen has a light central band with a dark heart mark that is more obvious in females.

OCCURRENCE Members of this genus are primarily spiders of northern latitudes and mountains. This species occurs from Alaska to the Maritimes and south to the northern tier of the United States. They have been found mostly in meadows and the shrub boundaries of openings in forests.

SEASONALITY Males: April through summer. Females: April through November.

REMARKS Previously known as *Tarentula aculeata*.

Alopecosa kochi (Keyserling, 1877)

Plate 52

IDENTIFICATION This is a medium-sized grayish spider with a broad light median band on the carapace that has a constriction just behind the head. The front legs of males are dark. The abdomen is pale with two rows of black spots on either side of the wide light median band. The heart mark is darker than the median band.

OCCURRENCE Most members of this genus are spiders of northern latitudes and mountains. This species extends from southern British Columbia to Vermont and south to California and Texas. It has frequently been captured in pitfall traps.

SEASONALITY Males: February through November. Females: May through December.

REMARKS This species is among the most common wolf spiders in the southern California region. Previously known as *Tarentula kochii*.

Arctosa littoralis (Hentz, 1844) • *Shoreline wolf spider*

Plate 54

IDENTIFICATION This is a large, sand-colored wolf spider that matches the beaches where it lives. They range from light gray to tan. The legs are banded and the abdomen usually has a pattern of light spots. The eye region is typically black, forming a masklike marking. Dark-colored individuals have the center of the cephalothorax black with a broad light marginal band and a dark median band on the abdomen.

OCCURRENCE Most of the spiders in this genus are found in open areas near bodies of water. This species occurs throughout the United States as well as the southern edge of Canada. These spiders have been collected from shorelines along rivers, lakes, and oceans at night. During the day they hide under driftwood or other debris.

SEASONALITY Adults: February through November.

REMARKS This spider so closely resembles the sand where it lives that it can be difficult to see unless it moves. This species has often been found at night by searching for its eyeshines.

Arctosa rubicunda (Keyserling, 1877)

Plate 54

IDENTIFICATION This dark-colored, medium-sized wolf spider has a shiny, hairless cephalothorax. The central region of the cephalothorax is a bit lighter in color. The abdomen is dark with light areas near the front. The legs are not banded, but the femora are darker than the other leg segments.

OCCURRENCE Most of the spiders in this genus are found in open areas near bodies of water. This species is found in across most of the northern third of the United States and Canada. It occurs in open habitats such as fields, bogs, marshes, and deciduous forests. They have been collected under rocks in open areas and under logs in forests.

SEASONALITY Adults: mid-May through October.

REMARKS According to John Henry Comstock (1912), the cephalothorax has hairs that are so short and sparse that it looks polished.

Arctosa virgo (Chamberlin, 1925)

Plate 54

IDENTIFICATION This is a small, dark wolf spider with a dark shiny cephalothorax, banded legs, and a variegated abdomen. The pattern on the abdomen usually consists of a light area in the front and a series of light spots at the back.

OCCURRENCE Most of the spiders in this genus are found in open areas near bodies of water. This species occurs from Michigan to Tennessee and east to New Jersey and Virginia.

SEASONALITY Adults: May through August.

Geolycosa missouriensis (Banks, 1895)

Plate 48

IDENTIFICATION This large burrowing wolf spider is tan, brown, or dark gray. The cephalothorax and abdomen are densely hairy. The hairs on the cephalothorax of older individuals wear off and the head region may be shiny. The front surfaces of the chelicerae are often covered with light brown or yellowish hairs. The legs are stout and not banded. The bases of the front two pairs of legs are light below, as is the abdomen. The tibia, metatarsi, and tarsi of these legs are dark below.

OCCURRENCE This is the most widely distributed species in the genus. This species occurs from New York to Florida and west to Alberta and Arizona. They dig their burrows in sandy soil where there is some surface litter of vegetation. They have been collected in open habitats, sandy plains, hillsides, and open sandy woodlands.

SEASONALITY Males: autumn. Females: autumn to the following summer; eggs in spring or summer.

REMARKS Like all of the members of this genus, the females are restricted to their burrows. The males spend their early lives in burrows but when mature they wander in search of females. The burrows are shallow in the spring and much deeper in late summer. The burrow often has a silk-lined turret extending up from the ground composed of bits of vegetation and soil. The burrows may be in the open or under the trees.

Geolycosa pikei (Marx, 1881)

Plate 48

IDENTIFICATION This large burrowing wolf spider has a pale abdomen with a dark median band of varying width. The cephalothorax is dark gray, with a light median band, face, and marginal band. The legs are plain dark gray above. The undersides of the front legs in females are pale at the base and dark from the tibiae to the tips. In males the undersides of legs I and II are black. The abdomen has dusky markings below in the female and a black Y-shaped mark in the male.

OCCURRENCE This species is an eastern one, occurring from Connecticut to Georgia. They typically build their burrows near the seashore in beaches and dunes. Some have been collected further inland.

SEASONALITY Males: August and September. Females: all year; eggs in spring.

REMARKS The spider remains inactive in winter at the bottom of the open burrow. This spider does not usually build a turret.

Geolycosa rafaelana (Chamberlin, 1928)

Plate 48

IDENTIFICATION This is a large, sooty-black burrowing wolf spider. The male is darker than the female. The chelicerae have light-colored hairs. The undersides of legs I and II are completely black, including the femora.

OCCURRENCE This is a western desert species. They have been found in shrub deserts and desert grasslands, most often in dunes and sandy areas.

SEASONALITY Males: spring and summer. Females: all year.

REMARKS The burrow entrance does not usually have a turret.

Geolycosa turricola (Treat, 1880)

Plate 48

IDENTIFICATION This is a large, dusky-brown burrowing wolf spider. The male is paler. The undersides of the patellae, tibiae, metatarsi, and tarsi of legs I and II are black in both sexes. The underside of the cephalothorax is light brown or tan, not black. The abdomen has a black median band that does not extend to the sides of the abdomen.

OCCURRENCE This species occurs from New England south to Florida and west to Ohio and Tennessee. It is a spider of open fields and grasslands. It has often been found in hilly areas.

SEASONALITY Males: spring and summer. Females: all year.

REMARKS This species builds a conspicuous turret around the entrance to the burrow, often extending up from the soil surface a centimeter or more. The turret is composed of silk and bits of grass or other dry vegetation. Previously known as *Lycosa nidifex*.

Geolycosa wrighti (Emerton, 1912)

Plate 48

IDENTIFICATION This is a large, light brown burrowing wolf spider. The carapace often becomes worn and loses its hairs, becoming shiny brown. The undersides of the tibiae, metatarsi, and tarsi of legs I and II are black, but not the patellae. The underside of the cephalothorax is pale, but the chelicerae are dark. The underside of the abdomen is black, extending to the sides.

OCCURRENCE This species occurs in the Great Lakes region, with additional records from further west in Canada and the southern Great Plains.

SEASONALITY Adults: all year; females with young in summer.

REMARKS This species does not usually build a turret.

Geolycosa xera McCrone, 1963

Plate 48

IDENTIFICATION This large, burrowing wolf spider is entirely covered with light gray hairs. The general body color typically matches the background sandy substrates where they live. The femora of legs I and II are black below, in the male III and IV are also dusky below. In females the patella I and II, tibia I, and base of tibia II are black below. In males all of the patellae, tibia I, and the bases of tibia II–IV are black below. There is variation in the extent of black under the legs.

OCCURRENCE This is one of several distinctive forms found in Florida. They are currently treated as subspecies. The *Geolycosa xera* complex is present in the central dunes and dune scrub areas from central Florida.

SEASONALITY Adults: all year; two breeding seasons, spring and fall.

REMARKS At least 10 species of *Geolycosa* have been found in Florida. *Geolycosa pikei* is more common in northern Florida and the coastal areas of the state. John McCrone (1963) has written an interesting article on the history of the dune areas and the *Geolycosa* that occur in them.

Gladicosa gulosa (Walckenaer, 1837)

Plate 52

IDENTIFICATION This species is a medium-sized to large brown wolf spider. It has a light median band on the cephalothorax that is constricted just behind the head region. The abdomen is variegated brown with a series of small dark spots with light centers. The legs have indistinct banding.

OCCURRENCE This is a common species in deciduous forests across the eastern half of the continent. It is replaced by related species in the southeastern states.

SEASONALITY Adults: late autumn through June; females with eggs April through June.

REMARKS The courtship sound may be audible from a few paces away. The male produces this stridulation by vigorously flexing his palps. He makes additional vibration signals by bouncing his abdomen against the dry leaves of the forest floor. The stridulation sound resembles a soft purring or humming when heard at close range.

Gladicosa pulchra (Keyserling, 1877)

Plate 52

IDENTIFICATION This is a medium-sized to large wolf spider. There are several color forms; some individuals are brown and similar to *Gladicosa gulosa*. Another color form is much grayer. This variable species can be found in all shades of color between these two extremes. All of the varieties have a light median band on the cephalothorax that is constricted behind the head region.

OCCURRENCE *Gladicosa pulchra* is a southern species extending from Texas to south Florida and as far north as Ohio and Long Island, New York. This species has been found on the trunks of trees in the fall. During the spring and summer they have been collected on the forest floor and in burrows or under rocks.

SEASONALITY Adults: autumn through spring; females with eggs through early summer.

REMARKS Micky Eubanks and Gary Miller (1993) have suggested that females climb trees to avoid predators on the ground. Males also climb trees, probably to find females.

Hesperocosa unica (Gertsch and Wallace, 1935)

Plate 49

IDENTIFICATION This is a small, dark slate–colored wolf spider. There is a pale median band of white hairs on the cephalothorax. The abdomen is also dark with a light band of hairs in the center of the upper surface. The legs are dark without banding.

OCCURRENCE This species has been found in arid regions of Arizona, New Mexico, and western Texas. It has been captured in pitfall traps in shrublands and desert grasslands.

SEASONALITY Adults: April through July.

Hogna antelucana (Montgomery, 1904)

Plate 54

IDENTIFICATION This is a large, light brown wolf spider that resembles *Hogna lenta* (Plate 55). It differs from this species by having the dark central band on the abdomen extending further back and lightly banded legs. There is a distinctive dark band on the distal portion of the tibia of the hind legs. The central light band on the cephalothorax is constricted behind the head region. The sides of the head region below the posterior eyes are usually dark, not light as in *Hogna lenta*.

OCCURRENCE This is a southern species; it has been found from western Florida to California, north to Kentucky and Utah.

SEASONALITY Nothing is published.

Hogna aspersa (Hentz, 1844) • *Tiger wolf spider*

Plate 55

IDENTIFICATION This is a large, dark wolf spider. It has distinctive orange color between the dark banding on the legs. There is a narrow light line between the eyes and often two small light eyebrow marks behind the posterior lateral eyes. The underside of the cephalothorax is dark brown or black. Underneath, the abdomen is pale in the center and spotted with irregular dark blotches at the sides.

OCCURRENCE This species is northeastern. It has been found between southern Ontario and Florida, west to Nebraska. This is a burrowing species whose burrows have usually been found in forested areas but occasionally in the open.

SEASONALITY Males: August through October. Females: all year; with eggs and young spring to summer.

REMARKS The burrow is relatively straight and usually has a turret of silk-bound bits of vegetation extending above the ground surface. The large entrance may be more than 30 mm in diameter. According to Willis Gertsch (1979), this spider sometimes constructs a lid to the burrow entrance, similar in form to that of a waferlid trapdoor spider.

Hogna baltimoriana (Keyserling, 1877)

Plate 55

IDENTIFICATION This is a large, brown wolf spider that is somewhat similar to a group of related species, particularly *Hogna lenta*. In most individuals there is a relatively narrow light band extending down the center of the cephalothorax from just behind the posterior median eyes to the back of the head region, wider in the thoracic region. There is also usually a narrow marginal light band that extends up toward the eyes in the front that looks like a light cheek mark. There are light lines radiating away from the center of the thoracic region toward each leg. The posterior lateral eyes are as large as the posterior median eyes and facing sideways. The abdomen has variable markings. The legs are not banded but have darker tips to the segments. The sternum and patellae are black underneath. There is also a black line under the femora. The abdomen has a black rectangular spot from behind the epigastric furrow almost to the end. This feature contrasts with *Hogna lenta* that has a completely black underside of the abdomen.

OCCURRENCE This species occurs from New England and Ontario, west to Montana, and south as far as Texas and Louisiana. It has not been recorded from the extreme Southeast. It is a burrowing species that seems to prefer sandy soils.

SEASONALITY Males: autumn. Females: all year.

REMARKS The burrows of this species have a large entrance (greater than 30 mm) but usually lack a well-defined turret.

Hogna carolinensis (Walckenaer, 1805) • *Carolina wolf spider*

Plate 55

IDENTIFICATION This is the largest wolf spider in North America north of Mexico. They are variable in color but usually have both a wide light median band and wide light marginal band on the cephalothorax. The broad median band usually extends only as far forward as the back of the posterior median eyes. The ventral surface of the abdomen, cephalothorax, and coxae of the legs are black. In the female there are also black spots under the ends of the femora. In the male there are black areas under the ends of the femora, patellae, tibiae, and the entire undersides of the metatarsi and tarsi. The legs are not banded above. The chelicerae are usually covered with light orange or yellow hairs.

OCCURRENCE This is a widespread species across the United States. It has been recorded from southern Ontario. This is a spider of open areas, fields, grasslands, pastures, meadows, and deserts. They are a burrowing species, but they wander away from the burrow at night and rarely during the day to hunt. Adult males also wander in search of females. They have also been found under rocks during the day.

SEASONALITY Males: late summer and autumn. Females: all year.

REMARKS The burrows of this species have a large entrance (greater than 30 mm) usually with a turret constructed of silk, grass blades, and small twigs. If threatened they will rear up, raising their front legs and spreading the chelicerae. This display is usually a bluff, and they are not known to bite unless provoked.

Hogna frondicola (Emerton, 1885)

Plate 55

IDENTIFICATION This is a medium-sized to large light brown wolf spider. Most of the body is pale tan. The sides of the cephalothorax are a contrasting dark brown. There are dark spots at the front of the abdomen that seem to extend the dark bands of the cephalothorax. The legs are not banded. The underside of the cephalothorax is black. There is an irregular dark stripe under the abdomen in males that may or not be present in females.

OCCURRENCE This species has a broad range, extending from Newfoundland to the Yukon Territory and south to Arizona and New Mexico but has not been recorded from the Southeast. This species has been found in both forests and open habitats. It has often been found running over the leaf litter. It does not usually build a burrow.

SEASONALITY Adults: summer until the following spring.

REMARKS The female may build a shallow silk-lined burrow as a nest when she has an egg sac.

Hogna georgicola (Walckenaer, 1837)

Plate 54

IDENTIFICATION This is a large, brown wolf spider. It is similar to *Hogna helluo* (Plate 55). The female is usually dark with a narrow light central line on the cephalothorax, narrower between the eyes. The legs are banded. The male is paler with less distinct banding or no banding on the legs. The ventral surface of the abdomen has three rows of dark spots that may appear as longitudinal lines, converging near the spinnerets.

OCCURRENCE This species has been recorded from southern New England west to Kansas and south to Texas and Georgia. It has been found in deciduous forests foraging among the leaves at night and hiding under logs in the daytime. It has also been found in burrows.

SEASONALITY Nothing is published.

REMARKS Previously known as *Lycosa riparia* or *Allocosa georgicola*.

Hogna helluo (Walckenaer, 1837)

Plate 55

IDENTIFICATION This is a large, dark brown wolf spider. It is similar to *Hogna georgicola* (Plate 54). The legs are not banded. The female is dark brown with a thin light median band on the cephalothorax. This band is narrowest between the eyes and sometimes with yellowish hairs. The abdomen of the female is dark with an indistinct dark heart mark that has a black margin. The underside of her abdomen is covered with an irregular pattern of small black spots. The male is much paler in color with a distinct light median band and a light marginal band on the cephalothorax.

OCCURRENCE This species occurs in the Northeast from New England and Quebec, west to Colorado, and south to Texas and Florida. This is a common and widespread species in agricultural fields and adjacent forested areas. It prefers moist areas.

SEASONALITY Males: May through September. Females: all year; eggs in spring and summer.

REMARKS This spider sometimes digs a shallow burrow without a turret. They also live in cracks in the soil and under rocks or logs. They hunt mostly at night. They often wander into buildings during the autumn.

Hogna lenta (Hentz, 1844)

Plate 55

IDENTIFICATION This is a large, light brown wolf spider with a dark mark over the heart region. The cephalothorax has a light median band, narrowing between the eyes and a light marginal band. There are sometimes light eyebrow marks behind the posterior lateral eyes. The legs are not banded. The markings of the male are more distinct. The underside of the abdomen is black behind the epigastric furrow. In this species, unlike the similar *Hogna baltimoriana*, the black ventral mark extends in the center as far as the front of the abdomen.

OCCURRENCE This species is mostly southeastern. It has been found as far north as Kentucky and southern Virginia. It has most often been found in moist, shady woods in the South and also drier forests in the North.

SEASONALITY Adults: in the North, summer and autumn; in the South, all year.

REMARKS The burrow has a relatively large diameter and may have a thin door similar to that found in some burrows of *Hogna aspersa*.

Pardosa distincta (Blackwell, 1846)

Plate 50

IDENTIFICATION This is a small to medium-sized wolf spider. The female is a yellow brown with alternating dark and light longitudinal bands on the cephalothorax. The abdomen of the female has a dark margin to the heart mark and a pair of dark bands on either side of the center of the abdomen. On the underside the abdomen has two parallel dark stripes that may converge at the back. The male is darker brown with dark brown femora and palps.

OCCURRENCE This species has been recorded across the northern tier of states and provinces from Nova Scotia to Alberta south through the Rocky Mountain States to Arizona and New Mexico. This is one of the most common spiders in the Rocky Mountain region. It is an abundant resident of grassy fields. It has also been collected in conifer forests, beaches, orchards, meadows, and clearings in woods.

SEASONALITY Males: April through August. Females: April through October; with eggs from June through October.

REMARKS According to Charles Dondale and James Redner (1990), this spider has been found at elevations up to 4,270 m in the mountains of Colorado.

Pardosa lapidicina Emerton, 1885

Plate 51

IDENTIFICATION This species is a small to medium-sized gray wolf spider. There is considerable color variation from dark reddish or chocolate brown to slate gray or black. The abdomens of females usually have light spots. The legs are usually banded.

OCCURRENCE This species has been found in the northeastern two-thirds of the United States, the Maritimes, Ontario, and Quebec. They have been found in rocky areas, talus slopes, clay banks, and frequently along the rocky shorelines.

SEASONALITY Adults: March through November; females with eggs in summer.

REMARKS This species often has coloration that matches its habitat.

Pardosa milvina (Hentz, 1844)

Plate 51

IDENTIFICATION This is a small gray or light brown wolf spider. The coloration is variable but the pattern is relatively consistent. The light median band on the cephalothorax is constricted behind the head region. The abdomen usually has a light heart mark followed by a series of paired spots farther back. The legs are banded. The males have black in the eye region and conspicuous black tarsi on the palps.

OCCURRENCE This spider is common in the eastern half of the United States as well as in southern Ontario and Quebec. It has most often been found on the ground in moist habitats, including swamps, marshes, meadows, pastures, and the muddy edges of ponds and creeks. They have also been found in wooded areas. They are particularly common in lawns and a variety of agricultural fields.

SEASONALITY Males February through August. Females February through November; with eggs from April through September.

REMARKS The males have often been found in the open during the day, constantly waving their black palps as they search for females.

Pardosa moesta Banks, 1892

Plate 51

IDENTIFICATION This is a small distinctive wolf spider of the genus *Pardosa*. Female *moesta* have a shiny brown cephalothorax, with an indistinct paler median area of short hairs. The cephalothorax of males is similar with a thin marginal band of white hairs. The abdomen is much lighter with a variegated pattern. The legs are yellowish in the female. In the male the palps as well as the femora of legs I and II are dark brown.

OCCURRENCE This is a common species from Alaska to the Canadian Maritime Provinces and south to Colorado and Tennessee. It has commonly been found in open habitats, including fields, meadows, bogs, lawns, and among debris on beaches. There are also records from deciduous and coniferous forests.

SEASONALITY Males: May through August. Females: May through October.

REMARKS This spider, particularly the adult male, is one of the few wolf spiders that is recognizable at a glance.

Pardosa saxatilis (Hentz, 1844)

Plate 51

IDENTIFICATION This is a small, brown wolf spider. Like *Pardosa milvina*, this species has a broad light median band on the cephalothorax that is constricted behind the head region. The abdomen has a light heart mark but usually lacks additional light areas behind. The male differs from male *Pardosa milvina* by having a much darker cephalothorax and dark femora of the legs. The male also has black tarsi on the palps as well as dark palpal femora. The palpal femora are pale in *P. milvina*.

OCCURRENCE This species has a northeastern distribution but has been recorded as far west as Minnesota and Nebraska. It is common in open grassy areas. It has also been found in marshes, bogs, beaches, and deciduous woodlands.

SEASONALITY Males: March through November. Females: May through September; with eggs in summer and autumn.

REMARKS This is one of the most numerous species captured in pitfall traps in fields.

Pardosa sternalis (Thorell, 1877)

Plate 50

IDENTIFICATION This is a small, dark-colored wolf spider. This spider is one of three similar dark-colored species found in the Northwest. The female has a light median band and irregular submarginal light bands on the thoracic region of the cephalothorax. The femora of the female are banded. The male is dark brown with a less distinct pattern similar to that of the female.

OCCURRENCE This species has been recorded from Kansas and Nebraska west to Alberta and south to California and Texas. It has been found around the margins of lakes and reservoirs, rivers, creeks, prairie wetlands, meadows, and lawns. According to Charles Dondale and James Redner (1990), it occurs in mountains at elevations up to 3,670 m.

SEASONALITY Males: March through September. Females: April through September; with eggs in June.

REMARKS This is one of the most numerous small wolf spiders in the Rockies and Great Plains.

Pardosa xerampelina (Keyserling, 1877)

Plate 51

IDENTIFICATION This is a medium-sized wolf spider with variable coloration. In this species the base color can range from nearly black to gray or light brown. Two color varieties are illustrated: a light brown form female and a gray form male. On the cephalothorax the submarginal light band is either replaced by a few light spots or is absent. The legs are usually banded although this is sometimes inconspicuous. There is usually a pattern of darker spots around the center of the abdomen.

OCCURRENCE This species is most common in the northern part of its range. It occurs from Alaska to Newfoundland south to Oregon and New Mexico in the West and West Virginia in the East. It has been found in short grass fields, cultivated crops, marshes, bogs, along roadsides, and open deciduous woodlands.

SEASONALITY Males: April through July. Females: April through October; with eggs May through October.

REMARKS This variable species is most common in the northern portion of its range.

Pirata insularis Emerton, 1885

Plate 49

IDENTIFICATION This is a small, yellowish-brown wolf spider. The abdomen has a series of light spots in along the top. The legs are pale, without banding.

OCCURRENCE *Pirata insularis* is found in Eurasia as well as in North America. Here the species occurs between northern British Columbia and the Yukon Territory to Newfoundland and as far south as Arizona and Florida. It has been found around the margins of lakes, ponds, streams, marshes, and bogs.

SEASONALITY   Males: in the North, May through August. Females: in the North, July through September. Adults: all year in the South.

REMARKS   This spider is adept at running across the surface of water. According to Charles Dondale and James Redner (1990), most Canadian specimens have been collected in sphagnum bogs.

Pirata minutus Emerton, 1885

Plate 49

IDENTIFICATION   This is a tiny black wolf spider. There is a light median gray band along the center of the cephalothorax split by a dark-colored mark in the head region. There is a thin marginal band of white scale-like hairs. The abdomen is dark with two parallel lines of white spots. The legs are dark and unbanded. In good light the femora of the legs in males may appear iridescent.

OCCURRENCE   This species has been collected from Newfoundland west to Saskatchewan south to Arkansas and South Carolina. It has been found in spruce-fir forests, woodlands, fields, meadows, marshes, swamps, bogs, and lawns.

SEASONALITY   Adults: May through September; females with egg sacs in summer.

REMARKS   The egg sacs of this species are often white, contrasting strongly with both the dark color of the spider and surrounding vegetation (see Fig. 2, lower).

Pirata piraticus (Clerck, 1757)

Plate 49

IDENTIFICATION   This is a small, yellowish-brown wolf spider. There is a wide submarginal light band on the cephalothorax. The legs are pale, without banding. The abdomen has a distinct light heart mark that often has an orange tint and is surrounded by a thin line of white hairs in unworn specimens. Beginning at the back of this mark and extending to the end of the abdomen are two lines of small white spots.

OCCURRENCE   This species is widespread in Eurasia as well as in North America. It is widely distributed from Alaska east to Newfoundland and south to California and South Carolina. This spider has been found in moist habitats at the edges of lakes, ponds, swamps, sphagnum bogs, and streams.

SEASONALITY   Adults: May through September; females with egg sacs June through September.

REMARKS   These spiders spend much of their time among low vegetation. Females with egg sacs move into the sun, presumably to warm the eggs.

Rabidosa punctulata (Hentz, 1844) • *Dotted wolf spider*

Plate 50

IDENTIFICATION   This is a large, light brown wolf spider. The cephalothorax has a distinctive pattern of alternating light and dark brown stripes. The abdomen has a dark median stripe without light spots. The underside of the abdomen has an irregular pattern of large brown spots.

OCCURRENCE This species has been found between Massachusetts and eastern Kansas and south to Texas and northern Florida. It is replaced by a similar species, *Rabidosa carrana*, in peninsular Florida. *Rabidosa punctulata* has been found in tall grasses and weedy habitats. They hunt mostly at night.

SEASONALITY Adults: September through October. Mating takes place in November. Adult females overwinter to lay eggs the following March.

REMARKS According to David Reed and Amy Nicholas (2008) in a study in Mississippi, the females carry their egg cases for 30 to 40 days before the young emerge. The young spend another one or two weeks riding on her back, then disperse.

Rabidosa rabida (Walckenaer, 1837)

Plate 50

IDENTIFICATION This is a large, light brown wolf spider. The cephalothorax has a distinctive pattern of alternating light and dark brown stripes. The abdomen has a dark median stripe with a series of paired light spots at the back. The front legs of adult males are black. The underside of the abdomen is pale.

OCCURRENCE This is an eastern species found from Massachusetts west to Minnesota and south to Texas and Florida. This spider is usually found in open habitats such as pastures and prairies, even open woodlands. It is most common in tall grasses, where it climbs in the vegetation at night. This spider also hunts in the grass during the day. This species seems to be more common than *Rabidosa punctulata* in the northern part of their overlapping ranges.

SEASONALITY Adults: June through October; females with egg sacs July through October. In the southern parts of the range, where both species are found in the same areas, *Rabidosa rabida* matures and mates in the summer while *Rabidosa punctulata* mates in the autumn.

REMARKS Jerome Rovner (1980) observed that this spider sometimes wraps the prey with silk after it has been immobilized.

Schizocosa avida (Walckenaer, 1837)

Plate 53

IDENTIFICATION This is a medium-sized, light brown wolf spider. The cephalothorax has both light median and marginal bands. The median band abruptly narrows between the posterior eyes, forming an odd reverse arrowhead. There is a broad light band on the abdomen enclosing a dark heart mark at the anterior end. The legs are not banded.

OCCURRENCE This species is eastern, occurring from Nova Scotia west to the 100th parallel and south to Texas and Florida. A closely related species, *S. communis*, shares much of the range of this species. The range of *S. avida* is complementary to that of *S. mccooki; avida* in the east and *mccooki* in the west. *Schizocosa avida* has been found in fields, pastures, meadows, and lawns. This spider has also been found under rocks and logs during the day.

SEASONALITY Adults: May through August; females with egg sacs in the summer.

REMARKS Subadult females dig a shallow burrow, often near the edge of a rock or log, and line it with silk. When mature, they will lay eggs in this secluded spot.

Schizocosa bilineata (Emerton, 1885)

Plate 53

IDENTIFICATION This is a small, light brown wolf spider. This species is a straw-yellow spider with a brown cephalothorax that has wide light median and lateral bands. The abdomen is light brown with a thin dark line around the heart area and rows of light spots on either side of a pale median band. The legs are unbanded. In the male there are conspicuous black brushes on the tibiae of the front legs.

OCCURRENCE This is an eastern species that has been found from New England west to Iowa and eastern Kansas south to eastern Texas and Georgia. It lives in tall grass, meadows, and beach grasses. It has also been reported from unkempt lawns.

SEASONALITY Males: May through July. Females: May through September; with egg cases in summer.

REMARKS The black brushes on the front legs of the males make them much more conspicuous running through the grass than are the females in the same areas. This species is active during the daytime.

Schizocosa mccooki (Montgomery, 1904)

Plate 53

IDENTIFICATION This medium-sized to large wolf spider has a gray or light brown body with two dark lateral bands on the cephalothorax as well as a dark heart mark on the abdomen. There are light bands lateral to the heart mark as well as a series of paired white spots. The males are similar to the females except for the tips of the front legs are dark. The color pattern is similar to *Schizocosa avida*. The total length of this spider varies from 9.6 to 22.7 mm in adult females, with the smallest adult males being only 9.1 mm long.

OCCURRENCE This is a western species, occurring from western Lake Erie south and west to California. The range of *S. mccooki* is complementary to that of *S. avida*; *avida* in the east and *mccooki* in the west. *Schizocosa mccooki* have been found in a variety of habitats, including lake edge dunes, sagebrush, pinyon pine woodland, grassland, and other open habitats.

SEASONALITY Males: June and July. Females: June through August.

REMARKS Unlike most members of the genus *Schizocosa*, some populations of this species dig a burrow. During courtship, the male makes drumming sounds with his palps.

Schizocosa ocreata (Hentz, 1844) • *Brushlegged wolf spider*

Plate 53

IDENTIFICATION This is a medium-sized, brown wolf spider. There is a broad light median band on the cephalothorax and abdomen. The dark areas on the sides of the cephalothorax and abdomen are more distinct, nearly black in adult males. The front legs of males are black with dense brushes of setae on the tibiae.

OCCURRENCE This is an eastern species that occurs from New England west to Nebraska and south to Texas and Florida. In the northern part of the range, this species is common in the leaf litter of upland deciduous forests, forest edges, and open fields near woods. In the southern part, it is more common in bottomland hardwood forests along rivers. Individuals are found foraging both day and night.

SEASONALITY Adults: April through October; females with eggs July through September. Over-winters as subadult and matures in early spring.

REMARKS This species is one of the most common wolf spiders in eastern North America and has been the subject of many scientific studies on behavior and ecology. For example, George Uetz and Andrew Roberts (2002) published a review of courtship behavior of this and related species by studying the response of spiders to video playback.

Schizocosa rovneri Uetz and Dondale, 1979

Plate 53

IDENTIFICATION This is a medium-sized brown wolf spider. There is a broad light median band on the cephalothorax and abdomen. The dark areas on the sides of the cephalothorax and abdomen are more distinct in adult males. The adult males lack the dark front legs and brushes of the similar *S. ocreata*. Females of these two species look nearly identical.

OCCURRENCE This is an eastern species that has been confused with *Schizocosa ocreata* until recently. The species occurs in bottomland deciduous forests and floodplains in the northern parts of its range. In the southern part it has been found in upland forests. It is known from Delaware west to Illinois and south to Arkansas, Mississippi, and North Carolina but may have a more extensive range.

SEASONALITY Adults: spring. According to Geraldine Denterlein and George Uetz (1979), they reproduce later in the season than *S. ocreata*.

REMARKS The absence of brushes on the legs and features of the courtship behavior of the males of this species serve to prevent hybridization with *S. ocreata*.

Schizocosa saltatrix (Hentz, 1844)

Plate 53

IDENTIFICATION This is a medium-sized, brownish-gray wolf spider. There is a broad light median band on the cephalothorax and abdomen. The median band on the cephalothorax contrasts strongly with the dark, nearly black lateral bands. The abdomen usually has thin dark lines around the heart area and a series of dark chevrons toward the back.

OCCURRENCE This is an eastern species. It has been found between Nova Scotia and Colorado, south to Texas and the panhandle of Florida. This species has been found in hardwood forests.

SEASONALITY Adults: March through August; females with eggs May through July.

REMARKS Jerome Rovner (1975) demonstrated that the stridulatory sounds produced by males in courtship are the result of rubbing structures at the joints during flexing of the palp, not "drumming" of the palps as had previously been thought.

Sosippus mimus Chamberlin, 1924

Plate 32

IDENTIFICATION This wolf spider is identified by its habit of building a funnel-shaped web. It is one of seven species in this genus sharing this odd habit. The cephalothorax has a pair of orange or whitish marks on the sides of the head region with a black mark between. The eye region is black. On the top of the head region there are three thin orange or whitish lines. The abdomen is dark in the center with a series of small white spots on either side of the dark mark. At the back the black mark is broken into a series of chevrons. The spinnerets are relatively long for a wolf spider.

OCCURRENCE This species and most of its relatives are southeastern. There is a similar species in southern California. They build their funnel-shaped webs extending out from some sort of protected retreat near the ground. The retreat may be in folded vegetation, between rocks, or other debris. This spider has been found in areas of sandy soil in woodlands or open fields. They have often been found in disturbed areas.

SEASONALITY Adults: January through October; females with eggs in spring and summer.

REMARKS The web is a spreading sheet extending from a tubular retreat. The retreat portion may form a tube under a log or other debris. This tube is 20 to 30 cm long with an opening at the end to permit escape.

Trabeops aurantiacus (Emerton, 1885)

Plate 49

IDENTIFICATION This is a tiny wolf spider. The eyes are particularly large; the posterior median eyes protrude over the face. The cephalothorax has dark shading that fades toward the center. There is a distinct light submarginal band. The abdomen is orange or light brown. The femora of the front legs are dark.

OCCURRENCE This is an eastern spider. It has been found from Nova Scotia west to Wisconsin and south to Mississippi and Florida. This spider has been found in both conifer and deciduous woods, swamps, near streams, ponds, and lakes.

SEASONALITY Adults: May through August; females with eggs in summer or fall.

REMARKS The silk of the egg sac is so thin that the eggs within it are visible.

Trochosa terricola Thorell, 1856

Plate 52

IDENTIFICATION This is a large, light brown wolf spider. The cephalothorax has a broad light brown median band with the sides being dark brown or black. At the margin there is a light brown band. Behind the eyes the pale median band is wide and has two parallel dark lines on either side of the midline. The legs are not banded or with faint rings. The abdomen is variegated gray brown with a light heart mark.

OCCURRENCE This is a widespread species present in Europe as well as in North America. They have been recorded from Nova Scotia west to Alaska and south to California and South Carolina. This species has been collected from forest edges and fields at night. During the day it has been found under rocks.

SEASONALITY Adults: all year.

REMARKS Occasionally adult individuals have been captured that have no eyes. Because they grew to mature size, it seems clear that this spider can hunt successfully without vision.

Varacosa avara (Keyserling, 1877)

Plate 50

IDENTIFICATION This is a medium-sized brown wolf spider. This species has a dark brown cephalothorax with a wide pale median band and indistinct light submarginal band. There are sometimes faint dark lines within the light median band. The abdomen is mottled dark brownish-gray. The legs are faintly banded.

OCCURRENCE This species has been found from southern Quebec west to Wisconsin and Nebraska south to Texas and the panhandle of Florida. This is a nocturnal species. During the day it has most often been collected under rocks in fields or at the edge of woods.

SEASONALITY Males: September through April. Females: October through July; with eggs in spring and summer.

REMARKS Previously known as *Trochosa avara*.

FAMILY MECICOBOTHRIIDAE · *Midget Funnelweb Tarantulas*

This family is native to the Americas and contains eight species. Six of the eight are found in western North America. The midget funnelweb tarantulas have long, widely separated, posterior spinnerets. The abdomen has two hard plates at the front, sometimes fused. Most of our species build small horizontal sheets with tube-shaped retreats. These webs are hidden under rocks, decomposing wood, or debris. During its life, perhaps of several years, the spider continues to add silk to the web, extending the sheet as well as building new layers. The largest species (*Megahexura fulva*, Plate 3) builds a larger exposed sheet web with a funnel-shaped retreat that could be confused with the web of a funnel weaver (Agelenidae).

Hexura picea Simon, 1884

Plate 3

IDENTIFICATION This is a small to medium-sized mygalomorph. The cephalothorax is shiny brownish with dark streaks from the eyes to the fovea. The fovea is a short longitudinal slot. The abdomen is purplish brown. This spider has huge chelicerae, but they lack the digging rastellum of the trapdoor spiders. The abdomen has one fused tergite at the front. The posterior spinnerets are long and flexible. The other four spinnerets are small.

OCCURRENCE This spider occurs in western Washington and Oregon. This forest species lives within a web of sheets and tubes in the deep coniferous duff.

SEASONALITY Males: summer through November. Females: all year.

Megahexura fulva (Chamberlin, 1919)

Plate 3

IDENTIFICATION This is a medium-sized to large mygalomorph. The cephalothorax is brown with large dark chelicerae. The legs are lighter. The abdomen is light orangish or pinkish brown with two tergites at the front, but only one is easily visible. The posterior spinnerets are long, with flexible tips.

OCCURRENCE This is a Californian species occurring from El Dorado and Alameda Counties south to the Mexican border. This spider lives in among the leaf litter in the Sierra Nevada and Coast Ranges.

SEASONALITY Males: wander in cool months. Females: all year.

REMARKS This spider builds a web composed of sheets and tubes, usually within deep leaf litter sometimes with an exposed funnel.

FAMILY MIMETIDAE · *Pirate Spiders*

This family is found worldwide, with 156 species. There are 18 species described from North America north of Mexico, but there are at least 10 more undescribed species. These spiders are distinguished by a unique arrangement of spines on the tibiae and metatarsi of the first two pairs of legs. The arrangement is a regular repeating pattern of long, stiff, and slightly curved spines. Between each pair of long spines there are two to five similar but shorter curved spines that increase in length gradually until the next large spine (see Fig. 33B). Pirate spiders do not build webs. These spiders prey primarily upon other spiders. They approach the web and pluck the strands, probably to draw the resident into the open, then attack. After consuming the resident spider, they may also eat her eggs. Some species are sit-and-wait predators; these also prey on other spiders.

Ero canionis Chamberlin and Ivie, 1935

Plate 38

IDENTIFICATION This is a small dark mimetid. The cephalothorax is pale, usually with a dark brown rim including the eye area, and a series of dark spots extending back from the eyes. The abdomen is usually variegated dark in the front. There are two or four humps along the margin of a light region in the back. The legs are marked with dark rings. The femora, tibiae, and metatarsi of the front legs are mostly dark.

OCCURRENCE This species is sparsely distributed across southern Canada and the northern United States. It has been recorded as far south as Texas.

SEASONALITY Adults: late summer through spring.

REMARKS This spider hunts other spiders. It is a sit-and-wait predator that hangs under a leaf awaiting the approach of prey.

Mimetus puritanus Chamberlin, 1923

Plate 38

IDENTIFICATION This is a large mimetid. The cephalothorax is pale with a dark region of three variable lines that converge at the center of the thoracic region. The abdomen has a dark dorsal surface with two transverse white spots that divide the front from the back. The legs often have heavy rings at the ends of the tibiae of the front legs. There are dark spots on the ventral surfaces of the femora of the front four legs.

OCCURRENCE This species occurs from New England, Ontario, and Quebec, west to North Dakota, and south to Texas and northern Florida. They have been found on low vegetation, fences, and walls.

SEASONALITY Adults: April through July.

REMARKS These spiders eat other spiders, including the common house spider (*Parasteatoda tepidariorum*).

Reo eutypus (Chamberlin and Ivie, 1935)

Plate 38

IDENTIFICATION This is a medium-sized mimetid. The cephalothorax is pale with a dark area surrounding the eyes. There are often three dark lines extending back from the eyes and converging near the center of the thoracic region. There are two pairs of dark spots near the edges of the sternum. The abdomen is often light with a pattern of small dark spots. Some individuals have a dark abdomen; others have a dark T-shaped mark. The legs are light with scattered small spots. There are rings at the ends of the femora and patellae of legs I and II.

OCCURRENCE This species has been found from central California and Nevada south into Baja California. This spider is found resting near the ground or under rocks. It forages in low vegetation.

SEASONALITY Nothing is published.

FAMILY MITURGIDAE • *Prowling Spiders*

Currently this worldwide family includes 355 species. Many species now included in this family used to be classified with the Clubionidae. They share the conical spinnerets, claw tufts, and the habit of building a silk retreat or sac. They have eight eyes in two regular rows of four. In North America north of Mexico we have 12 species. The prowling spiders are nocturnal wandering spiders. During the day they rest in a silk cocoon in a folded or rolled leaf, or under a rock on the ground. Most of the species in this family in North America are ground hunters. *Teminius* and *Syspira* could be confused with wolf spiders except for the arrangement of their eyes. The two species in the genus *Cheiracanthium* found here are active runners over vegetation at night. They are similar in appearance to spiders in the genus *Clubiona* of the family Clubionidae. They probably build a new silk cocoon before dawn each day. They are sometimes common in buildings.

Cheiracanthium inclusum (Hentz, 1847) • Agrarian sac spider

Plate 37

IDENTIFICATION This is a large, pale-yellow or yellow-green sac spider. It is one of two similar species. The other (yellow sac spider; *C. mildei*) is introduced from Eurasia. The cephalothorax is darkest in the eye region, and the tips of the chelicerae are dark brown, nearly black. The abdomen has a distinct heart mark. The legs and palps are not banded but the tips of the tarsi are dark.

OCCURRENCE This species is widespread in Central America, throughout the United States, and into parts of southern Canada. The introduced species (*Cheiracanthium mildei*) is also widespread and appears to be replacing *C. inclusum* in some areas and coexisting in others (Hogg, Gillespie, and Daane 2010). The agrarian sac spider is common in the foliage of trees and shrubs as well as in herbaceous vegetation. It has also been collected in urban and suburban yards, orchards, and field crops. It is one of the most common spiders inside buildings. Outdoors it builds its daytime sac in a folded leaf; indoors it often builds the sac at the intersection of wall and ceiling.

SEASONALITY Adults: early spring through November.

REMARKS This spider is a constantly active runner at night. Robin Taylor and Woodbridge Foster (1996) demonstrated that this active lifestyle is fueled by the consumption of plant nectar, which is high in sugar. The spiders eat the nectar from both flowers and other nectaries on plants. Members of this genus are known to occasionally bite humans. These bites may be painful but are usually not serious.

Strotarchus piscatorius (Hentz, 1847)

Plate 45

IDENTIFICATION This species is one of two similar spiders in this genus that are pale green. The eyes are in two parallel rows of four. Unlike the similar species of *Cheiracanthium*, the anterior median eyes of *Strotarchus* are distinctly larger than the other eyes. In both *Strotarchus* and *Cheiracanthium* the face region and jaws are darker in color than the rest of the cephalothorax. There is a narrow, distinct dark heart mark. The spinnerets of *Strotarchus* extend beyond the end of the abdomen. The male has extraordinary extensions of the palps (Plate 45) that make them easy to recognize.

OCCURRENCE This is an eastern species, but another similar species occurs in the West. Unlike *Cheiracanthium*, this spider has frequently been found on the ground. The retreat is a silken sac under a rock. Retreats are also found in vegetation, on trees, or in folded grass blades, but *Cheiracanthium* is far more common in these habitats.

SEASONALITY Adults: spring through August.

REMARKS According to Robert J. Edwards (1958), this spider acts dead when disturbed and does not flee. Previously known as *Marcellina piscatoria*.

Syspira longipes Simon, 1895

Plate 45

IDENTIFICATION This is a relatively large, pale-colored ground hunting spider that could be confused with a wolf spider except for the fact that it has eyes arranged in two parallel rows of four. This spider is tan or gray in color with dark markings.

OCCURRENCE This species is found in southern California and into Mexico. There are four similar species spread across the Southwest from Texas to California. These are nocturnal ground-inhabiting species of open, primarily arid, habitats. During the day they have been found under rocks, logs, or other debris.

SEASONALITY Adults: all year.

Teminius affinis Banks, 1897

Plate 45

IDENTIFICATION This species is relatively large and dark reddish-brown with a wide pale band on the cephalothorax. The abdomen has a variable pale band down the center, surrounding a dark heart mark. There are sometimes paired pale spots on the posterior third of the abdomen. It has relatively long, conical posterior spinnerets that typically extend beyond the end of the abdomen. The eyes are arranged in two regular rows of four at the front.

OCCURRENCE This species is found from Oklahoma, New Mexico, Texas, and western Louisiana. There is a similar species in Florida. These are ground hunters.

SEASONALITY Adults: March through September.

REMARKS Previously known as *Syrisca affinis*.

FAMILY MYSMENIDAE • *Dwarf Cobweb Weavers*

This family is primarily tropical, including about 90 species. Six species are known from North America north of Mexico. These are all tiny, inconspicuous spiders. They live near the ground, typically in the leaf litter. They build small webs in spaces within the litter. The webs are unusual among orb-weaving spiders because of the extensive three-dimensional structure. The spiders spin radial lines that are not in the same plane as the main orb to such an extent that the web resembles a space-filling web of the Theridiidae. Some species in this family are not known to build their own webs; rather, they visit the webs of other spiders and steal prey.

Microdipoena guttata (Banks, 1895)

Plate 20

IDENTIFICATION This is a tiny spider. The abdomen is usually dark with white spots. The spinnerets are under the center of the abdomen. The spider builds a space-filling web.

OCCURRENCE This species is found in the southeastern North America and well into the tropics. The spiders live in spaces within the leaf litter, low vegetation, or debris near the ground. The webs are built in small spaces within this habitat or in hollow logs.

SEASONALITY Adults: June through October.

FAMILY NEMESIIDAE • *Wishbone Spiders*

The members of the Nemesiidae, the wishbone spiders, look much like tarantulas and are distributed worldwide. There are five species known from North America, and with the exception of one that has been recorded in Nevada, are all from California. The wishbone spiders have dense scopulae on their tarsi, but on close inspection the tips of their legs lack the conspicuous claw tufts present in true tarantulas. The tips of their tarsi bear three claws, instead of the two found in tarantulas. The wishbone spiders are generally smaller and more slightly built than true tarantulas. These spiders are velvety hairy, sometimes appearing iridescent.

The wishbone spiders live in burrows. Their common name is derived from the fact that they often build forked burrows with two openings at the soil surface that would resemble a wishbone if viewed in cross section (Main 1976). Most people encounter males that wander in search of females, or individuals that have been flooded out of their burrows. They occasionally wander into houses. Such a large spider reacting defensively, or even biting when harassed, is frightening to some people. In reality, they are not dangerous to humans or pets.

Brachythele longitarsis Simon, 1891

Plate 4

IDENTIFICATION This is a very large mygalomorph. This spider resembles a tarantula; it has scopulae on the tarsi and metatarsi without claw tufts at the ends. There are three claws. This spider is covered with dense silky hair, shorter and more uniform than a tarantula. The fovea is usually a transverse slot. This spider may appear silvery, bluish, or brown. The males are often lighter in color and have a spur near the end of the first tibiae.

OCCURRENCE This spider occurs in northern and central California and western Nevada. This species has been found in a variety of habitats including oak grasslands, woodlands, and coniferous forests.

SEASONALITY Males: wander during the rainy seasons. Females: all year.

REMARKS This spider is also known as *Calisoga longitarsis*. The males, and even occasionally females, may wander into houses. When cornered, they rear up in a threatening defensive posture. The burrow has no lid and the silken lining extends outward a distance from the entrance.

FAMILY NESTICIDAE • *Cave Cobweb Spiders*

The cave cobweb spiders of the family Nesticidae are similar in appearance to cobweb weavers (Theridiidae) but are usually limited to caves or cavelike environments. Like the cobweb weavers, these spiders possess a comb of curved serrated bristles on their fourth tarsi. These diagnostic features are not readily visible without a dissecting microscope. Members of both families attack by wrapping their prey with sticky silk. Unlike most theridiids, cave spider females typically carry their egg cases with their chelicerae or attached to their spinnerets. There are more than 200 species worldwide, and 38 species are known from North America north of Mexico. Some of the species that are restricted to caves have become pale and lost their eyes.

Eidmannella pallida (Emerton, 1875)

Plate 24

IDENTIFICATION This is a small pale spider. Some individuals of this variable spider, such as the one illustrated in Plate 24, are pale and unmarked. Others show a dusky edge to the carapace, at the edges of the head region, and around the eyes. In a few populations the abdomen has dark chevron markings. Individuals in some cave populations have reduced eyes or no eyes.

OCCURRENCE This species is widespread in North America. Most records are for the East. They have been found in dark places, under rocks, and in caves.

SEASONALITY Adults: all year.

REMARKS The female is often found near the egg sac, presumably guarding it. Sometimes she has been observed carrying it attached to her spinnerets.

Nesticus silvestrii Fage, 1929

Plate 24

IDENTIFICATION This is a small variable spider. Some individuals have dusky markings as illustrated. Other individuals are uniformly pale. Individuals in some cave populations have much longer legs.

OCCURRENCE This is a western species. There are similar species in this genus throughout the East. This species occurs from British Columbia south to central California. They have been found in dark areas, under rocks, among tree roots, and in caves.

SEASONALITY Adults: all year.

REMARKS A related species (*Nesticus cellulanus*) has been introduced into North America, where it has occasionally been found in buildings.

FAMILY OCHYROCERATIDAE · *Midget Ground Weavers*

Most of the members of this family are tropical. There are nearly 160 species described worldwide. Only two are known from north of Mexico, and these are from southern Florida. Midget ground weavers may take the same ecological role as primary predators in the leaf litter in the tropics that members of the family Linyphiidae perform in higher northern latitudes. The midget ground weavers build small sheet webs in spaces within the litter. Some species in this family are known from caves, but our species have only been found near the ground among the leaves. They possess only six eyes in three groups near the front of the head.

Theotima minutissima (Petrunkevich, 1929)

Plate 82

IDENTIFICATION This tiny spider could easily be confused with either a midget cave spider (Leptonetidae), or longlegged cave spider (Telemidae). The legs are proportionately shorter than either of those. The six eyes are grouped in three pairs, together at the front of the head. Immatures are pale and darken with each molt.

OCCURRENCE This species is known from southern Florida and further south.

SEASONALITY Adults: probably all year.

REMARKS Females guard their clutch of about five eggs, carrying them in their jaws. After hatching, the spiderlings continue to be held until they disperse. This spider can reproduce parthenogenetically; the eggs develop and hatch without being fertilized (Edwards, Edwards, and Edwards 2003). Rarely, male spiders have been discovered with females, but it is unclear if they are of this species.

FAMILY OECOBIIDAE · *Flatmesh Weavers*

A total of 101 species of flatmesh weavers have been described worldwide. Only eight species are known north of Mexico. These small pale spiders are inconspicuous. The cephalothorax protrudes to a point in the front. The posterior median eyes have a peculiar kidneylike shape. There is a distinctive brush of spines on the anal tubercle. They build a thin, silk retreat consisting of two flat sheets. The spider rests between these two sheets, standing on the lower one. They add a series of silk lines radiating away from the retreat, which act as trip lines.

Flatmesh weavers often feed on ants. When a potential prey disturbs a trip line, the spider rushes out to attack. The attack consists of rapidly circling around the ant spreading a thin sheet of silk. As the spider continues to run around the prey and add silk, it becomes a wall then a tentlike enclosure. The spider runs around, first one direction then the other. This behavior is accomplished with the rear of the spider facing the prey and wagging the abdomen back and forth. They have a brush of extraordinary bristles on the anal tubercle that are used to spread the capture silk emitted from enlarged posterior spinnerets. Periodically the spider will turn, approach the prey, and attempt to bite. They often bite at the base of an antenna. Ants are often well-defended prey with powerful jaws; the movements of the spiders are furtive. If the ant begins to extract itself, the spider will return to spinning a stronger wall of silk.

Oecobius navus Blackwall, 1859

Plate 82

IDENTIFICATION This small grayish spider has a distinctive pattern of dark spots on the cephalothorax. The abdomen is variegated with both dark spots and bright white spots. The posterior median eyes are strange, somewhat kidney shaped. There is a point at the front of the cephalothorax. These spiders can hide in the open by flattening their body to the substrate and remaining motionless.

OCCURRENCE This species is cosmopolitan, probably because it thrives in buildings. It has been transported worldwide to such an extent that no one is certain where it originated. In North America it has been found across the continent, with most records in the South. Other species occur in the Southwest and the central part of the States. They live on the ground, the bark of trees, and among rocks. Many records are from buildings, frequently greenhouses. I have often found them occupying the small depressions on the surface of concrete block walls.

SEASONALITY Adults: all year; in summer outdoors.

REMARKS This and several similar species are considered among the most common human-associated spiders. They often prey on small ants.

FAMILY OONOPIDAE · *Goblin Spiders*

These tiny spiders are relatively poorly known, many species have yet to be formally described. Most of the known species live among leaf litter or under rocks. This family is primarily a tropical one with 455 species. Nearly 40 species are known from North America north of Mexico, mostly from southern states. They possess six eyes in a group at the front of the head. Most species are yellowish or orange; others are pale or purplish in color. The abdomen is sometimes covered with hard scutes. They move with jerky movements and can run rapidly if disturbed. Some species hop. They do not build a capture web but do construct silk molting chambers.

Escaphiella hespera (Chamberlin, 1924)

Plate 80

IDENTIFICATION This is one of two species chosen to represent a large group of goblin spiders, most of which are tropical. This spider has six eyes, two in front and a group of four behind. Other small orange spiders found on the ground, primarily in the family Caponiidae, have different eye arrangements. The abdomen of *Escaphiella* is covered by hard scuta. In the female they are paired left and right and present a clamlike appearance. In the male they are dorsal and ventral. Where soft parts of the abdomen are visible, they are white.

OCCURRENCE Found in the Southwest. This spider has been collected near the ground, usually in leaf litter or under rocks.

SEASONALITY Adults: all year.

Orchestina saltitans Banks, 1894

Plate 82

IDENTIFICATION This is one of two species chosen to represent a large group of goblin spiders, most of which are tropical. It is a tiny pale spider with all six eyes arranged in a group at the front of the head. The abdomen is occasionally marked with purplish. The femora of the rear legs are much fatter than the others.

OCCURRENCE According to B. J. Kaston (1978), this species occurs from New England south to Georgia and west to Missouri. It has been collected by sweeping low vegetation. Isolated populations have been accidentally introduced into buildings in a variety of localities, including Seattle, Washington.

SEASONALITY Adults: late spring through December.

REMARKS These spiders move with jerky motions, including remarkable jumps. They have often been noticed indoors, walking on walls or when hanging from a silk line.

FAMILY OXYOPIDAE · *Lynx Spiders*

The 405 lynx spider species are found worldwide. In North America north of Mexico we have 18 species. They possess an extraordinarily high clypeus and tall chelicerae. The eyes are arranged in a distinctive pattern. The anterior median eyes are small. Behind them the remaining six eyes are spread out in a hexagonal shape at the top of the head region (see Fig. 34). The legs are armed with numerous long, stiff, and conspicuous spines. The abdomen usually tapers gradually to a point at the posterior end.

These spiders get their name from the catlike way they stalk their prey. When threatened, they are capable of running and jumping rapidly through dense vegetation to escape.

Hamataliwa grisea Keyserling, 1887 · *Bark lynx spider*

Plate 38

IDENTIFICATION This is a large lynx spider. This species is covered with gray or light tan setae. There are sometimes faint dark bands at the back of the abdomen.

OCCURRENCE This spider occurs in the south from Florida to California. It forages among the stems of the low branches of trees and shrubs, including cacti.

SEASONALITY Adults: May through October.

REMARKS This species is less likely to jump or run when approached than the others in the family.

Oxyopes salticus Hentz, 1845 • *Striped lynx spider*

Plate 38

IDENTIFICATION This is a medium-sized lynx spider. This species can be distinguished from the others in our area by the thin black lines extending from the small anterior median eyes to the middle of the chelicerae. There are also thin black lines on the ventral surfaces of the femora. The female is often light green with dense white setae covering much of the dorsal surface with the exception of brown or black stripes. The male has an orange cephalothorax with conspicuous black palps and an iridescent silvery abdomen.

OCCURRENCE This species is found throughout the East and along the West Coast from Oregon south. It is absent from the Rocky Mountains, the Great Basin, and the northern Midwest. They have been found in tall grasses, prairies, old fields, and other herbaceous vegetation.

SEASONALITY Adults: spring and early summer.

REMARKS The name *salticus* refers to this spiders' habit of jumping frequently.

Oxyopes scalaris Hentz, 1845 • *Western lynx spider*

Plate 38

IDENTIFICATION This is a medium-sized lynx spider. This species has broad, dark lines below the anterior median eyes. Extremely variable in color, the cephalothorax and abdomen have a light band above and dark sides. The legs are usually banded rather than striped, but if they have stripes they are brown and relatively thick. The male is less iridescent than in the striped lynx spider. He is sometimes marked like the female or plain as illustrated in Plate 38. His legs often have thin dark lines on the dorsal surface of the femora and tibiae.

OCCURRENCE This species is found across the United States and southern Canada. It is more common in the West than the striped lynx spider. Western lynx spiders have been found in the low branches of shrubs and trees as well as in herbaceous vegetation.

SEASONALITY Adults: spring and early summer.

REMARKS According to Nicholas Hentz (1875), this spider is a frequent victim of predatory wasps.

Peucetia viridans (Hentz, 1832) • *Green lynx spider*

Plate 38

IDENTIFICATION This is a large green lynx spider. The body is usually bright green with conspicuous black spots on the legs. The abdomen often has a pattern of white chevrons. Some western individuals are green with red femora, red lines on the cephalothorax, and a pattern of red-and-white chevrons on the abdomen.

OCCURRENCE This spider occurs in the Southeast as far north as Virginia and across the southern United States to California. In California and Arizona it shares its range with the similar *Peucetia longipalpis*. The green lynx spider forages in low vegetation.

SEASONALITY Adults: summer.

REMARKS This species is so common in agricultural fields that it is considered an important natural biological control of pest insects. The egg sac is round, usually greenish in color, and covered with points. The female usually remains with the egg case, presumably to guard against predators or parasites.

FAMILY PHILODROMIDAE · *Running Crab Spiders*

This family includes 535 species worldwide. There are 96 species known from North America north of Mexico. Most of these are medium-sized light brown or gray spiders. Many species hold their legs in a laterigrade posture at rest. The eyes are at the front of a narrow cephalic region. The thoracic region of the cephalothorax is broad and often as wide as it is long. They lack the conspicuous eye tubercles present in the similar crab spider family (Thomisidae). The legs differ from those of the crab spiders by the presence of claw tufts and scopulae, and the lack of raptorial spines.

Apollophanes margareta Lowrie and Gersch, 1955

Plate 70

IDENTIFICATION This is a medium-sized running crab spider. The body is variegated brown. The head region is lighter in color. The abdomen has a light area at the back. Under magnification the posterior lateral eyes are further behind the posterior median eyes so that this row is recurved.

OCCURRENCE This is a western species that occurs from the southwestern corner of the Yukon Territory to southern California and east to New Mexico at elevations up to 2,600 m. They have been found in junipers or in other shrubs in montane woodlands. They have been found in low vegetation near the ground.

SEASONALITY Adults: June and July.

REMARKS It is one of the few spiders that is characteristic of arid juniper woodlands.

Philodromus cespitum (Walckenaer, 1802)

Plate 70

IDENTIFICATION This is a medium-sized light brown or gray running crab spider. These are very flat spiders. The body is usually light brown with a distinct white marginal band on the cephalothorax. The abdomen has a dark heart mark. The color varies from gray to brown or orange-brown. The legs are usually lighter than the body and only sparsely marked with spots at the base of some spines.

OCCURRENCE This species has a wide distribution in the interior of North America, generally in the North. It is also found in Eurasia. It has been found in a variety of habitats in herbaceous vegetation, grasses, shrubs, and trees.

SEASONALITY Adults: June through September.

REMARKS These cryptic spiders are rarely noticed but appear in sweep and beat samples.

Philodromus rufus Walckenaer, 1826

Plate 70

IDENTIFICATION This is a small, reddish running crab spider. There are four varieties or subspecies of *Philodromus rufus*. The cephalothorax has a broad light center and dark edges. The abdomen is also lighter in the center with the exception of the darker heart mark. The legs are pale and not banded.

OCCURRENCE This species is found throughout the Northern Hemisphere. In North America north of Mexico it is absent only in the Southeast. This spider has been found in many habitats, including broadleaf and coniferous forests. They occur on the foliage of the trees, understory herbaceous vegetation, and grasses.

SEASONALITY Adults: May through August.

Philodromus vulgaris (Hentz, 1847)

Plate 70

IDENTIFICATION This is a medium-sized gray running crab spider. These are flat spiders. The body is gray and marked with small spots and blotches. Some individuals are covered with a dense coating of pale gray setae. The second legs are only slightly longer than the first. The end of the abdomen is often white.

OCCURRENCE This species ranges across southern Canada from Nova Scotia to Alberta. Further south it is primarily eastern, occurring as far south as Texas and Florida. This species has been found on tree trunks, the foliage of conifers, and on wooden structures including fences and houses.

SEASONALITY Adults: April through November; subadults in winter.

REMARKS These are well-camouflaged spiders and are often overlooked until they move.

Thanatus vulgaris Simon, 1870

Plate 70

IDENTIFICATION This is a medium-sized running crab spider. The body is pale with a distinct dark heart mark. The legs are all of similar length and not usually held in a typical laterigrade posture.

OCCURRENCE This species has a mostly southwestern distribution in North America and is also found in Eurasia. It may have been accidentally introduced here. There are scattered additional records from the east, probably vagrants. This spider is often found inside buildings. The spiders in this genus are more terrestrial than most crab spiders, often found on rocks.

SEASONALITY Adults: all year.

REMARKS This spider is considered a tramp. It is often encountered in cars, trains, or ships.

Tibellus oblongus (Walckenaer, 1802)

Plate 70

IDENTIFICATION This is a medium-sized running crab spider. Members of the genus *Tibellus* often hold their legs with four forward and four backward, grasping thin twigs or grasses. The color matches the straws where they are found. The body has longitudinal stripes that enhance the camouflage. In this species there are two black spots near the back of the abdomen.

OCCURRENCE This is the most widespread member of the genus in North America. They are found in herbaceous vegetation usually in open areas. This species has often been found in prairies and grasslands, where it has been collected by sweeping.

SEASONALITY Adults: May through September.

REMARKS This spider usually freezes in position unless disturbed. When it does move, it escapes with lightning quickness.

Titanebo albocaudatus (Schick, 1965)

Plate 70

IDENTIFICATION This is a small and pale running crab spider. These small spiders have very long second legs, often twice as long as the first legs. In this species the body varies from nearly white, as shown in Plate 70, to darker and flecked with spots. The back half of the abdomen is white and without spines. The dark heart mark on the abdomen has lateral extensions forming a cross shape.

OCCURRENCE This is a spider of the arid Southwest. They have been found in grasses and shrubs in grazed areas as well as in arid grasslands and creosote bush shrubs.

SEASONALITY Adults: all months except September and October.

FAMILY PHOLCIDAE • *Cellar Spiders, Pholcids*

This family has 870 species worldwide. There are 34 species in North America north of Mexico. The Pholcidae are a group of spiders that has adapted well to human habitats. These spiders are common in buildings, particularly cellars and basements. The cephalothorax of cellar spiders is usually about as wide as it is long. Their extraordinarily long, thin legs give the group its alternate name, the daddylongleg spiders. They have unique flexible tarsi, usually held in a curved position. They are sometimes confused with harvestmen, but no harvestmen use silk. A spider in a web cannot be confused with a true harvestman or " daddylonglegs." Some species in this family possess only six eyes, but most have eight. Perhaps a candidate for the distinction "most common spider in buildings" is the longbodied cellar spider (*Pholcus phalangioides*). Most of the diversity of this family is found among spiders in the Southwest, where a number of different species have been found in dark recesses as well as caverns and caves.

The cellar spiders have an unusual defensive behavior. When disturbed, they move rapidly in their web, flexing their legs so that the body gyrates in a circular motion. This renders them difficult to see and perhaps difficult to capture by their principal predators: wasps. Cellar spider females carry the egg case in their jaws. The eggs are held together by a thin silken net; individual eggs are easily seen. The palps of adult males are very large and conspicuous.

Artema atlanta Walckenaer, 1837

Plate 23

IDENTIFICATION This is a relatively large light brown cellar spider with an oval or round abdomen. The cephalothorax has a dark mark surrounding the fovea, which is a large circular pit. There are also small dark marks at the base of each leg. The abdomen is round with a plain-colored heart mark and a series of paired black spots or bands on either side. The male has large black structures on the outer edge of the chelicerae.

OCCURRENCE This is a spider that has been transported around the world by humans. Most of the scattered North American records are from the West, but it could be expected anywhere. It has usually been found in protected spaces in buildings.

SEASONALITY Adults: all year.

REMARKS This is our largest member of the family Pholcidae.

Crossopriza lyoni (Blackwall, 1867)

Plate 23

IDENTIFICATION This is a medium-sized to large cellar spider. The abdomen has a distinctive point at the back, so that it looks angular from the side. The abdomen also has a silvery-white reticulated pattern of lines. There is a dark area at the back of the head region as well as a dark median spot on the thoracic part of the cephalothorax.

OCCURRENCE This is a spider that has been transported around the world by humans. Most of the worldwide records are tropical, but in North America it has been found in both southern and central states. Small populations have been found in buildings.

SEASONALITY Adults: all year.

REMARKS This species appears to be expanding its range in North America.

Holocnemus pluchei (Scopoli, 1763) • *Marbled cellar spider*

Plate 23

IDENTIFICATION This is a medium-sized cellar spider. The abdomen is long, similar in shape to that of the common longbodied cellar spider. The color is different: it has a pattern of silvery-white lines forming a reticulated pattern, the source of its common name. The underside of this spider is dark. The females have swollen palps, resembling those of immature males.

OCCURRENCE This is a Eurasian spider, introduced into North America. It has been most often found in the southwestern states. Unlike many cellar spiders, this species is often found outdoors with their tangle webs built in shrubs around buildings. They often eat other spiders and come to dominate debris, such as woodpiles. They have also been found inside.

SEASONALITY Adults: all year indoors; spring and summer outdoors.

REMARKS The webs of this species may form a sheet or dome where the spider hangs in the tangled web. Adult females with eggs and recently hatched young have been found in dome-shaped circular webs. Elizabeth Jakob (2004) has published a number of studies describing the group-living behavior of this species.

Pholcophora americana Banks, 1896

Plate 23

IDENTIFICATION This is a small, dark-colored spider. The cephalothorax is reddish brown. The abdomen is dark gray, sometimes with small black spots. The legs are plain brown. Proportionately, the legs are not as long as most of the spiders in this family.

OCCURRENCE This species has been found in the foothills and mountains in the West, from British Columbia south to California and east to Colorado and New Mexico. A similar species is known from Texas. This spider has been found in dark spaces under rocks, fallen tree bark, depressions in the ground, or cave entrances.

SEASONALITY Adults: May through September.

Pholcus phalangioides (Fuesslin, 1775) • *Longbodied cellar spider*

Plate 23

IDENTIFICATION This is a large gray spider with impossibly long legs. The cephalothorax and abdomen are light gray. The abdomen is considerably longer than tall. The cephalothorax in some individuals has a dark line down the center that splits into two spots in the thoracic region. The abdomen is usually pale gray or has dark markings on the upper surface.

OCCURRENCE The longbodied cellar spider is common in buildings worldwide. There are two other, smaller species in this genus in North America. In the northern parts of Canada as well as in Alaska it has been found only in buildings. In warmer areas this species is found in dark protected areas outdoors, including rocky overhangs and caves. It is common in houses where it builds a messy tangle web in upper corners near the ceiling, under shelves, in closets, basements, and cellars.

SEASONALITY Adults: all year.

REMARKS This spider often captures and eats other spiders, occasionally much larger ones. *Pholcus* approaches within a few body lengths; then, using its long legs, the spider draws silk from the spinnerets and throws it at the potential prey. It continues until the victim is thoroughly tangled. Only after this will the spider approach and bite the victim. If its web is disturbed, the occupant shakes the web violently, setting itself into a rapid rotation. Sometimes this behavior transforms the spider into an invisible blur.

Physocyclus californicus Chamberlin and Gertsch, 1929

Plate 23

IDENTIFICATION This is a small to medium-sized cellar spider. The cephalothorax has a dark Y-shaped mark in the center. The abdomen is round and light gray in color with paired black spots. The palps of the adult male are nearly as large as the entire cephalothorax.

OCCURRENCE This species has been found in California. Records are from the Sierra foothills, and Coast Ranges. This species has often been collected from dark spaces under rocks and other debris on the ground. It has also been found in buildings. There are three other species in the genus that occur as far east as Florida.

SEASONALITY Adults: all year.

Psilochorus hesperus Gertsch and Ivie, 1936

Plate 23

IDENTIFICATION This is a small cellar spider. The cephalothorax is pale gray with a Y-shaped darker area at the junction of the head and thoracic regions extending down the center of the thoracic region. The abdomen is round with a narrow dark heart mark. The remainder of the abdomen is variegated bluish gray with light spots and indistinct dark blotches.

OCCURRENCE This species has been recorded from the Northwest. It has been collected from southern British Columbia, Washington, Idaho, Oregon, and California. They have been observed in tangle webs under rocks, logs, or other objects. They have been found in webs among boulders on rocky slopes. They have also been found in buildings.

SEASONALITY Adults: February through September.

REMARKS The adult males have robust curved spikes extending forward on the jaws.

Spermophora senoculata (Duges, 1836) • *Shortbodied cellar spider*

Plate 23

IDENTIFICATION This is a small nearly white spider. The legs are, proportionately, not nearly as long as in most other cellar spiders. This species has only six eyes in two clusters of three. The pale abdomen is sometimes marked with paired gray spots.

OCCURRENCE The shortbodied cellar spider was introduced into North America from Eurasia. It has been found in temperate parts of North America. It occurs mostly in buildings. It often builds its tangle web under furniture.

SEASONALITY Males: late spring and summer. Females: April through December.

REMARKS Females are sometimes seen carrying their egg case in their jaws as they move from one area in the house to another.

FAMILY PIMOIDAE • *Large Hammockweb Spiders*

The 25 species of large hammockweb spiders have a scattered distribution in the Northern Hemisphere. There are about 15 species in this group in North America north of Mexico, all occurring in the West, ranging from western Montana to central California and north to coastal Alaska. As their name implies, these are generally larger spiders than sheetweb weavers (Linyphiidae).

This group has only recently been recognized as distinct from sheetweb weavers. The differences are mostly subtle features of the male palp. Most of these species are large (7–9.7 mm) in comparison to typical sheetweb weavers (1–8 mm). They build flat or slightly sagging sheet webs, usually close to the ground, typically in humid areas. The webs of large hammockweb spiders are often attached along one edge to a tree trunk or fallen log. They are sometimes found on buildings or other domestic debris. The spiders hide out of view near the web during the day. The retreat is near the web in a crack or crevice, or perhaps under leaves or bark. When active, like their relatives the sheetweb weavers, pimoids hang below their webs. They attack prey by grasping them from below, through the sheet. The webs of pimoids can be large, sometimes over three feet across. The individual species of large hammockweb spiders are difficult to distinguish. For more information about the members of the family Pimoidae, consult Gustavo Hormiga's 1994 monograph.

Pimoa altioculata (Keyserling, 1886) • *Large hammockweb spider*

Plate 77

IDENTIFICATION This is a reddish-brown spider, relatively large for a sheet web weaver. There is considerable variation in the extent of dark brown markings on the abdomen. The specimen illustrated in Plate 77 is a relatively light-colored example.

OCCURRENCE This species is a spider of humid coniferous forests of the Pacific Northwest. They have often been reported in cool canyons and near lakes or the ocean. The relatively large sheet webs of these spiders are often found near the bases of trees, stumps, or rocky areas. They are occasionally captured in buildings. The daytime retreat may be under bark, leaves, in a hole or crack.

SEASONALITY Adults: all year.

REMARKS The family Pimoidae are similar to linyphiids but are considered different because they possess distinctive features of the male palp that are only visible using a microscope.

FAMILY PISAURIDAE · *Nursery Web Spiders, Fishing Spiders*

This family is worldwide in distribution with more than 300 species. There are only 14 species known from North America north of Mexico. Most species are large, some extremely so. They appear somewhat like wolf spiders and also have a hunting lifestyle. The posterior eyes of the nursey web spiders are arranged in a recurved row and usually do not resemble the trapezoid characteristic of the wolf spiders (Lycosidae). Nursery web spiders frequently hunt from vegetation, the trunks of trees, or even from the surface of water. Unlike wolf spiders, they are rarely found on bare ground.

The alternate name for this group, the fishing spiders, refers to the ability of some large species (*Dolomedes*) to capture small fish, tadpoles, and large invertebrate larvae from ponds and slow-moving streams. They are capable of movement on water and can either skate over the surface of the water or plunge into the water to capture their prey. If disturbed, it is not unusual for them to climb down emergent aquatic vegetation and hide underwater.

The members of this family display extended parental care. The females carry the egg case under their body, holding it with their jaws. When the young are ready to emerge from the egg sac, she builds a nursery for the young. The female usually employs a folded leaf or other structure as a roof and then fills a space with a sturdy tangle of threads, placing the egg sac near the center. She stands guard near the nursery. When the young emerge, they remain in the nursery for a week or more, molt, then disperse.

Dolomedes albineus Hentz, 1845

Plate 57

IDENTIFICATION This is a large fishing spider. In this species the head region is distinctly higher than the thoracic region. The color of the cephalothorax in *albineus* is variable but often includes large areas covered with white. This may be the entire carapace or just the sides. The eye region is usually dark. The abdomen usually has light sides, often white in the front. The female has a dark-edged pale stripe on the underside of the abdomen. There is a contrasting dark heart mark. The legs are usually distinctly banded. Some individuals are green.

OCCURRENCE This species has been found in the southeastern states from Virginia west to Oklahoma and south into Texas and Florida. It has usually been found near slow-moving streams or still water.

SEASONALITY Adults: all year.

Dolomedes scriptus Hentz, 1845

Plate 56

IDENTIFICATION This is a large fishing spider. The female has an irregular light band around the cephalothorax. There are three light lines on the head region, one median, and one curved one extending back in a curve from the inside edge of the posterior lateral eyes. She also has four dark W-shaped black marks at the back of the abdomen. The male has a conspicuous wide white or cream band around the entire cephalothorax. He also has light around the outside edge of his abdomen. The legs have indistinct banding.

OCCURRENCE This species occurs from Nova Scotia west to Manitoba and south to Texas and northern Florida. It has most often found near actively flowing streams.

SEASONALITY Males: May through July. Females: late May through September.

Dolomedes tenebrosus Hentz, 1844

Plate 56

IDENTIFICATION This is a large fishing spider. This species is most often shades of medium brown. Both sexes have a dark area around the eyes and extending forward and covering most of the clypeus, except for a light spot in the center at the front edge. There are often two dark triangles just in front of the fovea. The broad lateral light bands around the carapace of the female are variably marked with light lines and dark edges. The abdomen is brown with variable markings, but there are usually three W-shaped marks at the back. In the male the marginal light bands around the cephalothorax and abdomen are often distinct. The legs are often distinctly banded.

OCCURRENCE This species occurs from Nova Scotia west to Manitoba and south to Texas and the panhandle of Florida. They have been found at night on tree trunks, particularly those with bases submerged in water. This species is the one member of the genus that has often been found well away from water. It wanders into buildings, including houses.

SEASONALITY Males: spring. Females: all year; with egg sacs in summer and early autumn.

REMARKS There is a similar species, *Dolomedes okefinokensis* in southern Georgia and Florida that is even larger, easily among the largest araneomorph spiders north of Mexico. According to James Carico (1973), these two species are indistinguishable without using a microscope.

Dolomedes triton (Walckenaer, 1837) • Sixspotted fishing spider

Plate 56

IDENTIFICATION This is a large fishing spider. The female is variably colored but usually has a dark brown or greenish-brown cephalothorax with a white submarginal band. The width of this band varies among individuals. Her abdomen is uniform greenish brown above with white sides and three or four pairs of distinct white spots. The legs are greenish brown with white lines or spots. The male is much smaller than the female and has similar coloration except that the white bands on the carapace and abdomen are broader.

OCCURRENCE This species has a broad range between Maine and British Columbia. There are records for Nunavut. It is sparsely distributed in the northwestern United States but occurs throughout the East. It is common throughout Florida. This is our most aquatic *Dolomedes*, usually found on still water near the edges of ponds or lake margins. They have frequently been found perched head down in vegetation near the water surface. This spider also skates over the water surface.

SEASONALITY Adults: all year.

REMARKS This spider is similar in appearance and habits to the famous *Dolomedes fimbriatus* of Europe.

Dolomedes vittatus Walckenaer, 1837

Plate 56

IDENTIFICATION This is a large fishing spider. Most of the females of this species are plain brown with two distinct dark triangles in front of the fovea. The brown marginal band on the carapace is relatively uniform in color. The abdomen has dark margin and a series of five pairs of small light spots within dark transverse spots. In the male the pale marginal band extends into they clypeal area. The male also has a light marginal band on the abdomen, absent in females. The markings on the legs are often indistinct but may form bands.

OCCURRENCE This uncommon species occurs from New England to the Great Lakes region and south to western Texas and northern Florida. Like *Dolomedes scriptus*, this species is most often found near actively flowing streams. *D. vittatus* seems to prefer smaller, shaded streams.

SEASONALITY Males: spring. Females: all year; with eggs during the summer and early fall.

REMARKS Females were previously known as *Dolomedes urinator*.

Pisaurina mira (Walckenaer, 1837) • *Nursery web spider*

Plate 57

IDENTIFICATION This is a large nursery web spider. This is a light brown or orange spider. Most females have a well-defined dark stripe just wider than the eye area down the middle of the cephalothorax. On the abdomen there is a dark stripe that may have a wavy boundary. Both of these dark bands may have a light rim with a thin dark margin. On some individuals the bands are obscure. The legs are not banded.

OCCURRENCE This is an eastern species occurring from New England and southern Ontario west to Minnesota and south to Texas and Florida. This common and widespread species has been found in fields, shrubs, and forest understory vegetation.

SEASONALITY Adults spring to early autumn.

REMARKS The nursery web is placed in low vegetation near the boundary between a field and woods. This spider frequently holds its front legs extended in pairs.

Tinus peregrinus (Bishop, 1924)

Plate 57

IDENTIFICATION This is a medium-sized fishing spider. This spider looks like a small version of *Dolomedes*. The spider is light brown with darker brown markings. The cephalothorax is dark in the center and tan on the sides. Some individuals have a thin white line down the center. The abdomen is light on the sides with a dark folium that encloses light and dark areas.

OCCURRENCE This species is known from the vicinity of the Colorado River from Arizona, Nevada, and California as well as southeastern Arizona, and southern Texas. It has been found near water.

SEASONALITY Adults: summer.

FAMILY PLECTREURIDAE · *Spurlipped Spiders*

This is a relatively small family of spiders; most species have been found in arid and semiarid areas, woodlands, and coniferous forests in the southwestern region of North America as well as in Cuba. Worldwide there are about 30 species, 16 species are found north of Mexico. These spiders have a high clypeus and tall chelicerae that slant forward, giving these spiders a distinctive face.

The family name is derived from distinctive extensions of the mouthparts (spurs on the endites) only visible with magnification. These spiders construct loose sheets or tangle retreats under a rock or other debris on the ground. The web may extend a short distance out from the retreat. The spiders hunt at night in the vicinity of their retreat.

Kibramoa madrona Gertsch, 1958

Plate 34

IDENTIFICATION These are dramatic dark-colored spiders with a deep reddish cephalothorax and chelicerae. The clypeus is high. The abdomen is dark gray with a lighter area over the heart region. The legs are spiny and both longer and thinner than those of *Plectreurys*. The patellae are reddish like the cephalothorax. Some individuals are lighter, dusky yellow.

OCCURRENCE This species is found in California. Other species in the group are from California, Baja California, Nevada, and Arizona. Most records are from arid and semiarid habitats but also woodlands. They have been found in loose tangle webs under rocks or other litter on the ground.

SEASONALITY Adults: all year.

REMARKS Males have been found wandering in the open at night.

Plectreurys tristis Simon, 1893

Plate 34

IDENTIFICATION This is a large dark plectreurid. The chelicerae are long and combined with the high clypeus they give the face a tall appearance. The legs are stout and densely hairy but lack the longer spines present in *Kibramoa*. The males have a conspicuous spur on tibia I, which they use in mating.

OCCURRENCE This is an arid or semiarid zone spider from California, Nevada, Idaho, Utah, and Arizona. They have been found in deserts, dry woodlands, and coniferous forests up to 2000 m. They build a loose web under rocks or other debris on the ground.

SEASONALITY Adults: probably all year.

REMARKS The males have been found wandering in the open at night.

FAMILY PRODIDOMIDAE · *Prodidomids*

This is primarily a family of the Southern Hemisphere, only two species occur in the region treated in this book. They are probably related to the ground spiders (Gnaphosidae). Like the ground spiders, they possess unusual spinnerets; the anterior pair consist of one segment, are cylindrical, and well separated. The anal tubercle is large and sometimes visible between the spinnerets when the spider is viewed from above. The eight eyes have an unusual arclike arrangement. All but the anterior median eyes are reflective. The posterior median eyes are oval in shape. These spiders are not well known; most have been found under rocks on the ground during the day. They hunt at night. There are a few records from buildings.

Neozimiris pubescens (Banks, 1898)

Plate 45

IDENTIFICATION This small pale-yellow spider has an eye arrangement (see Fig. 31) unlike any other except *Prodidomus*, which is red in color. The anterior spinnerets are relatively large and cylindrical, often visible from above. The anal tubercle is large and projects out between the spinnerets when viewed from above.

OCCURRENCE The spiders in this genus are primarily found in tropical America. This species, the sole representative north of Mexico, is known from the Southwest. They have been found on the ground, under rocks or debris.

SEASONALITY Adults: January through April.

REMARKS These small spiders are rarely encountered; little is known about them.

Prodidomus rufus Hentz, 1847

Plate 45

IDENTIFICATION This is a distinctive small, reddish-brown spider. The eight eyes form an arc (see Fig. 31). The only other spider in this region with this eye arrangement is the pale-yellow *Neozimiris*. The jaws are widely spread and relatively large.

OCCURRENCE The spiders in this genus are primarily found in the Southern Hemisphere. This species, the sole representative of its genus north of Mexico, is known from the Alabama, California, Louisiana, and Texas. They have been found on the ground, under rocks or debris, and occasionally in buildings.

SEASONALITY Adults probably all year.

REMARKS This species is likely a tramp that has been accidentally transported. The population in North America may have been introduced. It is also known from Asia, New Caledonia, South America, and scattered islands around the world.

FAMILY SALTICIDAE · *Jumping Spiders*

This is the largest spider family in the world with more than 5,300 species named. There are 315 species known for North America north of Mexico. Perhaps remarkably, considering the diversity of the family, jumping spiders are all easy to recognize by their distinctive eye arrangement (see Fig. 18). The anterior median eyes are huge and face forward. Next to these on a relatively flat face are the anterior lateral eyes, also fairly large. The posterior eyes are in such a curved row that it appears to be two distinct rows. The posterior median eyes are usually small, sometimes difficult to see on the side of the head area. At the back of the head region are the posterior lateral eyes, which are usually about as large as the anterior lateral eyes and may face sideways or backward.

The jumping spiders are intriguing because of their intelligent behavior. They are the favorites among many spider observers because of their habit of turning to look at us and even following our movements. The pair of large anterior median eyes give them a recognizable face. In addition, they are mostly diurnal, colorful, and active in open areas where they are easily spotted. Several species exhibit complex and flexible learned behavior that they use in hunting prey.

One of the most interesting aspects of the behavior of jumping spiders are their elaborate courtship rituals. Because jumping spiders are diurnal and possess color vision; they frequently employ movement and colorful body parts in their displays. The male is usually the primary actor. When he encounters a female, he will approach and begin a series of ritualized movements. These movements are the result of adaptations that make distinctive features of his coloration or body form conspicuous to the female. He may also use vibratory signals to enhance the effectiveness of his display. If the female is receptive, she will remain and sometimes even approach the male.

A moment in one such display is illustrated for *Habronattus viridipes* (Plate 62). The male of this species has green undersides and long fringes on his front legs. During part of the display he lifts his legs, exposing the green color to the watching female. In another part of the display, he lifts his third pair of legs to present the spots on his expanded patellae of those legs. Such displays serve to make his identity as a member of this species, and a potential mate, evident to the watching female. Many jumping spiders have complex courtship displays like this. An interesting study by Wayne Maddison and Marshal Hedin (2003) describes how courtship behavior and palp morphology have influenced diversification in the genus *Habronattus*.

Several species of jumping spiders have evolved an appearance similar to ants or the wasps called velvet ants (Edwards 1984). These ant mimics even adapt their behavior to resemble their ant models. They may run in a jerky fashion and wave a pair of legs like ants wave their antennae. The bodies of the most extreme ant mimic jumping spiders have constrictions or light bands that make them appear to have three body parts characteristic of an insect, like an ant. Some species have been observed in the same areas as ants, and sometimes among the ants in their foraging columns. This ant mimicry probably evolved as a defense against predators that avoid attacking the often well-defended ants and velvet ants.

Anasaitis canosa (Walckenaer, 1837) • *Twinflagged jumper*

Plate 58

IDENTIFICATION This is a small, dark-colored jumping spider. The sexes are similar with a thin white band around the abdomen that has two lines extending toward the center but not meeting. There may be green iridescence on the cephalothorax and abdomen. There is a white spot near the heart area on the abdomen. The cephalothorax has several white spots in the eye region, including elongated white spots behind the posterior lateral eyes. The femora and patellae of the palps are covered with conspicuous white setae.

OCCURRENCE This species has been found from South Carolina west to eastern Oklahoma and south to Texas and Florida. It has been found on the trunks of trees and fallen logs in a variety of woodland habitats, including pine stands, sand hills, and live oak hammocks. In open habitats it has been found on low vegetation in sandy shorelines, sedge meadows, and grasslands. This spider has frequently been found in houses.

SEASONALITY Adults: all year.

REMARKS Both sexes of this spider have the habit of constantly waving their conspicuous white palps in a circular motion. This spider eats ants. Previously known as *Corythalia aurata*.

Bagheera prosper (Peckham and Peckham, 1901)

Plate 58

IDENTIFICATION This is a small to medium-sized jumping spider. The male and female are very different. The male has extraordinarily long and parallel chelicerae, with long fangs. He is shiny bronze in color with a white band around the abdomen and on the sides of the carapace. The female has unremarkable chelicerae, and is brown to golden-brown with a thin covering of scattered white setae.

OCCURRENCE This spider has been found in central Texas. It has been found in a variety of habitats—oaks, cedars, mesquite, and riparian vegetation including both herbs and shrubs.

SEASONALITY Adults: April through October.

REMARKS Previously known as *Dendryphantes* or *Metaphidippus prosper*.

Eris militaris (Hentz, 1845) • *Bronze jumper*

Plate 60

IDENTIFICATION This is a medium-sized jumping spider. The female is light brown and spotted. The setae around the anterior median eyes are often orange. The male is bronze, with large divergent chelicerae. He has a broad white band on the side of the carapace behind the eyes. On the top he has one or two white spots. The abdomen of the male is brown with a white band around the edge.

OCCURRENCE This species has been found from Alaska to Nova Scotia and south to Mexico. It is absent from the arid Southwest. This is a spider of open sunny places. It has been collected in trees, shrubs, herbs, and tall grasses.

SEASONALITY Adults: early spring and summer.

REMARKS Adults sometimes overwinter in aggregations under bark.

Euophrys monadnock Emerton, 1891

Plate 58

IDENTIFICATION This is a small jumping spider. The male is distinctive, mostly black. The femora of legs III and IV are bright orange. The tarsi on all legs are pale yellow, as are the entire palps. The female is brown with a subtle pattern on the abdomen.

OCCURRENCE This is a northern species that has been found from Nova Scotia, across Canada to British Columbia, south to California in the West and Wisconsin, Ohio, and Pennsylvania in the East. It occurs in coniferous forests especially in sandy areas. This spider has been found among mosses and lichens.

SEASONALITY Adults: summer.

REMARKS The courtship display of the male involves his lifting and spreading the front legs. He also moves the third pair of legs, so that the orange femora are displayed. The fourth pair are positioned so that the orange femora are visible to the female. George and Elizabeth Peckham (1909) describe this display in their classic monograph on jumping spiders.

Evarcha hoyi (Peckham and Peckham, 1883)

Plate 61

IDENTIFICATION This is a small brown jumping spider. The male is brown with a light band behind the anterior eyes, a central white spot. There is a broad light band along the side of the cephalothorax narrow at the anterior lateral eyes, passing under the posterior eyes and broadest just behind the posterior lateral eyes. The abdomen has a broad white marginal band with two dark lines inside to the white band. The abdominal spots are variable. The female is mottled chestnut brown with a dark cephalothorax, darkest in the eye region and the posterior slope. The abdomen of the female has a subtle pattern of dark and light bands.

OCCURRENCE This is a northeastern species. There is a similar species in the boreal forests of the North and the West. The eastern species illustrated here has been found in low shrubs, herbs, and grasses.

SEASONALITY Adults: May through September.

REMARKS The scopula hairs of members of this genus, like many of the jumping spiders, have special tips that provide them with extraordinary clinging ability. This is why these little spiders have no difficulty walking up smooth surfaces including glass.

Ghelna canadensis (Banks, 1897)

Plate 59

IDENTIFICATION This is a small, dark-colored jumping spider. The female is dark brown and the male is reddish brown, both have a herringbone-like pattern on the abdomen. On the ventral surface of the abdomen are three longitudinal stripes.

OCCURRENCE This species is northeastern, extending west to Wisconsin and Kansas and south to Tennessee. Unlike similar small jumping spiders that usually forage in foliage, *Ghelna* is a ground-living genus, usually collected in pitfall traps or litter samples.

SEASONALITY Adults: June through August.

REMARKS This spider has been sparsely collected. They can be common on the ground under debris, particularly rocks or uneven concrete.

Habronattus agilis (Banks, 1888)

Plate 61

IDENTIFICATION This is a small brown jumping spider. The male has three white stripes on the top of the head above the anterior eyes. The two lateral white stripes continue through the posterior eye region to the back of the carapace where they meet a broad white lateral band. The clypeus of the male is brown. The abdomen of the male is light, nearly white with two broad dark bands on the upper surface. The male has prominent black tufts of setae on the tibiae of the front legs. The female is gray or pale tan with a light clypeus. She has small dark areas behind the anterior median eyes.

OCCURRENCE This species is restricted to the eastern seaboard from Massachusetts to North Carolina. This spider has been found in open areas of low vegetation or on the ground. It has also been found in marshes and under debris along the shore.

SEASONALITY Males: August through September. Females: June through September.

REMARKS During courtship the male holds his legs spread, displaying the dark fringes.

Habronattus americanus (Keyserling, 1884)

Plate 62

IDENTIFICATION This is a small and variegated brown jumping spider. The female is light brown with a light-colored face and a pattern of dark blotches on the cephalothorax and abdomen. The male has bright red brushes of long setae on his palps and the femora of his front legs. His clypeus is iridescent, usually appearing light blue with white margins above and below. The chelicerae of the male are also covered with long rows of iridescent scales. In some populations there is a crest of long setae over the anterior eyes. The legs of the male have tan margins and black stripes.

OCCURRENCE Has been found from Lake Superior west to British Columbia and south to California and Colorado. Usually found in open habitats on the ground in mountainous areas.

SEASONALITY Adult: spring and early summer.

REMARKS This is a highly variable species with distinctive male ornamentation in populations from different mountain areas.

Habronattus coecatus (Hentz, 1846)

Plate 62

IDENTIFICATION This is a small, light brown jumping spider. The female is variegated light brown with some dark spots on the abdomen. She has darker brown under each anterior median eyes. The male has a dark carapace with a bright red clypeus. The male also has a tan band just above the anterior eyes and tan stripes extending back from the posterior lateral eyes. There is sometimes a light band across the cephalothorax behind the posterior eyes that intersects these two stripes. In the male the undersides of the first legs are black, and the front edges of the tibiae of the third legs are lime green.

OCCURRENCE This species occurs from New England west to Illinois and Kansas and south to northern Florida and Texas. It is more common in the South. It has been found on or near the ground.

SEASONALITY Adults: March through November.

REMARKS The male begins his courtship by using his palps first to hide, and then suddenly expose, his red clypeus and white bases of the chelicerae. During parts of the dance he raises his front legs and displays their black ventral surface.

Habronattus decorus (Blackwall, 1846)

Plate 62

IDENTIFICATION This is a small brown jumping spider. The male and female are different. The female is pale brown or tan with two dark lines, or a series of large dark spots on the abdomen, converging at the spinnerets. The male has a dense covering of iridescent setae. His cephalothorax and legs are uniformly iridescent, either pale blue or dark blue. The top surface of the abdomen is iridescent rose or pink in color with a light band at the front.

OCCURRENCE This species occurs from Nova Scotia west to British Columbia and south to Oregon, Kansas, and northern Florida. There are isolated records as far south as Mexico. It has been found near the ground, in open fields, in dune grasses near beaches, and in tall grassy meadows.

SEASONALITY Adults: May through October.

REMARKS During courtship the male raises his abdomen and moves back and forth as well as turning around, displaying its iridescent rose color to the female.

Habronattus oregonensis (Peckham and Peckham, 1888)

Plate 62

IDENTIFICATION This is a small, mostly dark jumping spider. The female is tan on the top of the cephalic area between the eyes. She also has a light band at the front of the abdomen and a pattern of light spots and chevrons along the midline. The male is dark with a covering of light scales. His first legs are greatly expanded. The tibiae are bulbous and shiny with a fringe of dark blue-black setae. Males from the Northwest have shiny orange-brown scales on the body. In California and the interior the males have tan or brown scales on the body and long hairs on the chelicerae.

OCCURRENCE This species occurs from southern British Columbia east to western Montana and south to Arizona and southern California and Mexico. It has been found in a variety of open habitats, in low elevations as well as high in the mountains.

SEASONALITY Adults: spring and summer.

REMARKS The male spreads his enlarged front legs and shows their inner iridescent surface during his display.

Habronattus viridipes (Hentz, 1846)

Plate 62

IDENTIFICATION This is a small, light brown jumping spider. The male and female have similar patterns on the cephalothorax and abdomen, but they are much more distinct in the male. The male also has modified legs. The ventral surfaces of the front pair of legs are bright green, and the patellae of the third legs are expanded with a dark spot.

OCCURRENCE This species occurs from Vermont and Quebec west to Minnesota and south to Texas and the Florida panhandle. This spider is found on the ground or in low vegetation in open woodlands or fields. It also hides under rocks.

SEASONALITY Adults: April through summer.

REMARKS During courtship the male raises his front legs to display the green color and fringes of setae. He also positions his third legs so that the spots on the patellae appear as additional "eyes" from the female's perspective.

Hentzia mitrata (Hentz, 1846)

Plate 63

IDENTIFICATION This is a small jumping spider. The female is pale and covered with whitish scales. She usually has a series of reddish-brown spots down the midline of the abdomen. Both male and female have enlarged front legs, but the males are heavier and covered with white setae. The male is white below, and his clypeus is covered in a dense white mustache. The chelicerae are vertical and white in color, different from the dark projecting chelicerae of male *Hentzia pal-*

marum. The top half of the carapace and abdomen are orange or light brown. The color of the median abdominal band is most intense near the boundary with the white sides of the abdomen.

OCCURRENCE This species occurs from Massachusetts and Quebec west to Minnesota and south to Texas and Florida. This species is most often found on shrubs or the low branches of both deciduous and evergreen trees.

SEASONALITY Adults: all year.

REMARKS During courtship the male raises and spreads his legs and holds his abdomen up and tilted first to one side, then the other.

Hentzia palmarum (Hentz, 1832)

Plate 63

IDENTIFICATION This is a small jumping spider. The female is pale and covered with white- or cream-colored scales. Her abdomen has a dark series of spots, either brown or reddish down the center. The male is usually chestnut brown with a white submarginal band around the carapace and two parallel white lines on the abdomen. The front legs are much larger than the others. In the male they are dark brown. The male has a dense white mustache on the clypeus. He has long, dark, projecting chelicerae with long fangs. The chelicerae have a dense fringe of white setae on the outer edge. These chelicerae serve to distinguish this species from *Hentzia mitrata*.

OCCURRENCE This species occurs from Nova Scotia to Minnesota and south to Texas and Florida. It has primarily been found on the lower branches of trees and shrubs. In the northern part of the range it has also been found in herbaceous and grassy vegetation near the ground.

SEASONALITY Adults: all year.

REMARKS Some males have small chelicerae, similar in size to those of females.

Lyssomanes viridis (Walckenaer, 1837) • *Magnolia green jumper*

Plate 64

IDENTIFICATION This is a small to medium-sized jumper. This species is distinctive with bright grass-green color, an elongated abdomen, and long legs. There is an orange or red cap on the head region, usually surrounded by white. The anterior median eyes are conspicuous. The male has long divergent chelicerae with long fangs. His palps are also long. The anterior lateral eyes are farther back on the head than in most jumping spiders so that this spider seems to have four rows of two eyes each.

OCCURRENCE Most of the members of this genus are tropical; this species is found in the southeastern states. They have usually been found hanging under the leaves of herbaceous vegetation. Most records are from broadleaf woodlands and forests.

SEASONALITY Adults: spring and early summer.

REMARKS The anterior median eyes of these spiders are unusually large, even for a jumping spider. Internally these eyes have a tubular shape similar to a telephoto lens. The tube can move within the head. These eyes may appear green unless the retina is focused directly on you, then they appear black.

Maevia inclemens (Walckenaer, 1837) • *Dimorphic jumper*

Plate 64

IDENTIFICATION This is a small to medium-sized jumping spider. The female is much larger than the male, pale in color, with two longitudinal lines on the abdomen. These lines are most often red but sometimes black. She has a white clypeus. The males give this spider its common name because there are two distinct color varieties. One has a black body with pale legs. There are three extraordinary tufts of setae at the back of the eye region. The palps of this tufted form are black with a golden fringe of setae. The other form of male has a mottled-gray cephalothorax and legs. The femora of the legs have black lines on the underside. The palps are gray with gold-colored tarsi. The abdomen is gray with a series of large orange spots. All three of these spiders, the female and the two forms of males, are so different looking that it might easily have been called the trimorphic jumper.

OCCURRENCE This is an eastern species; relatives live in the West. They have been found in tall grasses, herbs, shrubs, trees, and on the walls of buildings in the sun.

SEASONALITY Males: April through summer. Females: all year.

REMARKS The behaviors of the two color forms of males are also dimorphic. The tufted form dances in a stilted upright stance, waving his front legs while held high. In a series of clever experiments, David Clark (1994) has shown that the elaborate visual displays of the tufted males are attractive to the females at a greater distance. The gray form males attract the females faster at closer distances. Thus each from of male has a particular advantage. (Clark 1994; Clark and Morjan 2001).

Marpissa formosa (Banks, 1892)

Plate 66

IDENTIFICATION This is a small to medium-sized jumping spider. The front legs are robust with thick tibiae with visible ventral spines. The male is black with white spots on the cephalothorax and abdomen. The female is bicolored, with a dark cephalothorax and light abdomen. There are two dark lines on the abdomen, which are usually red. The female has a white mustache of long white setae.

OCCURRENCE This is an eastern species. They have been found in understory shrubs as well as herbs and shrubs in open wet areas.

SEASONALITY Adults: all year.

REMARKS This is an agile and aware spider that is difficult to approach.

Marpissa lineata (C. L. Koch, 1848)

Plate 66

IDENTIFICATION This is a small jumping spider. The front legs are robust with thick tibiae with visible ventral spines. The female is larger and darker in color. The abdomen has four thin white lines, which are more conspicuous on the darker female.

OCCURRENCE This is an eastern species. It is found on the ground among the leaf litter in forests and open areas. It has often been collected in prairies. This spider has been found hiding under rocks or logs.

SEASONALITY Adults: summer.

REMARKS This species is sometimes the most common jumping spider in pitfall trap samples from grasslands and old fields.

Marpissa pikei (Peckham and Peckham, 1888) • *Pike slender jumper*

Plate 66

IDENTIFICATION This is a small to medium-sized jumping spider. This species is long and thin. The abdomen is more than three times as long as wide. The front legs are darker and more robust than the others and held forward and together when the spider is still. The male has a dark stripe on the abdomen, interrupted by pairs of white spots. The female is a sandy gray or tan. The abdomen of the female sometimes has three long and thin dark stripes.

OCCURRENCE This is a southern and eastern species. It has been found in tall grasses, prairies and old fields. This spider has been found in arid areas where other members of this genus are rare.

SEASONALITY Adults: in the northern part of the range, late spring through autumn; all year in Florida.

REMARKS The spiders cling to the long grass culms and are nearly invisible at rest.

Menemerus bivittatus (Dufour, 1831) • *Gray wall jumper*

Plate 66

IDENTIFICATION This is a medium-sized to large jumping spider. The cephalothorax is flat. The body is gray or tan with dark around the eyes and on the sides of the cephalothorax. In the male there are three long, black stripes on the abdomen; the central one is widest. The dark on the side of the carapace is interrupted by a light band of setae extending up under the posterior lateral eyes. In the female the abdomen has a darker upper surface with a variegated pattern and surrounded by a dark brown or black margin. The legs are banded.

OCCURRENCE This is a cosmopolitan species, mostly tropical. This spider been transported around the world by humans. It has been found in the southern United States from California to Florida. They have most often been found around buildings and bridges, often on walls in the sun.

SEASONALITY Adults: all year.

REMARKS The conspicuous males of this species can easily be distinguished from the similar-sized pantropical jumper by the fact that the gray wall jumper has black down the center of the abdomen; the pantropical jumper has a light stripe.

Messua limbata (Banks, 1898)

Plate 65

IDENTIFICATION This is a small jumping spider. The female is iridescent green or golden, with a darker cephalothorax. The male is dark with green iridescence and long divergent chelicerae. The male also has a white band on the side of the carapace and around the abdomen. They resemble the bronze jumper (*Eris militaris*).

OCCURRENCE This spider has been found along the Mexican border from Texas to California. They have been collected on the ground, in grassy areas, meadows, agricultural crops, or in low vegetation in woodlands. They have also been found on the low branches of trees near creeks.

SEASONALITY Adults: all year.

Metacyrba taeniola (Hentz, 1846)

Plate 64

IDENTIFICATION This is a small, dark jumping spider. The male is nearly black with two parallel lines of white spots on the abdomen. The female has the same lines as well as a white mustache on the clypeus and pale legs. The western form (*M. taeniola similis*) is illustrated in Plate 64. In the eastern form (*M. taeniola taeniola*) the femora (particularly femur IV) are darker, almost black. In some western females only femur IV is reddish-orange; the others are dark.

OCCURRENCE This species is widespread across the southern United States and south into Mexico, with two described subspecies. The eastern form has been found under bark or near the ground in deciduous forests. Near the western edge of its range this form is most often in wooded areas, but has also been collected under rocks in grassy areas. The western form is a desert inhabitant at low elevations and has been found in dry woodlands at higher elevations.

SEASONALITY Adults: all year.

REMARKS The adults overwinter in a cocoon under bark.

Metaphidippus manni (Peckham and Peckham, 1901)

Plate 59

IDENTIFICATION This is a small, dark jumping spider. There are two color forms. In the inland form (illustrated here) the male is a dark brown shiny spider with distinctive white spreading chelicerae and white bands on the face. Coastal form males lack the white bands on the face. Males have a white band around the abdomen. The female is lighter brown with a distinctive pattern on the abdomen of two longitudinal dark stripes separated by a paler center. In Arizona the female is often covered with yellowish setae. The legs are faintly banded in the male and less so in the female.

OCCURRENCE This is a western species, found from Vancouver Island to southeastern Arizona and west to the coast. This spider has been found in oak woodlands on oaks and shrubs at elevations up to 1,700 m.

SEASONALITY Adults: spring and summer.

Myrmarachne formicaria (DeGeer, 1778)

Plate 65

IDENTIFICATION This is a small to medium-sized antlike jumping spider. This spider resembles *Sarinda hentzi*. The cephalothorax is red-brown with a dark brown raised cap in the head area. The abdomen is light brown with one or more dark bands around the center. The legs are pale except for the metatarsi of the first legs. The male has extraordinary long chelicerae that extend directly forward. These have a flat top surface that is iridescent. The female has slightly enlarged black pedipalps, which she often holds in front of the chelicerae.

OCCURRENCE This species has been recently introduced into North America. It was first detected in Ohio in the summer of 2001. It has since been found in western New York and Pennsylvania. It has most often been noticed in sunny locations in low vegetation, on rocks or debris on the ground, or inside buildings near windows.

SEASONALITY Adults: summer and autumn.

REMARKS The spider moves with a jerky gait that resembles an ant. It also waves its front legs, almost like an ant's antennae. The coloration and peculiar body shape enhance this antlike appearance.

Naphrys pulex (Hentz, 1846)

Plate 61

IDENTIFICATION This is a small brown jumping spider. The cephalothorax has a triangular light mark behind the eyes, widest at the front. The eye region is nearly black in the male. The male is dark gray with a pale yellow or peach-colored clypeus and sides of the carapace under the anterior eyes. The palps of the male are pale at the base with a dark tarsus. The top of the abdomen is dark with a light central band enclosing a darker heart mark, and there is a broad white spot at the back. The female is brownish gray with a similar pattern on the abdomen and cephalothorax.

OCCURRENCE This is an eastern species. This spider has been found on tree trunks, rock walls, and in grass or leaves near the ground. It has often been found on the walls of buildings.

SEASONALITY Males mature in February and survive until October. Adult females have been found between March and October.

REMARKS This spider feeds on ants. Previously known as *Habrocestum pulex*.

Neon nelli Peckham and Peckham, 1888

Plate 58

IDENTIFICATION This is a tiny jumping spider with a compact body. The dark brown cephalothorax is so steeply sloped behind the posterior eyes that it almost appears to be square. The eyes seem large even for a jumping spider, particularly the posterior lateral eyes. The abdomen is mauve with a strong pattern of dark lines and central chevrons.

OCCURRENCE This species is eastern, occurring from Labrador and Newfoundland west to Manitoba, the Rocky Mountains, and south to Texas and Florida. This spider lives at ground level, in low vegetation, and around rocks and debris.

SEASONALITY Adults: February through late August.

REMARKS This species has often been found in a silk cocoon on the underside of a rock.

Paramarpissa albopilosa (Banks, 1902)

Plate 58

IDENTIFICATION These are medium-sized jumping spiders. The front legs have thick brushes underneath and are long and robust. The body is long and in the male there is a pale median band. The female is mottled gray.

OCCURRENCE This species occurs in Arizona and New Mexico. They have been found on the branches of shrubs such as Creosote and Mesquite as well as oaks and junipers.

SEASONALITY Adults: June through September.

REMARKS Bruce Cutler (1992) discovered that these spiders prefer to rest on branches rather than foliage. The females and immatures are nearly invisible when lying along Mesquite branches.

Paraphidippus aurantius (Lucas, 1833) • *Emerald jumper*

Plate 60

IDENTIFICATION This is a medium-sized to large jumping spider. The coloration is variable. The females and immatures usually have a green iridescence on the cephalothorax and abdomen. Some individuals have golden rather than green iridescence. The bands and spots around the abdomen are either orange or white. The male is darker; he has a dark clypeus without the white setal mustache of the female, and longer chelicerae. In good light the male shows green iridescence. The legs of males are dark; the first pair are longest and have black and white fringes of setae.

OCCURRENCE This is primarily an eastern species. It occurs from Delaware to Nebraska and south to Texas and Florida. This spider has been found in riparian trees and shrubs, and in open areas of shrubs, field herbs, and grasses. There is a population in southeastern Arizona that may represent a different species.

SEASONALITY Adults: late summer and autumn.

REMARKS This is one of the few common North American spiders that often appear bright green. This variable spider has been known by more than 20 different names, including *Eris aurantia*.

Peckhamia americana (Peckham and Peckham, 1892)

Plate 65

IDENTIFICATION This is a small, brown, antlike jumping spider. The abdomen is constricted near the middle, giving the illusion of a separate body section. The long and narrow body, with dark brown or tan coloration, is similar to a number of common ants. The spider moves with a distinctly antlike gait.

OCCURRENCE This species has been found in eastern deciduous forests and woodlands. There is a similar related species in the West. This spider has been found on tree trunks, understory vegetation in wooded areas, the walls of buildings, or on the ground.

SEASONALITY Adults: spring and summer.

REMARKS This species sometimes runs among foraging ants. These spiders often wave their secnd pair of legs, somewhat resembling ants wavng their antennae.

Pelegrina aeneola (Curtis, 1892)

Plate 59

IDENTIFICATION This is a small brown jumping spider. The female has a dark gray cephalothorax with scattered white setae and a prominent mustache of white setae on the clypeus. There is a weakly defined T-shaped pattern of white setae in the area on top of the head between the eyes. The abdomen has a light band flanked by two darker bands and within these are pairs of white spots. The legs are banded. The male is bronze-colored with a thin white rim around the abdomen. His cephalothorax is bronze in color, sometimes with a sparse band of white on the sides.

OCCURRENCE This is a western species. This species has most often been collected from the foliage of coniferous trees. It has also been collected from live oaks, shrubs, and ferns. There are records from riparian vegetation, marshes, and bogs. It occurs up to elevations of 3,000 m.

SEASONALITY Males: April through June. Females: April through September.

Pelegrina exigua (Banks, 1892)

Plate 59

IDENTIFICATION This is a small jumping spider. The female is a peppered-gray color with a dark central folium on the abdomen enclosing pairs of light spots. The legs are banded. The chelicerae of males are yellowish in color with distinct black spots on the interior surfaces. The male is bronze brown with two distinct color forms. The striped form has a lighter bronze color with a distinct white band around the abdomen. The males of this form have three distinct white patches of setae on the face. The plain form males are brighter red bronze with only the main white setal band from the anterior lateral eyes back to the end of the carapace. The three white spots on the face are absent.

OCCURRENCE This eastern species occurs as far west as Kansas, Arkansas, and Texas. They have been found mostly on conifer foliage. The have also been found on oak and walnut trees.

SEASONALITY Adults: May through October.

REMARKS The plain form is most common near the East Coast. The striped form is found further inland. Both forms occur together in some areas (Virginia, Maryland, and Arkansas). Even though the color forms of males look distinct, they share the same courtship displays.

Pelegrina galathea (Walckenaer, 1837) • Peppered jumper

Plate 59

IDENTIFICATION This is a small, gray jumping spider. The female is gray with a pattern of dark and light spots giving the species its common name. The male is bronze brown with distinctive white setal patterns on the cephalothorax and abdomen. The two white marks extending behind the anterior median eyes form a sort of V that gives his face a distinctive frown. There is a broad white setal band behind the anterior median eyes extending to the back of the carapace. The abdomen has a broad white setal band that may separate into spots at the back. There are four or five pairs of white setal spots, the third pair being oriented laterally and touching the white band.

OCCURRENCE This species occurs from New England west to Nebraska and Colorado and south to Texas and Florida. They have most often been found in herbaceous foliage of fields and prairies. They have also been collected from the lower branches of trees. *Pelegrina galathea* is an open field species; the similar species *Pelegrina proterva* is more common in wooded areas.

SEASONALITY Males: April through September. Females: May through autumn.

REMARKS This species is occasionally abundant in old fields and agricultural crops and may be important in the control of pest insects.

Pellenes wrighti Lowrie and Gertsch, 1955

Plate 67

IDENTIFICATION This is a small, dark jumping spider. The female is dark with light-colored palps and a series of light chevrons on the abdomen. The male can also be dark or have light brown on the top of the cephalothorax with white behind the posterior lateral eyes and along the rim. The abdomen of the male has white setae around the rim and in a line down the center. The abdominal color may be light brown, rust, or tan (western populations), or the pattern may resemble the dark female illustrated here (northern and eastern populations).

OCCURRENCE This spider has been found from Nova Scotia west to Manitoba and Wyoming and south to Kansas and Indiana. This species, like many in this genus, has been found in the mountains. They have been collected from the ground, sand, on top of rocks, or under rocks.

SEASONALITY Adults: May through September.

Phidippus apacheanus Chamberlin and Gertsch, 1929

Plate 68

IDENTIFICATION This is a large jumping spider. The chelicerae are iridescent green. The female is black with an orange top of the cephalothorax and abdomen. There is often a black stripe on the abdomen. The dorsal color is yellow, orange, or red. The underside is black. There may be small light spots or a light basal band on the abdomen. The male is similar but has solid coloration and is more often red-orange. The palps of the male are dark.

OCCURRENCE This species is found in most of the United States except the Northeast and far western states. It is most common from Nebraska to Utah south to Texas and Arizona. It has been recorded from dry grasslands, fields, and deserts at elevations up to 1,800 m. It has been found on shrubs, cacti, and other perennial vegetation, and along watercourses.

SEASONALITY Adults: all year.

REMARKS The female often lays her eggs under bark.

Phidippus ardens Peckham and Peckham, 1901

Plate 69

IDENTIFICATION This is a large jumping spider. The body is black with iridescent green chelicerae. The top of the abdomen in females is red or sometimes yellow with a black median stripe that starts about a quarter of the way back from the front and extends to the back. The top of the abdomen in the male is red. The palps of the male are dark, without white scales. The fringes on the front legs are short and alternate black and white.

OCCURRENCE This species occurs from western Kansas to southern Washington south to California and New Mexico. It is not known from Texas. This spider has been found in shrubs including mesquite and creosote as well as in grasses at elevations up to 2,500 m.

SEASONALITY Males: spring and summer. Females: all year.

REMARKS The name *ardens* refers to the glowing red abdomen.

Phidippus audax (Hentz, 1845) • *Bold jumper*

Plate 67

IDENTIFICATION This is a large, black jumping spider. The chelicerae are iridescent green. The cephalothorax is black with or without a white band from the posterior lateral eyes to the back of the carapace. The abdomen of females is black with three white or orange spots. Females may have a white band at the front of the abdomen tapering along the sides. The legs of females are black, with or without white bands at the ends of the segments. The abdomen in one form of male is black with a broad white band around the front, tapering along the sides. In these males there are usually white or orange spots on the dorsal surface. Another form of male has a more extensive white band on the carapace behind the posterior lateral eyes, a white band around the abdomen, widest at the front, and with a yellow, gold, or ocher dorsal surface surrounding black markings. Males have heavy front legs with fringes of white and black setae. The other legs usually have white rings near the joints. The dorsal surfaces of the femora and patellae of palps of the male are white. Juveniles may have orange spots that become white when they mature.

OCCURRENCE This species is most common in the East from New England and Ontario, west to South Dakota, Colorado, and south to New Mexico and Florida. It is less common in the West. It has been introduced into southern California. They are found in fields, prairies, suburban plantings, and low vegetation in open woodlands. This spider has often been found in houses. It sometimes builds its nocturnal retreat, a dense white silk cocoon, in mailboxes.

SEASONALITY Adults: April through late summer; subadults overwinter in a retreat, usually under bark.

REMARKS During the winter they may form aggregations in protected spaces. Females lay their eggs under the bark of trees or shrubs.

Phidippus cardinalis (Hentz, 1845) • *Cardinal jumper*

Plate 68

IDENTIFICATION This is a large jumping spider. The female is dark with rusty red or crimson on the top of the carapace and abdomen. Her chelicerae often have red at the base. There are usually two dusky lines on the abdomen, enclosing small white spots. The legs of the female are often suffused with red. The male is similar, but the red color on the cephalothorax extends down the sides. His chelicerae and palps are black. The legs of the male are dark with light rings around the bases of the tarsi and metatarsi.

OCCURRENCE This species is eastern, it has been found from New England west to Colorado and south to Texas and Florida. This spider has been found in fields, grasslands, shrubs, and the understory vegetation of woodlands.

SEASONALITY Adults: autumn; females may overwinter.

REMARKS As the spider matures, it changes from yellow to orange, then red. The egg sacs are placed under bark.

Phidippus carneus Peckham and Peckham, 1896

Plate 69

IDENTIFICATION This is a large jumping spider. The chelicerae are iridescent green. There are both plain individuals without white markings, or others with lateral bands on the cephalothorax, a white band around the abdomen and white spots. The female has a red pattern on the abdomen, usually enclosing a black central stripe. Females have either white or red spots on the abdomen. Rarely females are dark, without red. The male has two well-marked color forms. The plain form has a black cephalothorax, with or without a white lateral band, and a plain red abdomen. The "montivagus form" of male has extensive white markings on the abdomen and a broad white band on the sides of the carapace.

OCCURRENCE This spider has been found in Arizona, New Mexico, and a few localities in western Texas. Individuals have been collected from desert shrubs, including cacti, and oaks at elevations up to 2,100 m.

SEASONALITY Males: summer and autumn. Females: all year.

REMARKS The females deposit the egg sacs under rocks.

Phidippus clarus Keyserling, 1884

Plate 67

IDENTIFICATION This is a large jumping spider. The female is covered with tan, yellow, orange, or red scales. She has iridescent green chelicerae. There are two dark lines of spots on the abdomen that usually enclose light spots. There is a dark-margined light band at the front of the abdomen. On the pale underside of the abdomen there are two black stripes. The male is black with turquoise iridescent chelicerae. His abdomen has broad red bands surrounding a black central band and a white band at the front. The underside of his abdomen is pale with two black stripes. The palps are white above. The legs have white and black fringes.

OCCURRENCE This common species is found across southern Canada and most of the United States except the desert Southwest. This spider is found in old fields, prairies, agricultural crops, and other low herbaceous vegetation.

SEASONALITY Males: summer. Females: summer and autumn.

REMARKS The female constructs a conspicuous white retreat and egg case near the tops of plants (Edwards 2004).

Phidippus johnsoni (Peckham and Peckham, 1883) • *Johnson jumper*

Plate 68

IDENTIFICATION This large jumping spider is black with iridescent green chelicerae. The dorsal surface of the abdomen is red. The underside is black. In the female there is often a broad black stripe down the center of the abdomen starting about a third of the way back. The female often has white markings on the abdomen. These consist of a white band around the abdomen, widest at the front, and a series of white spots usually within the black median stripe or at the outer edge of the red area near the back of the abdomen. She may have a white band on the back of the cephalothorax. The male has a solid bright-red top of the abdomen.

OCCURRENCE This species has been found primarily in the West as far north as the Yukon and south into Mexico. It is common on the Pacific Coast. Its range extends eastward to Saskatchewan, Colorado, and Arizona. This spider has a remarkable ecological range from the coast up to elevations of 3,600 m. It has been found in suburban yards, beaches, fields, shrubs, chaparral, and coniferous forests.

SEASONALITY Adults: all year.

REMARKS The female places the egg sac under rocks or bark. When harassed, this spider can bite, but the bite is not considered serious. Previously known as *Phidippus formosus*.

Phidippus otiosus (Hentz, 1846)

Plate 69

IDENTIFICATION This is a large jumping spider. This is a variable species. The cephalothorax of the female may have a broad lateral band extending to the back, a narrow band, or no band at all. The sides of the carapace, face, and clypeus are white or yellow. The top of the abdomen of the female has a complex pattern of scales, either white or yellow. The underside of the abdomen is dark or dark at the back and light in the front with a black median stripe. The chelicerae in both sexes may be green, yellow, or orange. The male's carapace may be dark with a broad light band. The abdomen has a light band that tapers toward the back. The top of the abdomen is black with white spots. The fringes on the front legs are either black and white or, in Florida, black and yellow.

OCCURRENCE This species is southeastern. It has been collected from Maryland to Texas but is most abundant in Florida. It occurs in both hardwood and coniferous trees. It has been found in both dry and swampy habitats.

SEASONALITY Adults: autumn; females occasionally overwinter and survive into the next summer.

REMARKS According to G. B. Edwards (2004), the yellow coloration in males from Florida is probably due to hybridization with *Phidippus regius*.

Phidippus princeps (Peckham and Peckham, 1883)

Plate 69

IDENTIFICATION This is a medium-sized jumping spider. The female is light brown with only subtle markings. Her clypeus and cheliceral bases are covered with white scales. The ventral abdomen is black with white lateral stripes. The male is black with iridescent blue green. The femora and patellae of the palps have white scales. The top of the abdomen in the male is red with faint darker spots. Most of the fringes on the front legs are relatively short, but the ventral tibial fringes are long.

OCCURRENCE This is an eastern species occurring from New England to Saskatchewan and south to South Carolina and northern Texas. This spider has been found in fields and the forest understory.

SEASONALITY Males: spring. Females: spring through July.

Phidippus regius C. L. Koch, 1846 • *Regal jumper*

Plate 68

IDENTIFICATION This is a very large jumping spider. This species is variable in color. The cephalothorax of the female is usually light gray or white on the sides. Above, it may be smooth without scales or white, gray, brown, or orange. The clypeus is white, tan, or black. The chelicerae are iridescent green or purple. Two color patterns of the female are illustrated, but other females are dark like the male. The ventral abdomen is dark. The male is mostly black with white spots and bands. There are black and white fringes on the front legs. The iridescence on the chelicerae of males can be green, blue, or purple.

OCCURRENCE This species is found in the Southeast as well as the Caribbean islands. It occurs from Virginia to Mississippi but is most common in Florida. They have been found in open areas and fields as well as in shrubs at the edge of fields.

SEASONALITY Adults: all year; mating occurs in the autumn.

REMARKS This is one of the largest and most conspicuous jumping spiders; some females are 22 mm long.

Phidippus whitmani Peckham and Peckham, 1909

Plate 69

IDENTIFICATION This is a medium-sized jumping spider. The chelicerae are usually black or faintly iridescent blue. The female is usually shades of brown, nearly black. The top of the abdomen is often light brown but is occasionally yellow, orange, or red like the male. She has a light band around the abdomen, widest in front. The abdomen has two faint dark lines, usually with paired light tan or white spots. The underside of the abdomen is black with two white stripes near the center. The top of the head region of the female is reddish brown. The male is dark, thinly covered with light scales. The top of his cephalothorax is mostly red but is black from the posterior median eyes forward. The abdomen of the male has a pale band and a red dorsal surface, sometimes with paired light spots. The underside is black. The front legs of the male are covered with white scales including long ventral fringes.

OCCURRENCE This is an eastern species occurring from Nova Scotia west to Manitoba and North Dakota and south to Texas and northern Florida. This spider has been found in low herbaceous vegetation and the leaf litter in hardwood forests.

SEASONALITY Adults: spring.

REMARKS According to G. B. Edwards (2004), this is the only member of the genus *Phidippus* that is found primarily near the ground in forests.

Phlegra hentzi (Marx, 1890)

Plate 67

IDENTIFICATION This is a small, dark jumping spider. Both sexes have a dark abdomen with a white rim and a white line down the midline. They also have white lines on the cephalothorax extending from the posterior lateral eyes to the back. The male has a white or light blue clypeus, and the front of the cephalothorax between the eyes is covered with rust-colored setae.

OCCURRENCE This species occurs from Massachusetts and Ontario to Minnesota and south to Texas and Florida.

SEASONALITY Adults: summer.

REMARKS This species has recently been separated from *Phlegra fasciata*, which is from Eurasia.

Platycryptus undatus (DeGeer, 1778)

Plate 66

IDENTIFICATION This is a large, gray jumping spider. The cephalothorax and abdomen have a complex pattern of spots and chevrons. There is a lighter region in the center of the abdomen surrounded by darker gray. The cephalothorax is flat; the thoracic region is slightly lower than the head. The female has a white mustache on the clypeus. In the male this area is usually orange.

OCCURRENCE This is an eastern species. It occurs from Nova Scotia to Nebraska and south to Florida and Texas. This species has most often been found on the bark of trees and shrubs, wooden fences, and the walls of buildings.

SEASONALITY Adults: all year; in the northern part of the range they overwinter in aggregations under bark.

REMARKS The beautiful variegations on the body of this spider often resemble the lichen-covered bark where they live. Previously known as *Metacyrba undata*.

Plexippus paykulli (Audouin, 1826) • *Pantropical jumper*

Plate 60

IDENTIFICATION This is a large jumping spider. The sexes are different. The female is large and brown with a distinctive pattern of parallel lines and spots on the abdomen. She also has a long white spot on the back of the cephalothorax, sometimes with a dark center. The legs are brown without banding. The male is unmistakable. He has a black body with bold white stripes and white chelicerae. The legs are light brown with longitudinal stripes on most segments. The front legs are darker.

OCCURRENCE This spider has a worldwide tropical distribution. It has also been introduced into many subtropical areas. It is found in Georgia, Florida, the Gulf Coast, and Texas. They also show up further north and west, probably as accidental introductions. This spider has commonly been found on buildings and other artificial structures as well as at farms and in orchards.

SEASONALITY Adults: all year.

REMARKS The male and female are so different that it is easy to mistake them for two different spiders.

Salticus scenicus (Clerck, 1757) • *Zebra jumper*

Plate 60

IDENTIFICATION This is a small jumping spider. Both sexes have white stripes on a dark brown or black body. The alternating dark and light pattern is the source of the common name. Females often wave their white palps in front of the brown chelicerae. The males have large divergent chelicerae with long fangs.

OCCURRENCE This is a Eurasian spider that has been introduced in temperate areas around the world. It occurs across southern Canada and has been found as far south as northern California, Colorado, northern Texas, Ohio, and North Carolina. It has often been found on fences, buildings, and other artificial structures. They have also been found on tree trunks, rocks, and debris on the ground.

SEASONALITY Males: April through July. Females: May through October.

REMARKS This species is often found around houses. It is commonly noticed, perhaps because it is active in the open and near windows. This spider will approach almost anything that moves; it seems to be curious. Such behavior is sometimes considered either friendly or threatening, depending upon the temperament of the observer. Males engage in dramatic contests in which they interlock their spread chelicerae.

Sarinda hentzi (Banks, 1913)

Plate 65

IDENTIFICATION This is a small to medium-sized antlike jumping spider. This spider resembles *Myrmarachne formicaria*. The cephalothorax is red-brown, darker around the eyes, and higher in the head area. The abdomen is brown in front and nearly black behind. There is a constriction in the middle of the abdomen, whose appearance is enhanced by a band of light setae. The male has large diverging chelicerae with long fangs. The female has ordinary-looking chelicerae but enlarged pedipalps that make her look like a subadult male. This spider is extremely variable in color from orange to black.

OCCURRENCE This is an eastern species, but there is another member of the genus in Arizona and California. They have most often been found in low herbaceous vegetation and grasses. They have also been found on the ground and the walls of buildings.

SEASONALITY Adults: May through September.

REMARKS This spider resembles large brown ants in the genus *Formica*. It sometimes runs in the same areas as ants, and with the same jerky motion. It can run faster than the ants.

Sassacus cyaneus (Hentz, 1846)

Plate 63

IDENTIFICATION This is a small, dark jumping spider. These spiders have a smooth, shiny, compact body with short legs. They often look like tiny beetles. The body is black but it is covered with iridescent scales. The iridescence most often appears blue. The cephalothorax is flat and widest near the back. The cephalothorax is held close to the abdomen. This species is even more compact than *Sassacus papenhoei*; it also differs because it lacks the light band around the abdomen.

OCCURRENCE This species occurs from Massachusetts to Wisconsin and south to Texas and Florida. Most records are from the Eastern Seaboard. Individuals have been found on shrubs and small trees. They have also been captured in sweep samples in old fields and tallgrass prairies.

SEASONALITY Males: May through July. Females: May through August.

REMARKS Previously known as *Agassa cyanea*.

Sassacus papenhoei Peckham and Peckham, 1895

Plate 63

IDENTIFICATION This is a small, dark jumping spider. These spiders have a smooth, shiny, compact body with short legs. They often look like tiny beetles. The body is black but it is covered with iridescent scales. The iridescence can be purple, golden, green, or blue. The scales on the cephalothorax form a broad light band around the sides and face. The abdomen has a light band at the front that extends back along the sides almost all the way to the spinnerets.

OCCURRENCE This species has been recorded from Maryland south to South Carolina and west to British Columbia and California. It is absent from most of Canada, Georgia, and Florida. It is most common in the western states.

SEASONALITY Males: May through September. Females: May through November.

REMARKS According to Dave Richman (2008), the iridescence is usually pink or coppery on the cephalothorax and blue or green on the abdomen.

Sitticus concolor (Banks, 1894)

Plate 61

IDENTIFICATION This is a tiny jumping spider. This is a plain gray or tan-colored spider. The male has a white mustache and long dark palps. The female is nearly uniform gray. There is a herringbone pattern on the abdomen, which is usually obscured by gray scales. The abdomen is longer and less globular than most members of this genus. The legs are banded.

OCCURRENCE This is an eastern species. It has been found in tall grass and old fields near the ground. They have often been captured in pitfall traps.

SEASONALITY Adults: May through August.

REMARKS Previously known as *Sitticus cursor* or *Sitticus floridanus*.

Sitticus palustris (Peckham and Peckham, 1883)

Plate 61

IDENTIFICATION This is a small jumping spider. The male has a dramatic pattern of brown and white. He has white around the eyes. Below the thin white line under the anterior median eyes there is a gray band, then a thicker white line above the chelicerae. The palps are dark but the patellae and tibiae of the palps have tufts of white setae. The legs have white banding. The top of the cephalothorax has three narrow white lines. The abdomen has two prominent white spots at the back. The female is less strongly marked but has a similar pattern on the abdomen.

OCCURRENCE This species has a broad distribution across the northern part of the continent. This species has been collected in a variety of habitats near the ground, including fresh water and salt marshes, sphagnum bogs, herbaceous plants, grasses, rocks, and leaves on the ground. They have been found on the walls of buildings.

SEASONALITY Adults: early spring through August; some adults overwinter and become active early in the spring.

REMARKS The female of this species bears a remarkable resemblance to *Naphrys pulex* in size, coloration, and foraging behavior.

Synageles noxiosus (Hentz, 1850)

Plate 65

IDENTIFICATION This is a small, brown jumping spider that resembles an ant. The pattern of movement and postures enhance this effect. The head region is darker than the remainder of the cephalothorax. There are two light bands across the middle of the abdomen; the front band is wider. The abdomen is an elongated oval shape. In bright sunlight this spider appears dark brown in color, similar to medium-sized ants found in the same areas.

OCCURRENCE This species has a broad distribution across the southern half of the continent. Its range extends to southern Canada in the East. This spider has been found in a variety of open habitats, including chaparral, coastal scrub, mesquite dunes, grassy areas, open woodlands, sand dunes, meadows, and bogs.

SEASONALITY Adults: all year.

REMARKS This spider walks with a jerky antlike motion. According to Bruce Cutler (1988), spiders in this genus, as well as those in *Peckhamia*, habitually wave their second pair of legs in the air, thus appearing like the antennae of an ant. Using its second pair of legs is unusual; most antlike spiders wave their front legs.

Synemosyna formica Hentz, 1846

Plate 65

IDENTIFICATION This is a small, thin, antlike jumping spider. It has a long thin body with dramatic constrictions behind the head region of the cephalothorax, at the back near the pedicel, and at the front of the abdomen. The legs are pale and unmarked. There is variation in color; some are much lighter than the individual illustrated here.

OCCURRENCE This is an eastern species. It has been found in wetlands, and sunny places, on vegetation, rocks, or on the ground.

SEASONALITY Adults: May through September.

REMARKS This is the most extreme ant mimic among our spiders. The disguise is so perfect that I once had an entomologist friend hand me one that had been placed on an insect pin; he had discovered it in an insect collection among the specimens of ants.

Thiodina sylvana (Hentz, 1846)

Plate 64

IDENTIFICATION This is a medium-sized to large jumping spider. The female is pale-, white-, cream-, or straw-colored. The head has a pattern of darker markings around the head region, often with a dark central spot. The male is dark. One color form is nearly solid black with a white spot on the top of the cephalothorax as well as thin white lines below the posterior lateral eyes and at the very back where the thoracic region slopes steeply to the pedicel. In this form the abdomen is usually dark gray with two white stripes but is sometimes greenish. The other color form of the male has a deep maroon-red cephalothorax with similar white markings. The abdomen of the red form is usually greenish with two white lines.

OCCURRENCE This spider is one of two, primarily eastern, species but there is also a similar western one. This species occurs from southern New York west to Missouri and south to eastern Texas. This spider inhabits the understory vegetation in deciduous woods and forests as well as open fields near woodlands.

SEASONALITY Males: February through November. Females: March through December.

Tutelina elegans (Hentz, 1846)

Plate 63

IDENTIFICATION This is a small, iridescent-green jumping spider. The female is almost uniform iridescent green that may appear gold or even purple at different angles. There is a light band at the front of the abdomen. The chelicerae of the female are rose red. The femora of her front legs are black. The other light-colored leg segments have black lines down the centers of each segment. The male is also covered with iridescent scales that appear yellow-green or grayish except that the tip of the abdomen is even shinier. The male has conspicuous dense tufts of setae between his posterior median eyes and anterior lateral eyes; they tilt slightly toward the midline. Viewed from above, there often appears to be a dark line on the medial edge of the tufts. There are tufts of long dark setae on the ends of the tibiae of the front legs. These tufts extend out onto the metatarsi and tarsi as light colored tufts.

OCCURRENCE This species occurs from New England west to Minnesota and south to Arkansas and South Carolina. It has been found in shrubs, meadows, prairies, old fields, and tall herbaceous vegetation.

SEASONALITY Adults: late spring and summer.

REMARKS The male has the habit of waving his front legs incessantly as he walks in search of a female.

Zygoballus rufipes Peckham and Peckham, 1885 • *Hammerjawed jumper*

Plate 60

IDENTIFICATION This is a small jumping spider. The sexes are different in color, but they share a boxy-shaped cephalothorax. The female is relatively plain tan or brown often with a white band around the abdomen. Some females have spots on the dorsal surface of the abdomen. The femora of the front legs are dark brown. Some females are darker brown on the abdomen than the one illustrated. The male is brown with a white band around the abdomen. His legs are mostly yellow, but the femora of the first legs are orange or red. The chelicerae are relatively large and heavy.

OCCURRENCE This spider is mostly eastern, but there are isolated records from Arizona. They have most often been found in open areas on grasses, herbaceous vegetation, or small shrubs. In arid areas they have been found in riparian vegetation.

SEASONALITY Adults: all year.

REMARKS This spider frequently waves its front legs.

FAMILY SCYTODIDAE • *Spitting Spiders*

This family is composed of 158 species worldwide. Most of the diversity in this family is in the tropics. In North America north of Mexico we have seven species. The spitting spiders have only six eyes, arranged in three groups of two. The cephalothorax is enlarged in the back to house the huge modified venom glands. As the name suggests, these spiders subdue their prey by spitting. They approach a potential prey by walking slowly. When within range, the spider raises its fangs and spits a combination of venom and glue. This mixture pins the victim against the substrate and begins to poison it. The fangs actually vibrate during spitting, so that the stream of glue forms two zigzag lines covering a broad area. The spider then cautiously approaches and bites the prey, then backs off and waits. After some time the spider returns and begins to feed. Feeding occurs through one or several small bite holes; the remains of the prey are hollow but often appear intact.

The legs of spitting spiders are long and lack heavy spines. By far the most commonly encountered spitting spider in North America is *Scytodes thoracica*, a spider that has been transported inadvertently by humans around the world. It does not build a web and is sometimes common in or around human buildings. Some other species in this family build a loose space-filling web, usually under overhanging cover.

Scytodes thoracica (Latreille, 1802) • *Spitting spider*

Plate 76

IDENTIFICATION The cephalothorax has an odd shape, highest in the back, to accommodate the large venom and glue glands. The six eyes are arranged in three groups of two. The body and legs are light tan or yellow with bold black markings. The slow-walking behavior of this spider is also distinctive.

OCCURRENCE This species is found around the world. It is usually associated with buildings. These spiders are usually noticed as they walk slowly across an open surface.

SEASONALITY Adults: summer.

REMARKS The name refers to the unusual hunting method. These spiders approach their prey slowly. When near enough they spit, immobilizing the prey. Finding small prey glued to a wall or ceiling by sticky silk lines is a hint that there are spitting spiders nearby.

FAMILY SEGESTRIIDAE • *Tubeweb Spiders*

This family includes 120 species worldwide. There are seven species known in North America north of Mexico. They possess six eyes in three groups that are positioned near the front of the head region. The cephalothorax and abdomen are both longer than wide. Under ideal conditions the pair of tracheal spiracles located just behind the book lung openings can be seen using a magnifying lens. These spiders have the odd habit of resting with six legs extended forward and two backward.

The web and retreat give this group their name. The round tubular retreat is lined with silk and often positioned at the broken end of a branch, in other dead wood, or among debris on the ground (see Fig. 28, right). There are a number of trip lines radiating from this central tube. This arrangement is reminiscent of the often larger webs of the crevice weavers (Filistatidae, see Fig. 28, left).

Ariadna bicolor (Hentz, 1842) • *Tubeweb spider*

Plate 33

IDENTIFICATION This is a medium-sized spider with six eyes. The anterior median eyes are missing. The posterior median eyes are bright. The cephalothorax is brown, darker in the head region. The abdomen is gray with a satinlike appearance. The front two pairs of legs are the darkest, and the rear legs are lighter.

OCCURRENCE This species has been found from Maine to California and south to Florida and Mexico. There are few records from the Great Plains. This spider has been found under rocks and in cracks and crevices in wood. The retreat is a tubular silken structure including a few trip lines radiating from the circular entrance to the tube (see Fig. 28, right).

SEASONALITY Males: June through September. Females: May through September.

REMARKS The trip lines radiating from the tube serve as a signal to the spider that prey is nearby. When a line is disturbed, the spider rushes out to make a capture.

Segestria pacifica Banks, 1891 • *Western tubeweb spider*

Plate 33

IDENTIFICATION This is a medium-sized to large brown spider with six eyes. The anterior median eyes are missing. The posterior median eyes are bright. The cephalothorax is brown, darker in the front. The large, black, and shiny chelicerae can be extended forward when grasping prey. The abdomen is pale with a dark, distinctly lobed, median line flanked by scattered darker spots and blotches. The legs are light brown.

OCCURRENCE This species occurs from the southwestern coast of British Columbia south to Baja California. There are two other species with limited distributions along the West Coast. These spiders are most often found in tubular webs in wooden structures or under bark. They have most often been found in relatively humid environments.

SEASONALITY Adults: all year.

REMARKS The tubular retreats have often been found under the bark of *Eucalyptus* trees in California. This spider has scattered signal lines extending away from the tube that are used by the spider to detect prey.

FAMILY SELENOPIDAE • *Flatties*

This group has a worldwide distribution with 178 species. Much of the diversity is in Africa. In North America north of Mexico they are restricted to five southwestern species. The key distinctive characteristic of the selenopid crab spiders is the odd arangement of all eight eyes at the front of the wide face. Usually six are in what appears to be one row, the outermost being the smallest. Just behind and laterally to these are the last two eyes, that are often the largest and facing to the side.

The legs are held in a laterigrade posture. The body of these spiders is very flat, hence the group name. Due to this flat, laterigrade stance, they can hide in thin spaces. They have most often been found under rocks or between leaves. They wander at night and are very fast, making them difficult to capture.

Selenops actophilus Chamberlin, 1924

Plate 71

IDENTIFICATION This is a large, very flat, crablike spider. Six of the eyes are arranged in one row along the front of the very flat carapace; two others are behind at the sides. The abdomen is also flat. The body is light brown, with or without a variegated pattern of brown spots. The legs may be ringed or not.

OCCURRENCE This species is known mostly from California to West Texas and south into Mexico. It has been found under rocks in the desert. They forage on rocks and boulders at night.

SEASONALITY Adults: May and June.

REMARKS The flat body permits this spider to hide in thin cracks among rocks. The coloration closely resembles the granite rock surfaces where this species often forages.

FAMILY SICARIIDAE · *Sixeyed Brown Spiders, Recluse Spiders*

In addition to 13 North American species, this family includes an additional 110 species from the Mediterranean region, northern Africa, South America, and Central America. These are infamous spiders (see "Spider Bites"). This family includes the brown recluse and other related species whose reputation far exceeds their actual significance as a bite risk (Vetter and Bush 2002). These spiders are sometimes classified as a separate family Loxoscelidae. They possess only six eyes (see Fig. 29). Their bodies are shades of brown and have long, thin legs without heavy spines. They often hold these legs in a laterigrade posture. On the cephalothorax there is sometimes a dark central region, constricted in the middle. A dark line extends back to the end of the cephalothorax. This creates a mark that is said to resemble a violin body and neck. Recluse spiders build thin sheet webs, often under rocks, debris, or stored materials in buildings. One favorite haunt is a stack of unused cardboard. Because their bodies are flat, and the laterigrade habit, they can hide in narrow spaces. Occasionally they wander away from their retreat, mostly at night.

Loxosceles deserta Gertsch, 1973 · *Desert recluse*

Plate 76

IDENTIFICATION The six eyes are in three groups of two; the center group is nearest the front. This and three of the four related species from the arid deserts of the Southwest are uniformly pale in color. There is only a hint of the violin-shaped marking, if at all.

OCCURRENCE This species occurs in the deserts of southern California, southern Nevada, and western Arizona. These spiders have been found under rocks or other debris during the day, and rarely out in the open at night. Unlike several other species of recluse, these arid-zone spiders have not adapted to human habitations. When rarely found indoors, they are mostly in homes that are surrounded by native desert vegetation.

SEASONALITY Adults: all year.

REMARKS This species, like its arid-zone relatives, is rarely encountered.

Loxosceles laeta (Nicolet, 1849) · *Chilean recluse*

Plate 76

IDENTIFICATION The six eyes are in three groups of two; the center group is nearest the front. The head region is usually dark, but it is less like the shape of a violin than in the brown recluse. The Chilean recluse can be substantially larger. Females are usually darker and redder in color than the other recluse species native to North America; males are tan. The relatively long, thin legs are hairy but lack heavier spines.

OCCURRENCE The natural range of the species is in South America. This spider has established itself in a few cities in densely populated Los Angeles County but is restricted to commercial and historic buildings. They have not been found in homes. They are rarely encountered and have not spread far in the many decades that they have survived in Los Angeles.

SEASONALITY Adults: all year.

Spider Bites

It never fails that whenever I give a talk about spiders, someone in the audience will relate a story about a spider bite—theirs or a close friend's or a relative's. I politely listen and then ask a few questions. Usually the person relates finding a spot or wound, which develops into a painful sore or perhaps a swollen blister. More often than not, no spider was actually seen. If a spider was located, it was found sometime later and often not nearby. Why was spot called a spider bite? It is a widely held belief that spider bites are common. Even though there may be little or no evidence, the sudden appearance of a small wound is assumed to be the result of such a bite. In fact, however, evidence would indicate the opposite. Actual spider bites are rare.

There are a number of other medical conditions that cause small wounds, most prominently skin conditions related to bacterial infections. The increasing incidence of infections from antibiotic-resistant strains of staph bacteria, such as MRSA (methicillin-resistant *Staphylococcus aureus*), is an example of this phenomenon. Because these infections may develop into deep necrotizing wounds, which are superficially similar to the late stages of bite wounds from recluse spiders, they are frequently misdiagnosed as spider bites. A number of recent studies have demonstrated that spiders were not actually involved in most of these incidents.

Actual spider bites are usually the result of a defensive reaction of a spider being accidentally crushed against the skin. A common scenario is that when someone dons a piece of clothing that has been lying on the floor, or perhaps in the closet, they are bitten. A spider had moved into the item of clothing at the end of a night's wandering; later when the spider gets squished, it bites. Bites may occur in the yard, perhaps when a gardener picks up a handful of leaves with bare hands or thrusts a hand into a garden glove or into a storage area where the spider has taken residence. Only a few species of spiders are known to bite humans, and usually the bites have little medical significance. A small welt or painful spot may develop at the site of the bite, which heals within a matter of days. Some larger spiders can inflict a painful bite. Again, the bite is defensive. If you think about this from the spiders' perspective, there is little reason to bite a human. We are much too large to be a potential prey item. A spider's self-defense reaction usually involves a rapid retreat; spiders run and hide. Only when they have no choice will they bite.

If no spider was noticed when the wound occurred, or if there are other similar wounds, the culprit is not likely a spider. Even when they do bite, spiders almost never bite repeatedly. In the case of a bite where the spider was observed, you should seek medical attention. It is a good idea to bring the spider, even crushed, to the doctor's office. If the bite is serious, having the spider with you may help in determining the correct treatment. The physician can seek out a qualified arachnologist to identify the spider. For the few spiders that can cause dangerous injury, such as the black widows, an antivenin (antivenom) may be used. Antivenin, however, is only effective against the venom of the particular species of spider from which it was developed, making the expertise of the arachnologist invaluable. In the course of research and study, I have handled thousands of spiders; neither I nor any of my assistants have ever been bitten. Common-sense precautions, such as not handling dangerous species with one's bare hands, should prevent the unlikely event of a serious spider bite.

Loxosceles reclusa Gertsch and Mulaik, 1940 • *Brown recluse*

Plate 76

IDENTIFICATION The six eyes are in three groups of two; the center group is nearest the front. There is usually a dark violin-shaped mark on the front half of the cephalothorax. The "neck" of the violin is a dark line extending back from the cervical ridge to the back of the cephalothorax. The mark in the head region forms the "body" of the violin, its resemblance enhanced by a series of parallel rows of dark hairs, the "strings." The abdomen is without any contrasting markings and is usually pale in females, sometimes dark in immatures and adult males. The relatively long, thin legs are hairy but lack heavier spines. Any spider with conspicuous spiny legs and a patterned abdomen can be safely eliminated from consideration as this species.

OCCURRENCE The natural range of the species is restricted to a group of states surrounding Missouri and Arkansas. On occasion the spider is accidentally introduced outside this range, but then it is usually restricted to buildings. The spiders, befitting their name, hide in thin spaces under rocks, debris, or among stored materials. They build a thin sheet web that only occasionally extends out from the retreat area. Wandering individuals are found without webs. They have sometimes been found in cardboard boxes, perhaps explaining their accidental transportation to areas outside of the natural range. Many reports of this species outside of the natural range are the result of misidentifications or discovery of other related members of the genus.

SEASONALITY Adults: all year.

REMARKS A mythology has developed about the frequent occurrence of bites of this species resulting in slow-healing wounds. Actual bites are rare. Many wounds of unknown cause have been misdiagnosed as bites of the brown recluse but are often caused by bacterial infections and other skin conditions.

Loxosceles rufescens (Dufour, 1820) • *Mediterranean recluse*

Plate 76

IDENTIFICATION The six eyes are in three groups of two; the center group is nearest the front. This species often has a distinct reddish cast to the dark markings. The "violin" marking on the cephalothorax is more evenly rounded at the back and usually lacks the lines of dark hairs seen on the brown recluse. The long and thin legs are often dark in color.

OCCURRENCE Humans have transported this species, which is native to the Mediterranean region, to many places. It has been found in buildings worldwide. Rather than being widespread, this species is typically restricted to local, sometimes dense populations in particular buildings.

SEASONALITY Adults: all year.

REMARKS Perhaps because of the notoriety of the brown recluse, most individuals of the Mediterranean recluse are initially confused with that species. In areas outside of the natural range of the brown recluse, a recluse infestation in a building is more likely to be of the Mediterranean recluse.

FAMILY SPARASSIDAE • *Huntsman Spiders, Giant Crab Spiders*

This family includes some very large spiders with long legs. This family has a worldwide distribution and includes more than 1,000 species. There are only nine species known from North America north of Mexico. Despite their large size, these spiders' flat bodies permit them to hide in surprisingly small spaces. They have well-developed scopulae on their legs and are excellent climbers. For example, they easily cling to the ceiling in buildings. Their eight eyes are arranged in two regular rows, a feature that distinguishes them from the somewhat similar-appearing members of the families Selenopidae or Homalonychidae.

Heteropoda venatoria (Linnaeus, 1767) • *Huntsman spider*

Plate 71

IDENTIFICATION This is one of the giant crab spiders. This species is among the largest and the sexes are distinctive. Both sexes have a light or white clypeus. The female is light brown. The eye region is often dark. There may be an irregular pattern on the abdomen. The male is light brown but typically has the dark carapace pattern illustrated here. His abdomen may have a pattern of irregular dark blotches in the center and surrounding a light folium. On both sexes there are dark spots at the bases of some spines on the femora.

OCCURRENCE This is a tramp species that might be encountered anywhere in buildings, having been accidentally transported by humans. Outdoors they occur in southern states from Florida to Texas. It also occurs in urban areas in the Southwest.

SEASONALITY Adults: all year.

REMARKS This is the largest spider commonly found indoors in Florida with a leg span up to 10 cm (about 4 inches). The female grasps her flat egg sac with her palps and carries it under her body.

Olios giganteus Keyserling, 1884 • *Golden huntsman spider*

Plate 71

IDENTIFICATION This is one of the giant crab spiders. The body is light brown, tan or grayish. The cephalothorax is sometimes darker, particularly in worn individuals. The abdomen has a light heart mark with a brown or black rim forming a Y-shaped mark at the back. A series of dark spots extend the bar of the Y toward the spinnerets. The legs look almost as if the spider is wearing dark socks.

OCCURRENCE This species has been found from California, Nevada, Utah, Arizona, New Mexico, Texas, and northern Mexico. Found in arid areas in shrubs, rocks, or on the ground.

SEASONALITY Adults: all year.

REMARKS This spider occasionally wanders into buildings where its large size elicits consternation. This spider has no difficulty climbing on walls and even ceilings. This is not considered to be a dangerous spider.

FAMILY SYMPHYTOGNATHIDAE • *Dwarf Orbweavers*

This family contains the smallest spider (as an adult) in North America, *Anapistula secreta*, being only 0.2–0.5 mm long (Plate 20). It weaves a small horizontal orb. The orb is composed of a large number of radii and circular strands, which form a fine-mesh circular net. These spiders possess only four eyes in two groups. One remarkable feature, shared with ground orbweavers of the family Anapidae, is the absence of palps in adult females. This group is nearly worldwide in distribution but has been little studied. Only 42 species are known worldwide. *Anapistula secreta* is the only species known from North America north of Mexico; it has been found in the South.

Anapistula secreta Gertsch, 1941

Plate 20

IDENTIFICATION This species has the smallest adults of any spider in North America. It belongs to one of only two groups of spiders with adult females that have no palps. There are only four eyes in two groups at the side of the head. The body is pale, and the cephalothorax and legs are a bit darker than the abdomen. The small orb web has a fine screenlike mesh.

OCCURRENCE This species is widely distributed but rarely encountered. The true limits of its range are not known. Existing records north of Mexico are from the southern United States. Occurs in hollow logs, spaces in leaf liter, low vegetation, and caves.

SEASONALITY Nothing is published.

REMARKS Misting or dusting with cornstarch reveals the webs. When disturbed, the spider moves from the hub to the edge of the web. It has been suggested that pollen and spores, which fall on the web, may provide an important source of food for these tiny spiders.

FAMILY TELEMIDAE · *Longlegged Cave Spiders*

These spiders can be distinguished from other similar spiders such as the midget cave spiders and midget ground weavers by a peculiar zigzag-shaped ridge at the front of the abdomen, just above the pedicel. Unfortunately this feature is nearly impossible to see without a microscope. There is one genus in North America, *Usofila*, with four described species. There are many undescribed relatives in museum collections. Members of the family have been found in the West.

Usofila pacifica (Banks, 1894)

Plate 82

IDENTIFICATION This is a pale inconspicuous spider. The abdomen and legs often have a bluish iridescence. Four of the six eyes are arranged in line, the other two are behind the lateral eyes of this line. The females have a dense pattern of hairlike spines on the tarsi of the palps. There are three similar species known as well as a number of undescribed relatives awaiting description in museum collections.

OCCURRENCE This species occurs along the West Coast from coastal Alaska to Oregon. These spiders hang beneath a thin sheet web. They have been found near the ground among ferns fronds, in the leaf litter, or under rocks. They have also been found in caves, occasionally in large numbers.

SEASONALITY Nothing is published.

REMARKS Males and females may occasionally be found in the same web. The egg sacs are oval and usually placed at the edge of the web. Oddly, the abdomen of these spiders turns green when preserved in alcohol.

FAMILY TENGELLIDAE • *Tengellids*

This family has a tortured taxonomic history. Most members have at one time or another been placed in one of at least four other families. The characters that unite them are difficult to see without a microscope. One such feature is the presence of both claw tufts and three tarsal claws. Most spiders with claw tufts have lost the third (middle) claw. Tengellids are relatively large spiders with long, relatively stout legs. They possess a relatively long deep fovea. The spinnerets are conical and form a compact group. These are ground-living spiders, active on the surface at night but found hiding under rocks or other debris during the day.

Anachemmis sober Chamberlin, 1919

Plate 46

IDENTIFICATION This is a moderately large, plain brown- or reddish brown–colored spider. The abdomen is mottled brownish gray. There is an elongate fovea. The tarsi have claw tufts. This spider resembles *Liocranoides*.

OCCURRENCE This species occurs in California, but close relatives live in both California and Arizona, with ranges extending into Sonora, Mexico. This species has been collected from oak woodlands and coniferous forests at elevations from 70 to 2000 m. Individuals have been collected under leaves, rocks, logs, or other debris on the ground. It has often been collected in pitfall traps and occasionally in buildings.

SEASONALITY Adults: all year.

REMARKS According to Norm Platnick and Darrell Ubick (2005), this species can be very common. For example, on June 20, 1952, 95 males and 130 females were collected at Tanbark Flats, Los Angeles County.

Lauricius hooki Gertsch, 1941

Plate 46

IDENTIFICATION This is a medium-sized to large pale brown or tan spider with variegated markings. They often appear flat. Both the cephalothorax and abdomen are somewhat flattened. This effect is enhanced due to their posture at rest. They usually hold the legs laterigrade in one plane (Plate 46, female). The fovea is relatively long and deep. The tarsi have claw tufts.

OCCURRENCE This species has been collected in the mountains of Arizona and New Mexico at elevations between 1,500 and 2,100 m. They are found under rocks during the day but wander in the open at night.

SEASONALITY Adults: summer and autumn.

Liocranoides tennesseensis Platnick, 1999

Plate 46

IDENTIFICATION This is a medium-sized plain spider. They are ground active or cave species that range from pale tan to dark brown and gray ones. They have eight eyes in two regular rows. They possess claw tufts but also three claws. Sadly these claws are too small to be seen without a microscope, because this combination of characters is unique among spiders of eastern North America. Similar spiders in the genus *Anachemmis, Socalchemmis,* and *Titiotus* are all western, mostly Californian.

OCCURRENCE This species and others in the genus have most often been collected in the Appalachian region. Most are known from caves and cave-entrance habitats, but they have also been collected from pitfall traps in forests away from caves. They have also been found under rocks or logs.

SEASONALITY Adults: probably all year.

Socalchemmis dolichopus (Chamberlin, 1919)

Plate 46

IDENTIFICATION This is an unmarked species with long legs, a light orange or tan cephalothorax and legs, and a tan abdomen. There is an elongate fovea. The tarsi have claw tufts. This spider resembles *Titiotus.*

OCCURRENCE These spiders occur in southern California. This species has been collected in coastal sage, chaparral, desert shrublands, and woodland habitats in foothill and low elevations in mountains. They have mostly been collected during the daytime hiding under rocks, logs, or other debris. They wander on the surface at night. They have frequently been captured in pitfall traps.

SEASONALITY Adults: all year.

REMARKS Many individuals of this species have been captured in buildings.

Titiotus californicus Simon, 1897

Plate 46

IDENTIFICATION These are relatively large, ground-active spiders with pale coloration. This species is pale orange or tan in color. The abdomen is uniform pale. They resemble *Anachemmis* and *Socalchemmis.* There is an elongate fovea. The tarsi have claw tufts.

OCCURRENCE These spiders are from central California. This and related species occur in steep rocky ravines near montane streams. They live on the forest floor among the leaf litter. In addition to these habitats, they occur at lower elevations in both forests and grassy areas. They have been found wandering on the surface at night, hiding under rocks during the day. They have also been captured in pitfall traps.

SEASONALITY Adults: all year.

REMARKS These spiders have often been found in caves, occasionally in large numbers.

FAMILY TETRAGNATHIDAE · *Longjawed Orbweavers*

The majority of spiders in this family have long, thin bodies and legs. The most dramatic feature of the family are the long cheliceral bases and fangs, hence the name "longjawed." The first legs are the longest and the third legs are short. In most species the front two pairs of legs are held stretched out in front of the body, particularly when the spider is at rest. In this resting pose, with the body held close to a twig or branch, they are hard to see. The epigynum of females is usually a simple rectangular plate and it is not always obvious when a female is mature. The males have extremely long, thin palps. During mating, the pair tightly interlock their large jaws, and the function of males' long palps is revealed; they can reach the females' reproductive openings even while in this odd position.

Most of the members of the large genus *Tetragnatha* in this family build horizontal orbs, often over water. Emerging insects flying up from the surface become entangled. This makes these tetragnathids superb mosquito eaters. In addition to the horizontal orientation, most tetragnathid webs have an opening at the hub. One group, including the members of the genus *Leucauge*, builds webs at an oblique angle to the horizontal, often well away from the water. Their conspicuous webs and silvery and green colors make these spiders easy to notice. Members of one genus, *Pachygnatha*, are ground-living spiders and at least the adults do not build capture webs but wander among fallen leaves in search of prey.

Azilia affinis O.P.-Cambridge, 1893

Plate 13

IDENTIFICATION This is a brown spider with relatively large dark eyes. The abdomen is an oval shape, pointed at the back. Some individuals have a central hump on the abdomen near the back. The top of the abdomen has a wide dark band with several cream-colored spots that often form a cross-shaped mark in the center. The legs are conspicuously banded with dark brown.

OCCURRENCE This is a spider of the southeastern United States. They have been found in dark moist woods, ravines, shady areas near buildings, hollow stumps, wide cracks or depressions in the ground, and around the entrances to or within caves.

SEASONALITY Adults: all year.

REMARKS The web is either a vertical or horizontal orb. The spider hangs under the center of the web with the front four legs held together, slightly bent. When disturbed, they drop and run rapidly into hiding.

Dolichognatha pentagona (Hentz, 1850)

Plate 17

IDENTIFICATION This is a small dark orb-weaving spider with four humps on the posterior end of the abdomen. The cephalothorax is light with a dark boundary and dark bands demarcating the head region, including two that extend from the back of the head toward the eyes. The chelicerae are elongated. This species could be confused with *Colphepeira catawba*, but that species has grouped points on the abdomen.

OCCURRENCE This species occurs in the southeastern United States. It has been found in moist forests. *Dolichognatha* builds its nearly horizontal orb in shaded areas. Webs have been found near the base of fallen trees, in standing hollow logs, or among tree roots under soil banks.

SEASONALITY Adults: summer.

REMARKS The web is decorated with a vertical line of debris. The spider clings to the debris with the legs withdrawn during the day but moves out to the center of the web at night. The egg cases may be concealed within the debris line.

Glenognatha foxi (McCook, 1894)

Plate 12

IDENTIFICATION This is a small red, orange, or reddish-brown spider that builds a horizontal orb web about 5 to 10 cm (2 to 4 inches) in diameter. Females usually have a series of white and black smudges or spots on the abdomen. The males are nearly as large as the females, often brightly colored and have a conspicuous black palp.

OCCURRENCE This species is widespread in the East but extends to the West Coast only in the South. They have been found in a variety of wet habitats, including marshes, swamps, pond edges, wet fields, and lawns. They typically build their small orb webs within a few inches of the ground.

SEASONALITY Adults: spring and summer.

REMARKS The spider hangs from the undersurface of the web and drops to the ground if disturbed. Previously known as *Mimognatha foxi*.

Leucauge argyra (Walckenaer, 1841)

Plate 16

IDENTIFICATION This spider has a light brown cephalothorax with a tuning fork–shaped green mark extending down the center of the thoracic part of the carapace then splitting into two lines on the head. The legs are green. The abdomen is silvery-white with many dark markings, particularly on the sides and the back. On the top of the abdomen there is a dark centerline but the other lines form paired lobe shapes, progressively smaller toward the back of the spider. There are no red or orange spots on the abdomen. The somewhat similar basilica orbweaver has areas of yellow or golden markings on the top of the abdomen as well as a dome-shaped web.

OCCURRENCE Found in central and southern Florida and the tropics. The horizontal web is built in low branches of trees and shrubs in wooded areas as well as in mangroves.

SEASONALITY Adults: all year.

REMARKS There is a tuft of long and thin modified setae in two rows on the front side of the femurs on leg IV. These can sometimes be seen in photographs or with a magnifying glass. Previously known as *Plesiometa argyra*.

Leucauge venusta (Walckenaer, 1841) • *Orchard spider*

Plate 16

IDENTIFICATION This spider has a light brown cephalothorax with a dark median line and green legs. The most distinctive feature is the long oval abdomen that is silvery-white marked with a pattern of dark lines and colorful spots. In some individuals there are areas of yellow. Typically the dark lines include green and iridescent green. One dark green line forms a ring around the top of the abdomen. Within this ring is a single central line and two or three branching lines that angle backward. There may be a pair of bright red-orange spots at the back of the abdomen. There is usually a bow tie–shaped orange spot on the underside of the abdomen. The similar basilica orbweaver has obvious yellow or golden markings on the top of the abdomen as well as a dome-shaped web.

OCCURRENCE This species is widespread in the East but is also found on the West Coast and there are scattered records in between. This spider builds its web in low shrubs and the low branches of trees in moist forests and woodlands. It has often been found in suburban plantings and city parks.

SEASONALITY Adults: spring and summer.

REMARKS The web is usually built at an angle to the horizontal and has a tangle of threads below the orb. There is a tuft of long and thin modified setae in two rows on the front side of the femurs on leg IV. These can sometimes be seen in photographs or with a magnifying glass. The name *venusta* refers to its beauty.

Meta ovalis (Gertsch, 1933) • *Cave orbweaver*

Plate 20

IDENTIFICATION This is a long-legged brown spider with banded legs. The body varies from relatively light with contrasting dark marks to dark brown, nearly black. The cephalothroax usually has a central dark band, widest at the front and dark marginal bands. The abdomen has variable markings. The palp of the male has a large process at the base that extends perpendicular to the long axis of the palp. Because it lives in dark areas, the orb web is not always obvious. The general aspect of this spider, hanging in an inconspicuous orb among a tangle of threads, with a relatively teardrop-shaped abdomen, suggests a cobweb weaver, such as *Parasteatoda tepidariorum*.

OCCURRENCE This species occurs from Newfoundland to Minnesota and south to Arkansas and the southern Appalachian region. In addition to caves, this spider has been found in deep crevices in rocky areas, abandoned mines, culverts, cellars, as well as other shady spots.

SEASONALITY Adults: all year.

REMARKS The web is sometimes vertical but often nearly horizontal and has an open hub and associated tangle. There is usually a signal line extending from the hub to a rock wall or other surface nearby. Previously known as *Meta menardi*.

Metellina mimetoides Chamberlin and Ivie, 1941

Plate 18

IDENTIFICATION This is a medium-sized orb-weaving spider. The abdomen has a ridge or two small humps about a third of the way back at its widest point. The front legs are relative long, the jaws are large, but otherwise the spiders of this genus often resemble orbweavers more than they do typical longjawed orbweavers. The underside of the abdomen has a central dark line with white lines on either side. The males may be larger than the females but have a smaller abdomen.

OCCURRENCE This is a western species. It has been found from southern Alaska to Baja California east to Texas and Oklahoma. The nearly vertical web is usually built in dark, humid habitats. There are, however, some records from grasslands. Many of the records from the eastern part of the range are from caves.

SEASONALITY Nothing is published.

REMARKS The orb web has an open hub. One of the three similar species in this genus, *Metellina segmentata*, has been introduced from Europe and is now the most common one found around Vancouver and Seattle.

Metleucauge eldorado Levi, 1980

Plate 18

IDENTIFICATION The shape of this spider is intermediate between *Leucauge* and *Tetragnatha*. It has long legs, particularly the first two pairs that have dark bands at the ends of the segments. The abdomen has a dark folium that surrounds a light lobed central band.

OCCURRENCE This is our only species but there are several more in the Far East and Japan. Our species occurs in the Northwest Coast and the mountains of California. This spider builds its horizontal orb web between the rocks of streambeds.

SEASONALITY Adults: July and August.

Nephila clavipes (Linnaeus, 1767) • *Golden silk orbweaver*

Plate 16

IDENTIFICATION There is no other spider north of Mexico that could be confused with this species. The body of a mature female can be longer than 30 mm (1¼ inch) and the leg span nearly 10 cm (4 inches). The abdomen is yellow or orange with white dots and a red mark at the front. The cephalothorax is covered with silvery hairs. The legs (I, II, IV) have large black tufts at the ends of the femora and tibiae. The male is a dwarf. One or several males are occasionally found in or near the web of an adult female. The web is both large and sturdy and contains golden-colored silk lines that give the spider her name.

OCCURRENCE *Nephila* is a tropical group and our species occurs near the coast from South Carolina to Texas, common in Florida. There are occasional records further north and west, but these probably represent individuals that were accidentally transported by humans. They build their large distinctive webs in the lower branches of trees, often across trails.

SEASONALITY Males: May through September. Females: all year.

REMARKS There are tiny silvery spiders (*Argyrodes nephilae* and *Argyrodes elevatus*; Plate 25) that are occasionally found in the webs of *Nephila*. They take advantage of small prey captured in the web and are either small enough to avoid notice or agile enough to avoid being captured by the host. They look like drops of mercury or dew. The genus *Nephila* and a few related genera are sometimes placed in a separate family, the Nephilidae.

Pachygnatha furcillata Keyserling, 1884 • *Thickjawed orbweaver*

Plate 14

IDENTIFICATION This species is characterized by heavy spreading chelicerae, hence the name "thickjawed." The abdomen is pale tan with a brown or reddish-brown double folium. The outer edge and the center of this folium are darker than the band between them. The cephalothorax is reddish brown usually with a dark central stripe. Some have described the carapace as yellow. The legs lighter in color and are not banded.

OCCURRENCE This species is eastern and there are other species in the genus across most of the continent, but they are absent from the arid Southwest. They have been found on the ground, usually in humid habitats.

SEASONALITY Adults: probably all year.

REMARKS They are not found in webs but are found walking among the litter on the ground. There are scattered published reports of small webs built by young spiders of this genus in Europe.

Tetragnatha elongata Walckenaer, 1841

Plate 18

IDENTIFICATION This species is pale reddish brown and has a relatively dark central folium on the abdomen, usually with three sets of prominent lobes. The chelicerae are similar in length to the cephalothorax in females and longer than the cephalothorax in males. The head region is darker than the thoracic part of the cephalothorax.

OCCURRENCE This species is common throughout the East and there are scattered records from the West. The horizontal web is usually built in low branches above streams. This spider has only been found near water.

SEASONALITY Adults: early spring through autumn.

REMARKS The egg sacs are decorated with strands of greenish silk and attached to branches.

Tetragnatha guatemalensis O.P.-Cambridge, 1889

Plate 18

IDENTIFICATION This species is variable in color but usually has a lovely silvery reticulated pattern on the abdomen. The chelicerae are shorter than the cephalothorax.

OCCURRENCE This species has been found in the East with many records from Florida and southern Texas. There are scattered records from the southwestern states. The range extends into the tropics. Individuals have usually been found near the edges of bodies of water.

SEASONALITY Adults: May through September.

REMARKS The males of this species and others in the genus grasp the chelicerae of the female during mating. The males have projections near the ends of the chelicerae that interlock with the fangs of the female, holding them still. The palps are extremely long and reach under the female who curls her abdomen toward that of the male during mating.

Tetragnatha laboriosa Hentz, 1850 • *Silver longjawed orbweaver*

Plate 18

IDENTIFICATION This species is one of the smallest in the genus. It is usually silvery white, sometimes with yellow. The abdomen is shorter and more oval than the other species in the genus. The chelicerae are distinctly shorter than the cephalothorax.

OCCURRENCE This species has been found from Alaska to Central America and is among the most common members of the genus. They have often been collected by sweeping in fields and herbaceous vegetation away from water.

SEASONALITY Adults: late May through autumn.

REMARKS The webs are not as conspicuous as those of the other species probably because they build them in the dense, loose vegetation. The web is occasionally oriented vertically. The spider may drop to the ground if disturbed.

Tetragnatha versicolor Walckenaer, 1841

Plate 18

IDENTIFICATION This species is variable in color. It usually has a dark folium on the abdomen with a light center. Some individuals have an abdomen that is pale silvery white.

OCCURRENCE This species is found from Alaska to Central America. It is one of the most common members of the genus. The horizontal webs are built in the low branches of trees, shrubs, and even grasses near water.

SEASONALITY Adults: early spring through autumn.

REMARKS The egg sacs have a loose texture and a greenish color. Like other members of this genus, individuals may sometimes be found near the web with the legs stretched parallel to the twig where they are resting. In this pose they are hard to see.

Tetragnatha viridis Walckenaer, 1841 • *Green longjawed orbweaver*

Plate 18

IDENTIFICATION This species is small and light green in color. There is often, but not always, a red patch at the front of the abdomen. It is the most distinctive member of the group.

OCCURRENCE This species has been found along the East Coast and in the Southeast. They are a spider of pine forests and occasionally other conifers. The web is built in the low branches of pine trees.

SEASONALITY Adults: early summer.

REMARKS These spiders are inconspicuous because they rest among the green needles. Their long, thin green body blends well with this habitat. They are not easily dislodged by sweeping or beating.

FAMILY THERAPHOSIDAE • *Tarantulas*

The tarantulas are famous, probably because of their large size. The family, which contains nearly 900 species, is distributed throughout the tropics. There are currently 55 species recognized in North America north of Mexico, but there may be many more. Recent research on the genetics of tarantulas has revealed that some forms are actually composed of several separate genetic units. Some of these genetically distinct species look identical and cannot be identified morphologically. Even before this discovery, identification of species in this family was considered notoriously difficult because the genitalia are simple. The tarantulas that are native to North America north of Mexico occur west of the Mississippi River. One introduced species, *Brachypelma vagrans*, has become established in Florida.

These mygalomorph spiders build relatively large silk-lined burrows whose entrances are usually open. During the day the resident may spin a thin silk sheet over the burrow entrance. If the spider is inactive for a longer period, it may plug the burrow entrance and the thin silk cover can be hidden under debris. When open, the burrow entrance usually has a visible silk lining, which can be helpful in distinguishing these large burrows from those of other large invertebrates or mammals.

Male tarantulas are conspicuous during the mating season. These large spiders wander across the landscape, at times even during daylight hours. Because of their large size, they can be seen from a car as they attempt to cross roads. Tarantulas frighten some people, but our species are harmless to humans and pets.

Aphonopelma chalcodes Chamberlin, 1940 • *Desert blond tarantula*

Plate 4

IDENTIFICATION This is a large tarantula. The large size and hairy body including claw tufts are characteristic of our tarantulas. This species is light tan in color with darker brown femora. Males often have much darker, chocolate-brown or black legs. Some males have a beautiful slate-blue coloration. The abdomen of the male often has rust-colored hairs.

OCCURRENCE This species occurs in Arizona in the vicinity of Tucson.

SEASONALITY Males: wander in search of females after rain. Females: all year.

REMARKS These spiders wait near their burrow entrance at night. During inactive periods they may cover the entrance with silk or a soil plug. When threatened, tarantulas may use rapid movements of their hind legs to brush hairs from their abdomen into the air. This cloud of hairs can be extremely irritating to the eyes or nose of humans or pets. The bite of this spider is not serious.

Aphonopelma hentzi (Girard, 1852) • *Texas brown tarantula*

Plate 4

IDENTIFICATION This is a large tarantula. The large size and hairy body including claw tufts are characteristic of our tarantulas. This species has a tan cephalothorax and dark legs. The abdomen is dark brown, sometimes with rusty-colored hairs.

OCCURRENCE This species is found in Oklahoma, but similar related spiders are found in Arkansas and Texas. This species has been found in grasslands with scattered trees.

SEASONALITY Males: wander in search of females after rain, summer through September. Females: all year.

REMARKS This spider is a member of a poorly understood species group.

Aphonopelma moderatum (Chamberlin and Ivie, 1939) • *Rio Grande gold tarantula*

Plate 4

IDENTIFICATION This is a large tarantula. The large size and hairy body including claw tufts are characteristic of our tarantulas. This species is a light golden brown color with the exception of the patellae, metatarsi, and tarsi, which are dark brown or black.

OCCURRENCE This species occurs in the lower Rio Grande Valley of Texas. They have been found in Chihuahuan desert and desert grasslands with scattered shrubs.

SEASONALITY Males: wander in search of females in spring. Females: all year.

REMARKS The burrows have been found on the slopes of low ridges.

FAMILY THERIDIIDAE • *Cobweb Weavers*

The family Theridiidae is one of the most diverse worldwide with more than 2,200 species. The cobweb weavers are also found in abundance throughout our region. There are 234 species in 32 genera north of Mexico. The common house spider (*Parasteatoda tepidariorum*) is perhaps the most common spider in the world. This bold statement is based on the fact that this species lives in harmony with humans on every continent except Antarctica. Few other species are so cosmopolitan and abundant. It can be found both indoors and out. Most cobweb weavers build their webs under some sort of cover. This may be a rim on a rock face, a branch, or even a single leaf. Around buildings they often build under the eaves or furniture. Some members of the cobweb weaver family build scant webs consisting of only a few lines of silk. Among the cobweb weavers are the black widows of the genus *Latrodectus*. The widows have toxic venom but are actually reticent to bite (see "Spider Bites," on page 210). There are a many conspicuous and colorful members of this family.

Many cobweb weavers have a rounded abdomen tapering to a point at the spinnerets. This gives it a characteristic teardrop shape to the abdomen when the spider is hanging upside down in her web. Some members of this family have unadorned webs, others hang a leaf or other debris in the web and rest below this cover as a retreat. A few members of this family with modified body shapes, different from the typical teardrop shape, are illustrated on Plate 25. Cobweb weavers generally lack conspicuous heavy spines; they do possess a covering of thin hairlike spines. They also possess curved serrated bristles on their fourth tibiae. They use these bristles to draw out and fling sticky silk during a wrapping attack. The chelicerae are usually small and lightly built.

One interesting variation of the space-filling web type is the tanglefoot or gumfoot web of some cobweb weavers (see Fig. 22, right).

The webs of these spiders include a series of vertical, or nearly vertical, silk lines attached at their ends to the substrate below. The region near this attachment has two modifications. One is that the end is thin and fragile, prone to breakage. The other modification is the addition of a series of glue droplets. When a potential prey organism bumps up against such a line, it sticks to the glue. Struggling to escape, the prey breaks the line at the weak point, and the line, which is under tension, contracts and lifts the prey into the air. The prey now has no traction and hangs, struggling, helpless in the web. The resident spider usually rushes in and wraps the prey, completing the capture.

Anelosimus studiosus (Hentz, 1850)

Plate 28

IDENTIFICATION This is a small cobweb weaver. Some are shades of brown or tan; others have a red or even greenish tint to the abdomen. The cephalothorax usually has a dark margin and dark central area, but sometimes it is plain. The abdomen shows a central dark band with a wavy boundary and a thin white rim.

OCCURRENCE The species is found from southern New England to Texas and Florida. The webs of this species have been found in a variety of bushes and trees.

SEASONALITY Males: May through July in the North; all year in the South. Females: May through September in the North; all year in the South.

REMARKS This spider is considered social. The young spiders remain in their mother's web until maturity. All individuals may participate in capture and feeding on prey. The colonies occupy small compact tangle webs, often at the tip of a low branch in a tree or bush.

Argyrodes elevatus Taczanowski, 1873 • *Dewdrop spider*

Plate 25

IDENTIFICATION This is a small silvery spider. The cephalothorax is dark brown, appearing black. The femora, patellae, and ends of the tibiae of legs I and II are black. The abdomen is cone-shaped and iridescent silvery. The median line and much of the ventral part of the abdomen appear black.

OCCURRENCE This species has been recorded from the Southeast as well as southern California. The similar *Argyrodes nephilae* is known from Florida. This spider is usually found in the web of a larger orb-weaving spider, often the golden silk orbweaver (*Nephila*) or a garden spider (*Argiope*).

SEASONALITY Adults: June through October.

REMARKS This is a kleptoparasitic species; it steals food from its host. Sometimes it takes small prey that have been ignored by the larger host. At other times it wraps and removes larger prey before the host arrives to claim the catch. On rare occasions this species has been observed feeding on the host spider.

Asagena americana (Emerton, 1882) • *Twospotted cobweb weaver*

Plate 27

IDENTIFICATION This is a small cobweb weaver. This species has a dark chestnut-brown cephalothorax and legs. The abdomen is dark brown with two large white or cream-colored spots just behind the middle.

OCCURRENCE This species is widely distributed in North America. They have been found near the ground. It has been collected under rocks, logs, moss, leaves, and other debris.

SEASONALITY Males: May through July. Females: May through August.

REMARKS Males are much more frequently encountered because they often wander over the ground in search of females during the daytime. Probably for the same reason, these spiders have often been recorded as prey of mud dauber wasps. Previously known as *Steatoda americana*.

Asagena fulva (Keyserling, 1884)

Plate 27

IDENTIFICATION This is a small to medium-sized cobweb weaver. The cephalothorax is red or reddish-brown. The legs of females are red, banded with darker tips to the segments in males. The abdomen is red or brown with white spots or a variable-width white median band. There are also white marks on the sides that sometimes have a zigzag shape. In males there is a white arc across the front of the abdomen.

OCCURRENCE This species is most common in arid regions of the Southwest and into Mexico. Its range extends from Oregon to Florida. It has most frequently been found near the ground.

SEASONALITY Adults: March through August.

REMARKS Bert Hölldobler (1970) observed this species building its web near the entrance to harvester ant nests and capturing the ants. Previously known as *Steatoda fulva*.

Canalidion montanum (Emerton, 1882)

Plate 28

IDENTIFICATION This is a small cobweb weaver. The cephalothorax is yellow with a brown band down the center, wider in front as well as a narrow black marginal band. The abdomen has a pale band down the center that may be yellow or white and dark lateral areas with a marbled pattern of light areas. The legs are yellow with brown banding.

OCCURRENCE This species occurs in the North and the Rocky Mountains up to elevations of 2,600 m. This spider has been found in coniferous forests and meadows and fields adjacent to conifers.

SEASONALITY Males: June through August. Females: June through September.

REMARKS Previously known as *Theridion montanum*.

Crustulina sticta (O.P.-Cambridge, 1861)

Plate 81

IDENTIFICATION This is a small globular spider. The cephalothorax is dark reddish-brown with a rough warty surface. The eyes are in two regular rows on a head region that protrudes out in front of the relatively high clypeus. The round abdomen is usually pale, marked with dark blotches at the front in females. The species is variable and some individuals have a variety of light and dark markings on the abdomen. The legs are pale orange and unmarked.

OCCURRENCE This species is widespread across North America and Europe, usually found near the ground. This spider has been collected by sifting litter, or by searching among debris.

SEASONALITY Adults: summer; may overwinter as adults.

REMARKS The thin tangle web is rarely noticed.

Cryptachaea porteri (Banks, 1896)

Plate 24

IDENTIFICATION This is a small cobweb weaver with a mottled purplish-gray coloration. Some individuals are lighter in color. The abdomen has a small hump or tubercle at the back. The legs are banded.

OCCURRENCE This has been recorded from New York to Colorado and south to Texas and Florida. It has been found in webs under rocks, wood, or other debris on the ground. It has also been collected in caves.

SEASONALITY Adults: late summer through December.

REMARKS Some individuals lack the hump on the abdomen. Previously known as *Theridion porteri* or *Achaearanea porteri*.

Dipoena nigra (Emerton, 1882)

Plate 82

IDENTIFICATION This tiny, dark gray spider has a protruding head area with a very high clypeus. The body is round and compact. The legs are paler than the body. There is much geographic variation in this spider; western specimens are often lighter in color.

OCCURRENCE This species is found in southern Canada and throughout the United States. It has been collected in trees, shrubs, low vegetation, and on the ground.

SEASONALITY Adults: spring through autumn.

REMARKS This spider preys on small ants. This spider is not known to construct a capture web.

Enoplognatha marmorata (Hentz, 1850) • *Marbled cobweb spider*

Plate 26

IDENTIFICATION This is a small to medium-sized cobweb weaver. The cephalothorax and legs are shades of brown, darker near the joints. The abdomen is grayish with a brown central folium mark that has a variety of granular-looking silvery-white markings. These markings are also seen along the outer edge of the folium.

OCCURRENCE This species has been found throughout the United States and southern Canada. They have most often been found under rocks and other debris on or near the ground. It has also been discovered among leaves and in low vegetation.

SEASONALITY Males: June through September. Females: April through September; overwinters as a subadult.

Enoplognatha ovata (Clerck, 1757)

Plate 26

IDENTIFICATION This is a medium-sized polymorphic spider. There are three common color forms. One has a plain abdomen, one has a pair of red or pink lines, and the third variety has a red or pink median band. The cephalothorax is usually pale with a narrow dark rim and an indistinct darker median band. The ends of the tibiae usually have a dark band. There is a prominent white-bordered black line on the underside of the abdomen between the spinnerets and the epigastric furrow. The chelicerae of the male are large with long fangs.

OCCURRENCE This species was probably introduced into North America from Eurasia. They have been found between Nova Scotia and British Columbia south to California, Ohio, and New York. This species is usually found with small webs under leaves in shrubs.

SEASONALITY Adults: summer.

REMARKS The female has often been found guarding her bluish egg sac in a small web under a folded leaf. According to William Bristowe (1958), the plain color form is the most common, the two-banded form is next, and the broad red-banded form is least common. I have found the same relationship in Ohio.

Episinus amoenus Banks, 1911

Plate 25

IDENTIFICATION This is a small cobweb weaver. The body shape is distinctive. The abdomen is longer than wide and viewed from above appears to have a wedge shape, wider at the back. On the top of the abdomen at the back are two lumps at the corners that give it a truncated look. Below, the end of the abdomen tapers to a point near the spinnerets. The top of the abdomen and cephalothorax are dark brown, the sides are paler, and there is a distinct wavy boundary. The legs are banded.

OCCURRENCE This species is mostly southeastern, but there are records as far north as Ohio and Maryland. It has been observed in the low branches of trees and near the ground. At rest, the spider clings to small twigs and is nearly invisible. Specimens have often been obtained by beating or sweeping.

SEASONALITY Adults: summer.

REMARKS This spider builds a minimal web composed of a frame with two main sticky strands. These strands extend from support above, such as a low branch, to the ground or another branch. The spider hangs in the web holding the two vertical strands with its legs. When a potential prey brushes the strands, the spider attacks.

Euryopis funebris (Hentz, 1850)

Plate 25

IDENTIFICATION This is a small odd-looking cobweb weaver. This spider is black with an iridescent area covering the posterior third of the abdomen. This iridescent spot may appear silvery or golden depending upon the lighting. The shape is unusual for a cobweb weaver; the cephalothorax is nearly as wide as it is long, and the abdomen is pointed at the back.

OCCURRENCE This species is eastern but others can be found throughout North America. This spider has been found from New England and southern Canada, west to North Dakota, and south to Louisiana and Florida. This species has been collected from the low branches of trees and shrubs as well as among leaves on the ground.

SEASONALITY Adults: April through September.

REMARKS This spider waits with its legs outstretched as illustrated. It positions itself in an area where ants are likely to walk. When an ant brushes into it, the spider dashes out to capture the ant. The spider turns quickly and throws silk on the ant while circling around it. This behavior pins the ant in place. As the ant struggles the spider adds more silk. After the ant is immobilized, the spider moves in to bite. The spider then waits until the venom acts before feeding.

Hentziectypus globosus (Hentz, 1850)

Plate 24

IDENTIFICATION This is a tiny cobweb weaver. Despite its small size, this spider is distinctive. The abdomen is dark in front and has a pale area on the back with a black spot in the center. The cephalothorax is light brown with a black area around the eyes. The legs lack bands. The male is sometimes pale. There is considerable variation in the base color and pattern, but all individuals have the black spot.

OCCURRENCE This is an eastern species. It has been found between Maine and Minnesota south to Texas and Florida. The space-filling tangle is built among leaf litter, in hollow logs and tree stumps in forests.

SEASONALITY Adults: summer.

REMARKS The eggs are spindle-shaped, and several may be placed within the tangle web. Previously known as *Achaearanea globosus*.

Latrodectus bishopi Kaston, 1938 • *Red widow*

Plate 26

IDENTIFICATION   This is a large beautiful spider. The cephalothorax and legs are bright red-orange. The abdomen is shiny black with red, orange, or black spots on the back with white outlines. The typical widow hourglass mark on the underside of the abdomen is reduced to a triangle behind the epigastric furrow. The basic body shape is similar to the other widow spiders.

OCCURRENCE   This species occurs in Florida. They are most common in central and southern Florida. It has been found in sand pine scrub habitat. The webs are built in low vegetation such as small palmettos.

SEASONALITY   Adults: all year.

REMARKS   This spider has toxic venom, similar to the southern black widow. They are not aggressive and rarely bite. The egg case is creamy white.

Latrodectus geometricus C. L. Koch, 1841 • *Brown widow*

Plate 26

IDENTIFICATION   This is a light brown spider with a teardrop-shaped abdomen. The abdomen has an orange or red-orange hourglass-shaped mark on the underside. In the adult male the orange spot on the underside of the abdomen may be broken into two spots. The sides of the abdomen have a series of spots and streaks forming a distinctive pattern. The legs are light brown with dark near the joints.

OCCURRENCE   This species has been transported by humans around the world, and may occur anywhere. In the United States it is common in the South. They have been found in protected areas, under overhangs, in sheds, garages, and other buildings. They have often been found in large numbers under mobile homes.

SEASONALITY   Adults: all year.

REMARKS   This species has toxic venom, similar to that of southern black widows. They are timid and rarely bite. This species is not considered dangerous because the spider rarely injects enough venom to cause anything other than minor symptoms. The egg sac is round, pale tan in color, and covered with points.

Latrodectus mactans (Fabricius, 1775) • *Southern black widow*

Plate 26

IDENTIFICATION   This species is among the most recognizable of spiders. The shiny black abdomen of the female has a red hourglass-shaped mark on the underside. In this species the bottom portion of the hourglass is usually rectangular not triangular. The back of the abdomen can be plain black, or it may have a series of red spots, a red line, or red-centered white marks. There is usually a red dot at the end of the abdomen near the spinnerets. Two of the many color varieties are shown. The males and immature females are often more colorful with conspicuous red-and-white abdominal markings, but they also possess the red hourglass. Similar species include the northern black widow (*Latrodectus variolus*), which is more frequently marked with red on the top of the abdomen, and the hourglass mark is broken by a black gap in the center. In the western states, the western black widow (*Latrodectus hesperus*) probably replaces the southern black widow but these two species are easily confused.

OCCURRENCE This spider is common in the southern United States, but it has been reported as far north as southern Canada. It is not clear whether these northern records are of this species or a related one. Widows are usually found in their space-filling tangle webs under an overhanging ledge, under benches or other objects, in outbuildings, barns, and garages. Some are found in low vegetation in a variety of habitats. They are more common in warm and dry areas.

SEASONALITY Adults: all year.

REMARKS This is a dangerously toxic spider but it is not an aggressive animal (see "Spider Bites" on page 210). This spider is reticent to bite; considering how common they are, bites are very rare. Typically, the spider bites defensively if the web is disturbed. The bite often requires medical attention and the neurotoxic venom can, rarely, be fatal to small children if untreated. With a severe reaction, the bite may be treated with antivenom. It is important to bring the offending spider to the hospital so that the correct treatment can be determined. The egg sacs are round and grayish with a small point at the top.

Neospintharus trigonum (Hentz, 1850)

Plate 25

IDENTIFICATION This is a small to medium-sized cobweb weaver. This a distinctive triangular-shaped spider. The cephalothorax is orange-brown. The abdomen is variable in color, usually brown, orange, or reddish. The legs are similar in color to the cephalothorax and not banded. The male has a smaller abdomen. The male has an extraordinary cephalothorax with a two large projections from the head region, each with a tuft of hairs.

OCCURRENCE This is an eastern species. They have been collected from Maine to Wisconsin and south to eastern Texas and Florida. They are recorded from deciduous forests, coniferous forests, and woodlands.

SEASONALITY Adults: May through August.

REMARKS This spider sometimes builds its own web, but often it invades the webs of other spiders. It may live with the resident as a kleptoparasite, stealing prey. Often it kills the occupant and appropriates the web. The unusual egg case is shaped like an inverted urn. Previously known as *Argyrodes trigonum*.

Parasteatoda tabulata (Levi, 1980)

Plate 24

IDENTIFICATION This is a medium-sized cobweb weaver. The abdomen appears teardrop-shaped as the spider hangs upside down in its web, spinnerets uppermost. The cephalothorax is dark brown. The abdomen color is extremely variable, usually shades of light brown with dark streaks or blotches. The legs are darker at the joints. This species is easily confused with the common house spider (*Parasteatoda tepidariorum*).

OCCURRENCE The distribution of this spider poorly documented. There is speculation that it has been introduced into North America, but its origin is uncertain. It is often overlooked because of its similarity to the common house spider. There are records from southern Canada, and from Maine west to Wisconsin and south to Ohio. This species probably has a larger, undetected range.

SEASONALITY Adults: May through October.

REMARKS This spider hangs under a retreat composed of debris near the center of the web. The egg cases are tan and round. They are the same color as, but may be less teardrop-shaped than, the egg cases of the common house spider.

Parasteatoda tepidariorum (C. L. Koch, 1841) • *Common house spider*

Plate 24

IDENTIFICATION This is a medium-sized to large cobweb weaver. The abdomen appears teardrop shaped as the spider hangs upside down in its web, spinnerets uppermost. The cephalothorax is tan or brown. The abdomen color is extremely variable, usually shades of light brown with a mottled appearance. Some individuals are nearly black, others unmarked and pale. The legs are darker at the joints. Adult males often have a reddish coloration. This species is easily confused with *Parasteatoda tabulata*.

OCCURRENCE This is one of the most widely distributed spiders worldwide. It has been found throughout the United States and Canada, but it is less common in the center of the continent. Outdoors it is rarely found north of the conterminous United States, but indoors it is possible almost anywhere.

SEASONALITY Adults: all year in the South, usually absent outdoors during the cold months at high latitudes. Adults may live more than a year.

REMARKS The egg cases of this spider are tan and teardrop-shaped, point uppermost. Some females have been found in webs with up to four egg cases. When the young emerge, they remain together in a tight cluster for some time before dispersal. Previously known as *Achaearanea* or *Theridion tepidariorum*.

Phoroncidia americana (Emerton, 1882)

Plate 29

IDENTIFICATION This is a tiny cobweb weaver. This spider resembles a small lump of dirt or a bit of bark. The compact body, including the abdomen, is hard. The cephalothorax is brown and the eye region is extended into a prominent lobe. The abdomen is variegated pale or brown with many dark brown-colored pits at the bases of hairs. The abdomen also has a series of lumpy projections, paired at the back. The legs are pale with dusky bands.

OCCURRENCE This is an eastern species. It has been found between Nova Scotia and Indiana, south to Louisiana and Florida. It has been recorded from deciduous forests, coniferous forests, and in low branches at the edge of fields. This species has often been collected by sweeping or beating the low branches of trees.

SEASONALITY Adults: May through September in the North; all year in the South.

REMARKS These remarkable little spiders build a single-line sticky web. The spider hangs near one end of a silk line in the understory of forests. It has been speculated that it may use an odor to attract male flies as prey.

Platnickina alabamensis (Gertsch and Archer, 1942)

Plate 28

IDENTIFICATION This is a small cobweb weaver. The cephalothorax and legs are pale yellow. There is an indistinct dark central band and marginal band on the cephalothorax. The abdomen is dark above with white blotches in the center and sides. The underside of the abdomen is light with two dark lines running from the spinneret region forward then bending laterally near the center. The legs are unbanded.

OCCURRENCE Most of the records of this species are eastern, from New Brunswick west Wisconsin and south to Texas and Florida, but there is also a population in southern California. This spider has been found under bark, logs, and boards on the ground.

SEASONALITY Adults: April through August.

REMARKS Previously known as *Theridion alabamense*.

Rhomphaea fictilium (Hentz, 1850)

Plate 25

IDENTIFICATION This is a medium-sized cobweb weaver. The distinctive abdomen has a long extension projecting well beyond the spinnerets. This extension can appear nearly tubular, or triangular, but always tapering to a point. The tip of the abdomen is flexible and often bent or curved. The male does not have the odd projections on the head region that are characteristic of *Neospintharus trigonum*.

OCCURRENCE This species has been recorded from the East and far West. They have been found in low branches of trees, bushes, and grasses in forested regions.

SEASONALITY Adults: June through autumn.

REMARKS This spider is able to move the abdomen, bending or wriggling it. The purpose of this behavior is undetermined. The spider may build its own space-filling tangle or invade the web of another spider species, where it is known to kill the resident spider. Previously known as *Argyrodes fictilium*.

Robertus riparius (Keyserling, 1886)

Plate 80

IDENTIFICATION This is a small plain spider. They have brown, unmarked, somewhat shiny cephalothorax and legs with a gray abdomen. The chelicerae are large. *Robertus* somewhat resemble the larger, ground-living species of dwarf sheetweavers (Linyphiidae, Erigoninae). The clypeus in *Robertus* is only moderately high, about equal to the distance between the eye rows, lower than most sheetweavers.

OCCURRENCE This species has been collected primarily in the Northeast. There are scattered records further west in the literature, but this species has sometimes been confused with other similar-looking species of this genus that are found in the West. It has usually been found on the ground among the leaves in forests, under moss, logs, or other debris. They are not normally associated with a capture web but do build a small courtship web in low vegetation.

SEASONALITY Adults: all year.

REMARKS They are sometimes active in daytime. On one occasion I found a great number of individuals in the nest of a mud dauber wasp in Ohio. Previously known as *Ctenium riparius*.

Spintharus flavidus Hentz, 1850

Plate 25

IDENTIFICATION This is a small or medium-sized cobweb weaver. This distinctive spider has a pale body with bold markings on the abdomen. The abdomen is widest about a quarter of the way back, tapering to a point at the rear. The top of the abdomen has a black-bordered area from the wide point of the abdomen back to the tip. Additional thin transverse black lines are often present; one at the front, two in the middle, and one near the rear. The central region of the abdomen is marked with pink or red with cream or yellow spots. The cephalothorax is pale yellow to orange and unmarked. The anterior eyes and posterior lateral eyes are dark and arranged in a recurved line. The posterior median eyes are widely separated and just behind the posterior lateral eyes completing the arc. The legs are pale with red bands at the ends of the tibiae of legs I and IV.

OCCURRENCE This is an eastern spider with a range that extends into the tropics. It has been found from Massachusetts, Ohio, and Oklahoma in the North, south to Florida and Texas. It has been found in shrubs and understory vegetation at the edge of woods.

SEASONALITY Adults: late summer and autumn.

REMARKS This spider has often been found among a few strands of webbing under leaves. The capture web is minimal, somewhat like that of *Episinus*. *Spintharus* perches on a pair of lines that extend to a substrate that a prey item might walk on. If the prey brushes a line, it sticks and is attacked.

Steatoda albomaculata (DeGeer, 1778)

Plate 27

IDENTIFICATION This is a medium-sized cobweb weaver. The cephalothorax is dark brown. The abdomen pattern is variable; most individuals are white with a prominent dark brown central area with white spots, usually paired. In the West some individuals are nearly solid black. The legs are strongly banded with the light brown and dark brown areas of approximately the same size. Adult males have large robust chelicerae.

OCCURRENCE This species has a cosmopolitan distribution. In North America it is widely distributed in the West but absent from the Southeast. This spider lives near the ground, it has often been found under rocks, logs, or boards.

SEASONALITY Adults: May through October.

REMARKS This species eats ants.

Steatoda borealis (Hentz, 1850)

Plate 27

IDENTIFICATION This is a medium-sized to large cobweb weaver. This species has a dark reddish-brown cephalothorax and legs. The abdomen is either reddish-brown or purplish. There is a lot of variation in the abdominal color; some individuals are nearly black. The abdomen has a distinctive light colored T-shaped mark at the front of the abdomen. The carapace has a rough pitted texture, more prominent in males.

OCCURRENCE This species is found in the East and the North. It is replaced by a similar-looking species, *Steatoda hespera*, in the West. It has been found in low vegetation, under bark, logs, and rocks. It has often been found in and around buildings. Adult males often wander into houses.

SEASONALITY Adults: all year; egg cases in summer.

REMARKS The egg case is white and circular. The pink eggs are sometimes visible through the case. Both males and females have a rough surface with ridges on the front of the abdomen that forms a stridulating organ. The spider can make sounds by rubbing this structure against the carapace.

Steatoda grossa (C. L. Koch, 1838) • *False black widow*

Plate 27

IDENTIFICATION This is a large cobweb weaver. The cephalothorax and legs vary from pale brown to black. The abdomen is dark purplish-brown with light spots and bands, or completely black. The black variety is probably the source of the common name. The abdomen is oval rather than the teardrop shape of true widows. There is no red hourglass mark.

OCCURRENCE This species has been collected mostly in coastal states; it is particularly common in southern California. The web is built under overhanging protection among vegetation, rocks, furniture, and buildings.

SEASONALITY Adults: all year; females may live at least six years.

REMARKS This species sometimes preys upon black widows. This spider is considered harmless to humans.

Steatoda triangulosa (Walckenaer, 1802) • *Checkered cobweb weaver*

Plate 27

IDENTIFICATION This is a medium-sized cobweb weaver. The cephalothorax is reddish-brown. The legs are light brown or orange with dark brown bands. The abdomen is light brown with two rows of angular spots or bands of dark reddish-brown separated by white areas with a mottled appearance. In some individuals this pattern looks checkered.

OCCURRENCE This species may have been introduced from Eurasia. It is widespread in North America from New England to Oregon and south to central California, Texas, and Louisiana. It has been found under stones, culverts, and other cover. It has most often been found in buildings, where it may be abundant.

SEASONALITY Adults: May through October.

REMARKS The egg case is a fluffy white sphere. Individual females may construct between five and ten egg cases.

Theridion differens Emerton, 1882

Plate 28

IDENTIFICATION This is a small cobweb weaver. The cephalothorax is orange or orange-brown with a darker median band, widest in the eye region. There is a thin brown marginal band. The legs are orange with black banding. The abdomen is orange-brown with a broad light median band. This lobed band has a white margin and is orange in the center.

OCCURRENCE This species has been found between Nova Scotia and British Columbia south to central California and Florida. This spider builds its small cobweb in the tops of small herbs such as goldenrod, tall grasses, and the low branches of bushes or trees.

SEASONALITY Males: May through July. Females: May through September.

REMARKS This small cobweb weaver is among the most common small spiders in fields in the East. It has often been collected by sweep netting.

Theridion frondeum Hentz, 1850

Plate 29

IDENTIFICATION This is a small to medium-sized cobweb weaver. This spider shows remarkable variation in color. Most individuals are creamy white or yellow with distinct black bands on the ends of the femora, tibiae, and metatarsi. There is usually a thick dark line down the center of the cephalothorax, but some individuals have a completely black cephalothorax. The abdomen is pale or yellow. The abdomen may have no markings, a pair of lobed dark bands, or a completely black central area. In some individuals the abdomen is glossy white with two dull greenish or cream bands and a pattern of small black spots.

OCCURRENCE This is primarily a northeastern species, but there are scattered records in the West from Alberta and British Columbia to Arizona. It has been collected from bushes, herbaceous vegetation, and the low branches of trees. It is sometimes abundant in the understory of deciduous forests.

SEASONALITY Males: May through July. Females: May through September.

REMARKS This polymorphic species is sometimes found in the same areas as the polymorphic *Enoplognatha ovata*, creating an assemblage with extreme color and pattern variation. The female stays with her egg case in a loose web under a leaf until the young disperse.

Theridion murarium Emerton, 1882

Plate 28

IDENTIFICATION This is a small cobweb weaver. This species has a pale orange or cream-colored cephalothorax with a dark median band, widest in the eye region. There is a dark marginal band around the carapace. In some individuals the carapace is plain orange. The legs are light orange-banded with many thin black, reddish, or dark orange bands. The abdomen is mottled orange-brown with a light median band. This lobed median band has a white margin and orange center.

OCCURRENCE This spider is found throughout the United States and southern Canada. It has been collected on walls, fences, bushes, and the low branches of trees.

SEASONALITY Adults: February through October; may overwinter as an adult.

REMARKS The female has been found guarding her small light-colored egg case under leaves or a rock.

Theridula emertoni Levi, 1954

Plate 29

IDENTIFICATION This is a small cobweb weaver. The cephalothorax is pale yellow with a wide dark central band. The legs are pale yellow without bands. The abdomen has an odd shape, wider than long, with dark spots on the lateral lobes and a light spot in the center. Typically the base color is red or red-orange with a yellow central spot. The cephalothorax is sometimes completely black. Variations in abdominal color include a brownish base color with a white or cream-colored central spot. The male is entirely orange with a dark central band on the cephalothorax. He has an ovoid abdomen with a light yellow central spot.

OCCURRENCE This species occurs from Newfoundland to British Columbia in the North and from New England to Wisconsin in the South. They have been found in fields, bushes, and low vegetation in forests. The web is reduced to a sparse tangle under a leaf. This species has occasionally been found together in the same areas as *Theridula opulenta*.

SEASONALITY Adults: late May through August.

REMARKS The adults have sometimes been found together in one web. The female remains in her web with a small white egg case until the young emerge. Females have been observed carrying the egg case.

Theridula gonygaster (Simon, 1873)

Plate 29

IDENTIFICATION This is a small cobweb weaver. The cephalothorax is black. The legs are pale yellowish-green. The abdomen is usually wider than long with the lateral lobes more pointed than in the other two species in this genus. There is often a third point at the back of the abdomen. The abdomen is black with four or more yellow or white spots. The principal spots are arranged two in front and two behind usually farther apart. There are sometimes several smaller spots between these. The male has a light carapace with a dark central band. The abdomen of the male is dusky gray or blackish, often without a central spot.

OCCURRENCE This is primarily a tropical species with a worldwide distribution. In North America north of Mexico it has been found in southeastern Arizona as well as in southern Florida. This spider has been found in small webs under leaves in a variety of habitats.

SEASONALITY Adults: probably all year.

REMARKS Antonio Castineiras (1995) has observed that in Cuba this species is a frequent predator of a pest insect: the sweet potato whitefly.

Theridula opulenta (Walckenaer, 1841)

Plate 29

IDENTIFICATION This is a small cobweb weaver. The cephalothorax is black; the legs are pale yellow without bands. The abdomen is wider than long with lateral lobes. The color of the abdomen is dark bluish gray or nearly black. There is a small cream-colored or yellow spot on the center of the abdomen. The male is lighter in color than the female and has a dark band down the center of the carapace. The male has an ovoid abdomen with or without an indistinct pale central spot.

OCCURRENCE This species is primarily southeastern but there are scattered records from Utah and Oregon. They have been found in fields, bushes, and low vegetation in forests. The web is reduced to a sparse tangle under a leaf. This species has occasionally been found together in the same areas as *Theridula emertoni.*

SEASONALITY Adults: May through September.

REMARKS The adults have sometimes been found together in one web. The female remains in her web with a small white egg case until the young emerge. Females have been observed carrying the egg case.

Tidarren sisyphoides (Walckenaer, 1841)

Plate 28

IDENTIFICATION This is a medium-sized to large cobweb weaver. The coloration is extremely variable from dusky to light brown. Most often the cephalothorax is light brown with a dark median band widest in the eye region and a dark marginal band. The legs are light brown, banded with brown. The abdomen is light brown with dark mottling that often has an orange tint. There is a white line of variable width extending on the dorsal surface from near the spinnerets to the center of the abdomen. There are usually two other pairs of white lines on the sides. The male is a dwarf, growing to only a fraction of the adult female's size.

OCCURRENCE This is a tropical species that has been found in the southern and coastal states from North Carolina to California. The female builds a retreat under a curled leaf in her space-filling tangle web where she hides when not feeding and where she lays her egg case. These webs have often been found in rocky areas and on porches near buildings. This spider has been found in similar situations as *Parasteatoda tepidariorum.*

SEASONALITY Males: summer. Females: all year; egg cases in summer.

REMARKS As described by Jefferson Branch (1942), males of this species pull off one of their palps after molting into the subadult stage. He actually spins a small web, tangles one of his palps in it, then twists and pushes the palp until it breaks off. He later feeds on the remaining fluids in the palp. When the male molts into an adult, there is no trace of the palp. He has only one remaining, unusually large one. The male dies during mating and his body is removed and discarded, not eaten, by the female.

Yunohamella lyrica (Walckenaer, 1841)

Plate 28

IDENTIFICATION This is a small cobweb weaver. This species has a brown or grayish-brown cephalothorax with a lighter area behind the eyes. The legs are pale with brown banding that is faint in some individuals. The abdomen is brown or grayish-brown and has a whitish median band.

OCCURRENCE This is an eastern species. It has been found from Maine to Wisconsin and south to Texas. It has been collected from the understory of forests as well as on fences and mailboxes. This spider occasionally wanders into houses.

SEASONALITY Males: June and July. Females: June through August.

REMARKS Previously known as *Theridion lyricum.*

FAMILY THERIDIOSOMATIDAE • *Ray Orbweavers*

The name of this family is derived from the peculiar nature of their orb webs. The hub is modified so that groups of radii converge and each group is connected by one line, or ray, to a point where the spider waits, perched on a tiny silk platform. The spider, facing away from the orb, holds a silk line that extends from the hub out to a nearby attachment. The web is pulled into a cone shape and held under tension (see Fig. 25, left). This family has a worldwide, mostly tropical, distribution and includes 72 species. In North America there are only two species. The most common one builds it web near the ground in humid forest environments of the eastern states and provinces. The other species is found in the Gulf coastal plain.

Theridiosoma gemmosum (L. Koch, 1877) • *Ray orbweaver*

Plate 20

IDENTIFICATION   This small globular spider is usually identified by its unusual web. The web is a vertical orb pulled into a cone shape by the spider holding tension on a line that extends from the center of the hub to the substrate. The spider is relatively small and dark brown with a network of silvery-white markings on the abdomen. The cone-shaped orb web is about 10 cm (about 4 inches) in diameter and resembles an inverted umbrella in shape when the spider is hunting (see Fig. 25, left). A second species, *Theridiosoma savannum*, builds a horizontal web above leaf litter and has been found along the Gulf coastal plain.

OCCURRENCE   This species occurs in eastern North America, mostly in dark humid environments. The webs are built near waterfalls, creeks, wet rocky areas, moist forests near logs, and in low vegetation. Sometimes they build their web among the roots of fallen trees or under an eroded stream bank. Ray orbweavers usually hunt during the day.

SEASONALITY   Adults: spring and summer.

REMARKS   The spider sits upright in a tiny modified silk platform at the hub facing away from the orb. When a potential prey hits the web, the spider releases the tension on the web and it springs back ensnaring the prey. The distinctive egg cases are frequently more obvious than the spiders themselves. They are tan in color, about 3 mm in diameter, and attached to the substrate (branch, log, or rock) by a single silk line about 2 to 2.5 cm (1¼ inches) long. When empty egg cases are found, even during the winter, their presence reveals that this species occurs in the area.

FAMILY THOMISIDAE • *Crab Spiders*

These are among the most familiar spiders because of their habit of perching in flowers and ambushing arriving pollinators. The family is worldwide in distribution with 2,152 species. In North America north of Mexico there are 130 species. Half of these are in the genus *Xysticus*, with 67 species north of Mexico. Most crab spiders have compact globose bodies. The relatively short, thick legs are held in a laterigrade crablike stance. The first pair of legs is often armed with heavy raptorial spines that are used in prey capture.

One group of species is colorful and common in flowers. Some of these spiders are capable of changing color to match the background. Other crab spiders are variegated shades of brown and hunt in vegetation away from flowers or among debris on the ground. Some species are flat and live in cracks in the bark of trees.

Bassaniana versicolor (Keyserling, 1880) • *Bark crab spider*

Plate 72

IDENTIFICATION This is a small flat crab spider. The carapace of these spiders is very flat. The female is mottled brown. In contrast to most of the similar spiders of the genus *Xysticus*, the cephalothorax does not have much color contrast between the central region and the sides. The male has a black shiny cephalothorax. He also has black legs with some light rings on the ends. His abdomen is mottled with dark brown or black. The legs of spiders in this genus are proportionately longer than those of *Xysticus*.

OCCURRENCE This species occurs from New England west to Minnesota and south to Florida and Arizona. Other similar species are found throughout the United States and Canada. These spiders have been found under bark, on dead wood, under rocks, and in buildings.

SEASONALITY Adults: May through August.

REMARKS This spider's flat body and somber coloration closely matches the bark and wood where it is found. Previously known as *Coriarachne versicolor* or *lenta*.

Diaea livens Simon, 1876

Plate 74

IDENTIFICATION This is a medium-sized crab spider. The female has a green cephalothorax and legs. The abdomen is yellowish or green with a large red brown spot on the dorsal surface, rimed by white. The spot may be reduced to a red ring with a pale center. The male has a greenish or greenish-orange cephalothorax. The abdomen of the male has a variegated red-brown spot on the dorsal surface rimmed with white, or a series of small red spots on a pale background. The legs of the male are green with dark spots and pale or orange from the patellae to the tips.

OCCURRENCE This species has been found in California. This spider has been collected from the branches of oaks, usually in coast live oak habitat.

SEASONALITY Adults: December through June.

REMARKS Previously known as *Diaea pictilis*. The other species in the genus north of Mexico, *Diaea seminola*, was described from Florida and known only from the male. It has not been recorded since the original description.

Mecaphesa asperata (Hentz, 1847) • *Northern crab spider*

Plate 75

IDENTIFICATION This is a small to medium-sized crab spider. The members of this genus are generally spinier than either *Misumenoides* or *Misumena*. This is a particularly spiny crab spider with stiff spines on the legs, carapace, and abdomen. The cephalothorax is yellowish or pale and usually has two red or brown stripes extending back from the lateral eyes. The eye area is light-colored. The abdomen is pale or yellowish with red spots down the center, forming a V converging at the back. There are usually also heavier red stripes on the sides of the abdomen. The front legs are banded with red or merely darker at the ends.

OCCURRENCE This species has been found from southern Quebec west to southern British Columbia and south into the United States from Arizona to Florida. It has been found in fields and field edge scrub. They also occur in the low branches of trees in both deciduous and coniferous forests.

SEASONALITY Adults: early spring through autumn.

REMARKS The female lays her eggs in a folded leaf and stays nearby guarding them. Previously known as *Misumenops asperatus*.

Mecaphesa celer (Hentz, 1847) • *Celer crab spider*

Plate 75

IDENTIFICATION This is a small to medium-sized crab spider. The members of this genus are generally spinier than either *Misumenoides* or *Misumena*. The color is variable from pale greenish or yellow to individuals with dark markings. The cephalothorax usually has two wide dark stripes extending back from the lateral eyes. The female's abdomen has a few dark or black spots that converge near the rear, or is marked with red spots and lateral stripes. The male's abdomen has a dark spot at the front and two lines of dark spots forming a V that converge at the back. The legs of the female are only occasionally banded with red. The front legs of the male have distinct bands.

OCCURRENCE In Canada this species is known from southern British Columbia, Alberta, and Saskatchewan. It has been found throughout the conterminous United States but is less common in the Northeast. Most of the records of this species are from field habitats, including agricultural crops. Some have also been collected from the low branches of trees.

SEASONALITY Adults: early summer through autumn.

REMARKS Previously known as *Misumenops celer*.

Mecaphesa lepida (Thorell, 1877)

Plate 75

IDENTIFICATION This is a small to medium-sized crab spider. The members of this genus are generally spinier than either *Misumenoides* or *Misumena*. The carapace in the female is light green, pale in the eye area and the center of the cephalothorax. The abdomen is white or yellowish. There is often a pattern of red spots on the dorsal surface of the abdomen forming a V that converges at the back. There are also red spots along the lateral edges of the abdomen. The legs are not banded. The male has light orange carapace with broad red or brown stripes extending back from the lateral eye region to the back. His abdomen is white with two dark brown lines at the back. The first two pairs of legs of the male are long and have red bands.

OCCURRENCE This species occurs in the West from British Columbia south to Mexico. There are records as far east as Utah and Nevada.

SEASONALITY Adults: March through September; mountain records are from April and May.

REMARKS Previously known as *Misumenops lepida*.

Misumena vatia (Clerck, 1757) • *Goldenrod crab spider*

Plate 74

IDENTIFICATION This is a medium-sized to large crab spider. It resembles *Misumenoides formosipes*. The key difference, visible with a good hand lens, is a white ridge in the middle of the clypeus, absent in this species but present in *Misumenoides*. Both of these spiders have various color varieties. *Misumena vatia* is typically white or yellow with reddish or orange in the eye region.

The abdomen often has two pink or red bands along the sides. The male has a dark brown or reddish-brown cephalothorax, lighter in the eye region. His first two pairs of legs are long and dark reddish-brown. The rear legs are pale. The abdomen of the male is pale, yellow or white, with two reddish lines at the back.

OCCURRENCE This species occurs across southern Canada and the United States. It is most often found in fields and occasionally flowering shrubs. The spider waits at a flower for visiting pollinators that become prey.

SEASONALITY Adults: May through August; in the North this spider may take two years to mature.

REMARKS This spider can change color from yellow to white and back. This ability enhances their camouflage by helping the spider to match the flowers where it waits for prey. We know a great deal about this spider because of the careful experiments conducted by Douglass Morse and summarized in his book (Morse 2007).

Misumenoides formosipes (Walckenaer, 1837) • *Whitebanded crab spider*

Plate 74

IDENTIFICATION This is a large crab spider. It resembles *Misumena vatia*. The key difference, visible with a good hand lens, is a pale or white ridge (carina) in the middle of the clypeus on the whitebanded crab spider, hence its name. This spider is remarkably variable in color. The cephalothorax of the female is usually lighter in color in the center and darker, often greenish at the sides. The basic color of the cephalothorax and abdomen is yellow, white, or light pink. The abdomen usually has dark markings. These are sometimes two rows of spots, or bands. These marks often form a V-shape (open at the back). There may also be dark marks on the sides of the abdomen. The legs are sometimes dark near the joints. The male may have a green, yellow, brown, or orange cephalothorax. His abdomen is often a different color, yellowish or orange. The male has long dark front legs and pale hind legs.

OCCURRENCE This species occurs across southern Canada and the United States. It is usually found in fields or shrubby edges of fields. It waits at flowers for potential prey.

SEASONALITY Adults: summer through September.

REMARKS Like the similar goldenrod crab spider, the whitebanded crab spider can change color. Previously known as *Misumena aleatoria*.

Misumessus oblongus (Keyserling, 1880)

Plate 75

IDENTIFICATION This is a small to medium-sized crab spider. It can be distinguished from the species in *Mecaphesa* by having few spines on the cephalothorax and none on the abdomen (female). It can be distinguished from *Misumenoides formosipes* because it lacks the white ridge (carina) on the clypeus of that species. The female's cephalothorax is usually pale green without bands. Her abdomen is creamy white, only rarely with red marks on the sides. The legs are usually pale green without spots or rings. The male has short spines on the abdomen and sometimes a red margin on the cephalothorax.

OCCURRENCE This species is known in Canada only from Ontario but is found across the United States. It is more common in the southern states. This species has been found in fields and prairies.

SEASONALITY Adults: summer through early autumn.

REMARKS Previously known as *Misumenops oblongus*.

Ozyptila americana Banks, 1895 • *Leaflitter crab spider*

Plate 72

IDENTIFICATION This is a small crab spider. The cephalothorax has a broad light brown band enclosing the entire eye area and extending back from the anterior lateral eyes to the back of the carapace. The sides of the carapace are dark brown. The lateral eyes are larger than the median eyes. The abdomen is light brown with broken dark brown bands at the back. The front legs are blotched with brown on the femora and banded on the other segments. The rear legs are lightly spotted and banded.

OCCURRENCE This is an eastern species occurring from southeastern Canada and New England west to Iowa, and south to Virginia, Mississippi, and Texas. Other species occur throughout North America. This species has been found in protected areas near the ground, under bark, rocks, or logs. It has been recorded from forests, hawthorn thickets, fields, and swamps.

SEASONALITY Adults: May through October.

Synema parvulum (Hentz, 1847)

Plate 72

IDENTIFICATION This is a small crab spider. The female has a smooth orange cephalothorax lighter around the eyes. Her abdomen is bicolored with the anterior three-quarters yellow and the posterior one-quarter black. The male has a dark orange or brown cephalothorax and a brown or bicolored abdomen. In both sexes the front two pairs of legs are dark brown or black. The rear legs are light greenish or yellowish.

OCCURRENCE This spider has been found from New Jersey west to Kansas and south to northern Florida and New Mexico. It has been collected from low vegetation with a sweep net. It hunts in or near flowers.

SEASONALITY Adults: May through November.

Tmarus angulatus (Walckenaer, 1837)

Plate 72

IDENTIFICATION This is a small crab spider. The cephalothorax is highest at the posterior eyes and slopes forward, and with the chelicerae gives the spider a distinctive face. The abdomen is tallest at the squared-off posterior end. There is a prominent rounded point or tubercle above the spinnerets. The color is mottled-brown or grayish-brown.

OCCURRENCE This species occurs across the United States and southern Canada. This spider rests on twigs of shrubs and the small branches of trees.

SEASONALITY Males: May through August. Females: May through November.

REMARKS This spider rests on twigs with its four front legs extended. In this pose it is well camouflaged. Immature individuals have been captured in the forest canopy.

Xysticus alboniger Turnbull, Dondale, and Redner 1965

Plate 72

IDENTIFICATION This is a small crab spider. The cephalothorax is black or reddish-black with a thin light rim in the thoracic region. The abdomen is uniform gray. The legs are black with pale joints and tarsi.

OCCURRENCE This species occurs from southern Ontario to North Dakota and south to northern Florida. This spider has been collected in pitfall traps in forests and open habitats as well as from low vegetation.

SEASONALITY Adults: May through June.

REMARKS Previously known as *Synema bicolor*.

Xysticus auctificus Keyserling, 1880

Plate 73

IDENTIFICATION This is a small crab spider. This species is pale brown. The cephalothorax is light around the eye area and in a broad band toward the back. The sides of the cephalothorax are brown. There are two thin white lines on either side of the pale band. There are three distinct dark spots at the back edge of the carapace. The abdomen is pale cream in color with or without dark spots. The legs are tan with tiny brown spots.

OCCURRENCE This species occurs between southern Alberta and Saskatchewan west to states bordering the Great Lakes and south to Arizona and Georgia. It has been collected in and around fields in low vegetation or on the ground.

SEASONALITY Adults: May through September.

Xysticus californicus Keyserling, 1880

Plate 73

IDENTIFICATION This is a medium-sized crab spider. The cephalothorax is pale in the center and dark on the sides. The abdomen is pale with faint paired brown spots separated by light lines. The front two pairs of legs have three dark marks on the femora and dark lines along the tibiae. The hind legs are banded. The male is dark reddish brown with a pale median band of variable width on the abdomen. The femora and patellae of the front legs of the male are uniformly dark, nearly black.

OCCURRENCE This is a western species occurring from British Columbia to the Mexican border. It has been found both at low elevations near the coast and in the mountains.

SEASONALITY Adults: March through July at low elevations; May through July in the mountains.

Xysticus elegans Keyserling, 1880 • *Elegant crab spider*

Plate 73

IDENTIFICATION This is a medium-sized crab spider. The cephalothorax is brown with a lighter area in the middle that is sometimes outlined in white. There is a light tan band through the eye region. The abdomen is has brown spots outlined in white. The legs are brown, the femora and patellae are nearly black in males. The female is paler than the male.

OCCURRENCE This species is widespread across southern Canada and south to New Mexico and Georgia. This spider has been found on or near the ground, under leaf litter, logs, and rocks. It has also been found in low vegetation in both fields and forests.

SEASONALITY Adults: April through November.

Xysticus ferox (Hentz, 1847)

Plate 73

IDENTIFICATION This is a medium-sized crab spider. The female is variegated light brown with a pale band down the center of the cephalothorax, lightest in the back and flanked by darker brown. There are often two light lines within the thoracic dark bands. The abdomen is palest in the center and has two or three transverse light bands with black edges. The legs are light brown. The male is darker brown on the cephalothorax with a light line through the anterior eye region. The abdomen of the male is darker brown with a light rim and two or three white bands. The legs of the male have dark brown femora, patellae, and the tips of the tibiae. The metatarsi and tarsi are light.

OCCURRENCE This species occurs across Canada. It is primarily eastern in the United States south to Texas and Georgia. This spider is often found under cover on the ground or in low vegetation.

SEASONALITY Adults: May through September.

REMARKS This spider often hunts near flowers.

Xysticus pretiosus Gertsch, 1934

Plate 73

IDENTIFICATION This is a small crab spider. The eye region and central cephalothorax are tan, with the head region outlined in white. The sides of the cephalothorax are brown. The abdomen is light brown with thin brown spots at the back highlighted by white lines in front of each. The legs are tan with scattered spots, and the rear legs have rings at the ends of the segments.

OCCURRENCE This species occurs along the West Coast from British Columbia to the Mexico border. This spider has been collected from low vegetation including cultivated plants. It has occasionally been found in shrub habitats including chaparral.

SEASONALITY Males: May and June. Females: May through August.

FAMILY TITANOECIDAE · Rock Weavers

In North America north of Mexico this family is represented by only 4 species, but worldwide there are 44. Together, these four species occur throughout most of the continent except Florida. They are medium-sized spiders that build thin cribellate sheet webs usually under rocks. They look much like small members of the hackleband weavers (Amaurobiidae). The abdomen is usually without any pattern but may have an iridescent sheen. They are typically found in relatively drier habitats than hackleband weavers.

Titanoeca nigrella (Chamberlin, 1919)

Plate 34

IDENTIFICATION This species has an orange-brown cephalothorax and legs with dark femora. The legs are hairy, the femora densely so, but they lack heavy spines. The abdomen is dark gray above. In the other species there are a pair of light spots on the underside of the abdomen; in *nigrella* the abdomen is uniform dark below. In good light the abdomen frequently shows an iridescent sheen.

OCCURRENCE This species is found in the western half of the continent. Other species are found in the East as well as farther north. They have been found in thin webs under rocks. Unlike the somewhat similar members of the hackledmesh weavers (family Amaurobiidae), these spiders prefer drier habitats.

SEASONALITY Adults: May through October.

REMARKS These are cribellate spiders and their thin webs often have a faint bluish look.

FAMILY TRECHALEIDAE · *Longlegged Water Spiders*

This family contains 72 species, mostly from tropical America. There is only one species that occurs north of Mexico. These spiders are relatively large and use a laterigrade stance near desert streams in the Southwest. They have flexible tarsi that often appear curved. Like the somewhat similar nursery web spiders (Pisauridae), longlegged water spiders are adept at moving on the water surface. They may also climb down under the water surface by using emergent vegetation or rocks for traction.

Trechalea gertschi Carico and Minch, 1981

Plate 71

IDENTIFICATION This is a large crablike spider. The body is moderately flat with long legs held in a somewhat crablike stance. When waiting near the water, they occasionally adopt the posture illustrated, with the legs extended sideways. The body is mottled light brown or grayish brown. The posterior eye row is nearly straight, not as strongly recurved as in *Dolomedes*, which are otherwise superficially similar spiders. The tarsi are fringed, flexible, and often held in a curved position.

OCCURRENCE This species occurs in the Gila River drainage in both Arizona and New Mexico and south into Mexico. The spiders cling to rocks above the water, and they may even walk under the surface clinging to rocks.

SEASONALITY Adults: July through January.

REMARKS This spider runs over the surface of the water skillfully. The female carries the egg sac attached to the spinnerets.

FAMILY ULOBORIDAE • *Hackled Orbweavers*

The family Uloboridae is most diverse in tropical latitudes, relatively few species are found in the temperate zone. There are 16 species north of Mexico. The webs are typically horizontal orb webs, but some species build reduced webs. These spiders possess a cribellum and calamistrum and spin hackle-band adhesive silk for the sticky spiral part of the orb. Members of the Uloboridae are unusual because they possess no venom glands and do not use poison to subdue their prey. They wrap the prey in many layers of silk, forming a thick cocoon, and do not bite until they are ready to begin feeding. This use of silk may seem wasteful, but it is not because the spider eventually consumes the silk along with the prey. Like all spiders, hackled orbweavers spit out liquid onto the prey during feeding. This fluid includes digestive enzymes, which liquefy the prey. The spider then sucks up the resulting soup. The extensive wrappings of silk may assist in holding the gushy mass together during feeding.

Most of our species are annual, but some adult females are thought to survive into a second year. Hackled orbweavers are often tolerant of other individuals nearby. Members of the genus *Philoponella* form communal aggregations, with many individuals living in a large aggregation sharing the support frame strands around their individual orb webs. Individuals in these congregations often defend their own orb from others in the group. Many uloborid spiders add decorations, such as stabilimenta, to their webs. In the featherlegged orbweaver the adults often have a linear stabilimentum while the young construct a spiral-shaped one. A few other types of uloborids build reduced webs. The triangle weaver builds a web that resembles a pie slice–shaped portion of an orb. The Mexican stickspider captures prey using a single line stretched between shrubs.

Hyptiotes cavatus (Hentz, 1847) • *Triangle weaver*

Plate 19

IDENTIFICATION This spider has a small round body with a series of lumps on the abdomen, and with the eyes on raised tubercles on the cephalothorax. It has variegated markings that make it resemble a flake of bark or a bud. This spider can be identified by its triangular web. The web consists of a pie slice–shaped wedge of an orb web. Four radial strands are connected at one point to a single line that extends at one end to a support. At the other end each of the four radial strands are connected to a simple frame of silk. There are a series of hackled-silk strands stretched between the four radials in the same manner as a portion of a spiral orb in other hackled orbweavers.

OCCURRENCE This species occurs throughout much of the eastern half of the United States. The other species are distributed further north and in the western part of the continent. The webs are usually built among dead branches in the understory of forests.

SEASONALITY Adults: summer and autumn.

REMARKS The hunting spider waits, holding the web stretched under tension, near the end of the single line. The spider is actually holding onto the line close to the attachment point with its hind legs; the front legs are holding the taut line and a loop of slack line above the spider. When a potential prey strikes the web, the spider reacts by releasing the web so that it collapses somewhat, snaring it. The spider may jerk the web one or more times to further entangle the prey.

Miagrammopes mexicanus O.P.-Cambridge, 1893 • *Mexican stickspider*

Plate 19

IDENTIFICATION This is an unusual-looking spider. They resemble a small twig or stick. The body is thin and elongate. There are only four eyes in one row, where the cephalothorax is widest. The long first pair of legs are usually held extended forward and have dense setae that make them appear thick. The second and third pairs of legs are thin and much smaller. The fourth pair of legs are long and held extended backward parallel to the abdomen.

OCCURRENCE This spider is a member of a worldwide tropical group. This species occurs in Mexico and just extends into the Rio Grande Valley of Texas.

SEASONALITY Nothing is published.

REMARKS This spider builds the simplest of webs, a single line stretched between two shrubs. The strand is decorated with a patch of white flocculent silk that is attractive to potential prey who land there. The spider, resting near one end of the line, pulls it taut and holds the slack in a loop over the body. When an insect alights on the line, the spider releases the tension and the vibration causes the prey to become entangled.

Octonobus sinensis (Simon, 1880)

Plate 19

IDENTIFICATION This is a large hackled orbweaver that is tan in color with white bands near the tips of the femora of the legs. When hanging in the web awaiting potential prey, they hold their front legs separated widely or only meeting at the tips. Other hackled orbweavers hold their front legs together from the patellae to the tips. The second pair of legs are also held separated. From the side view there are four pairs of tubercles along the dorsal surface of the abdomen.

OCCURRENCE This spider is found in the southeastern part of the United States as well as eastern Asia. It has often been found near buildings and in greenhouses. It was probably accidentally introduced into North America from Asia.

SEASONALITY Adults: summer.

REMARKS Previously known as *Uloborus octonarius*.

Philoponella oweni (Chamberlin, 1924)

Plate 19

IDENTIFICATION This species is often found with the legs held in a folded, crouched position. There are either well-demarcated black and white areas on the abdomen as shown in the plate, or the spider is unmarked. The spider huddles in the center near the two stabilimenta. The webs are often built in shaded situations, and this combined with the posture makes the spider inconspicuous.

OCCURRENCE This is a species of the southwestern United States. It has been collected from shady areas in desert and mountain canyons.

SEASONALITY Adults: May through October.

REMARKS These spiders are sometimes found in communal webs with other individuals, often immatures. The females hold the egg case with their legs until the young emerge. Previously known as *Uloborus oweni*.

Siratoba referens (Muma and Gertsch, 1964)

Plate 19

IDENTIFICATION This is a small gray or light brown hackled orbweaver with a compact body. The front two pairs of legs are typically extended forward and held together. There is one hump just anterior to the center of the abdomen.

OCCURRENCE This species is known from Arizona and northern Mexico. There is one other member of this genus also known from the desert Southwest.

SEASONALITY Adults: late summer.

Uloborus glomosus (Walckenaer, 1842) • *Featherlegged orbweaver*

Plate 19

IDENTIFICATION This is a medium-sized spider with conspicuous tufts of hairlike spines on the tibiae of the front legs. This species is typically beige in color and has four conspicuous humps on the front half of the abdomen. While waiting at the hub, the first two pairs of legs are extended and held together. When feeding, the spider holds silk-wrapped prey close to the face using the palps; the front pairs of legs are held spread apart to either side.

OCCURRENCE This species is widespread in the eastern part of North America, other species in this group are found in the West.

SEASONALITY Adults: spring through late August.

REMARKS This species builds its horizontal web in low vegetation. There are often many individuals with webs nearby. The web may contain a stabilimentum in the form of a single line, the spider hanging underneath. Juveniles may build a spiral-shaped stabilimentum. The egg case is an irregular lumpy shape; the same color as the spider and often placed near the web.

Zosis geniculata (Olivier, 1789)

Plate 19

IDENTIFICATION This is a large hackled orbweaver with both black and white banding on the legs. The coloration varies from gray to brown or yellowish. The thoracic part of the cephalothorax is considerably wider than the head portion. There is usually a prominent hump near the front of the abdomen. The front pair of legs are typically held outstretched, together from the patellae to the tip. The second pair of legs are spread. The third pair of legs are held out to the side.

OCCURRENCE This is a tropical species, also known from the Gulf Coast region and Florida. This species is often associated with buildings, barns, and grain-storage facilities.

SEASONALITY Nothing is published.

REMARKS Many individuals often build their webs near each other. This species may decorate the web with a thin mat of silk at the hub or circles just around the central area.

FAMILY ZODARIIDAE · *Zodariids*

The five species of spiders in this family are rarely encountered in North America. There are two distinct genera. One, *Zodarion*, is represented by a small Eurasian species that has probably been accidentally transported by humans to our region. The other, *Lutica*, is represented by a group of species found along the coast of southern California. The family is most easily distinguished by the odd pattern of the spinnerets. The anterior spinnerets are greatly enlarged and extend beyond the end of the abdomen. The other spinnerets are reduced and sometimes difficult to see.

These are sand-loving spiders. They build retreats coated with sand grains. In *Zodarion* the retreat is a small globular one that has been found under rocks. In *Lutica* the spider constructs a silken tube within the sand.

Lutica maculata Marx, 1891

Plate 47

IDENTIFICATION This spider has long cylindrical posterior spinnerets. The cephalothorax is bulbous in the head region, highest behind the two rows of small, closely spaced eyes. The legs are without banding. The tarsi lack claw tufts. The abdomen is gray with a pattern of dark markings.

OCCURRENCE This spider occurs only in the coastal sand dunes of southern California, Baja California, and the California Channel Islands. It is one of four similar species in this region. According to Martin Ramirez (1995), it builds its sand tube near dune vegetation and leaf litter.

SEASONALITY Little is known; males have been observed wandering in search of females from May through November.

REMARKS The unusual burrow of this spider has a sand-covered extension at the surface. The burrow itself enters the sand at a 45° angle but may extend vertically down into the sand. They are approximately 15–30 cm deep. According to Ramirez, there may occasionally be a delicate sand-covered flap. The spiders typically capture their prey by lunging through the thin wall of the silk-lined sand tube to capture prey walking on the surface.

Zodarion rubidum Simon, 1914

Plate 47

IDENTIFICATION This is a small spider with an orange cephalothorax. The femora are the same color but the other leg segments are paler. The abdomen is gray, lighter on the sides. The anterior median eyes are much larger than the others. The anterior spinnerets extend beyond the end of the abdomen.

OCCURRENCE This is a European species, it has been found in Pennsylvania, Ohio, Quebec, Colorado, and British Columbia. These inconspicuous spiders may yet be discovered in other localities in North America. A similar undescribed species has been noted in California. During the day these spiders hide under rocks in a small silk-lined chamber that is covered with sand. If disturbed during the day, they run rapidly and hide. At night they emerge to hunt ants. They have often been captured in pitfall traps.

SEASONALITY Adults: summer.

REMARKS According to Paula Cushing and Richard Santangelo (2002), these little spiders eat exclusively ants. The spider dashes out and bites the ant, often on a rear leg, then waits until the ant is paralyzed before approaching again. In about half of the observed attacks, the spider bit the ant more than once. After the ant is immobilized, the spider carries the ant to a protected place and feeds.

FAMILY ZORIDAE · *Zorids*

This family contains 65 species distributed in Eurasia, Australia, New Zealand, and tropical America. Only two species, both in the genus *Zora*, are found in North America north of Mexico. The cephalothorax is nearly as wide as it is long and has two parallel dark longitudinal stripes. The anterior eyes form a nearly straight line, the posterior row is so strongly recurved that there appear to be three rows. The central four eyes form a trapezoid superficially similar to the eye arrangement of the Ctenidae. The most distinctive feature of the family is a group of paired spines on the underside of the tibiae and metatarsi of the first two pairs of legs. The western species lives near the ground, often in the leaf litter; the eastern species hunts in low vegetation.

Zora pumila (Hentz, 1850)

Plate 37

IDENTIFICATION This is a small, light brown spider. The cephalothorax has two longitudinal dark stripes as well as a dark marginal band. The posterior eye row is so recurved that there appear to be three rows of eyes. The tibiae of the front four legs are distinctly darker than the other segments. There are a series of paired overlapping spines on the ventral surfaces of the tibiae and metatarsi of the front four legs. The abdomen is spotted down the center. The sides of the abdomen have dark longitudinal bands or scattered dark spots.

OCCURRENCE This species is eastern, but there is a similar one in the West. These spiders have been found in both wooded and open habitats. They are active diurnal hunters either in vegetation or on the ground. They have most often been found hunting is small shrubs. This species has occasionally been found indoors.

SEASONALITY Males: April through July. Females: January through October.

REMARKS Females protect their egg case but do not build a retreat.

FAMILY ZOROCRATIDAE · *Zorocratids*

There are fewer than 50 species worldwide. Most species are Mexican, but related spiders have been described from Africa and Madagascar. This family is represented by six species of medium-sized spiders north of Mexico. Some taxonomists consider this a subgroup of the family Zoropsidae. The North American species build loose sheets or tangles of cribellate silk under rocks, logs, or among the cracks in lava fields. Little has been published about the biology of these spiders.

Zorocrates unicolor (Banks, 1901)

Plate 35

IDENTIFICATION This is a moderately large spider with a pale brown cephalothorax. The legs are similar in color to the cephalothorax and lack any markings. It has a gray abdomen. They are darker in the face region with robust jaws.

OCCURRENCE This species is known from Arizona, Texas, and Mexico. Most records are from southeastern Arizona. Individuals have been found in a loose cribellate sheet or tangle under rocks in desert canyons, oak, pine, and riparian woodlands.

SEASONALITY Adults: all year.

REMARKS Males have been captured wandering away from their retreats at night. Previously known as *Chemmis unicolor*.

FAMILY ZOROPSIDAE · *Zoropsids*

The composition of this family is somewhat controversial. Some believe that it should be combined with the Zorocratidae. Taken separately, the family contains 78 species worldwide. There is only one species known for North America north of Mexico and was introduced. This species (*Zoropsis spinimana*) is native to the Mediterranean region. This is a large, ground-living wandering spider. It frequently strays into houses. The general body form, size, and habits resemble a wolf spider. The eyes of this spider are in two rows of four with the posterior row recurved. The posterior median eyes are closer together than the length of the anterior row (see Fig. 20D). This contrasts with the eye arrangement of true wolf spiders (Lycosidae), where the posterior median eyes form a row nearly as wide as the anterior eyes (see Fig. 20B).

Zoropsis spinimana (Dufour, 1820)

Plate 46

IDENTIFICATION This large, tan-colored spider is occasionally mistaken for a wolf spider. The eyes are arranged differently. They form two rows of four; the posterior row is only slightly recurved. The abdomen has a distinctive pattern of spots around a dark heart mark in the front, and thin transverse lines in the back. These are cribellate spiders but the cribellum and calamistrum are often difficult to see without magnification.

OCCURRENCE This species is native to the Mediterranean region and was probably accidentally introduced into central California. The first records were from the San Francisco Bay region in the autumn of 1992. This spider seems to be slowly expanding its range in central California. It has often been found wandering in buildings.

SEASONALITY Adults: autumn, winter, and spring.

REMARKS This spider occasionally assumes a defensive posture when disturbed, raising its front legs and spreading its fangs. It is not actually aggressive and will run away if given a chance.

GLOSSARY

ABDOMEN The back portion of a spider's body, behind the pedicel.

ANAL TUBERCLE A fleshy lobe at the rear of the abdomen that bears the anus (anal opening) at the end of the digestive tract.

ANTERIOR EYE ROW (AER) The four eyes at the front of the head.

ANTERIOR LATERAL EYES (ALE) The outermost eyes in the front row.

ANTERIOR MEDIAN EYES (AME) The innermost eyes in the front row.

ANTERIOR SPINNERETS (AS) The spinning organs in the front row.

ARACHNID Any member of the class Arachnida.

ARACHNIDA (CLASS) The formal taxonomic category including the members of eleven orders in North America, including micro whipscorpions, spiders, whipspiders, vinegaroons, scizomids, hooded tick spiders, mites and ticks, harvestmen, scorpions, pseudoscorpions, and wind scorpions.

ARANEAE (ORDER) The formal taxonomic category to which spiders belong.

ARANEOMORPH A spider in the infraorder Araneomorphae. This group constitutes most spiders, more than 95 percent of the species in North America north of Mexico. One distinctive feature is jaws that oppose each other and move in a pinching action.

ASPIRATOR A device for catching spiders composed of two tubes and a chamber; this device is also called a pooter.

BERLESE FUNNEL An apparatus used to extract small animals from leaf litter that consists of a funnel over a collection container with a light or heat source placed above. The animals migrate down, away from the light, and drop into the collecting container.

BOOK LUNGS Respiratory structures on the underside of the abdomen.

BOSS A reinforced area at the lateral edge of the cheliceral base where it meets the carapace.

BRIDGE LINE The line of silk that spans a gap between the upper points of attachment of a spider's web. It is usually the first part of an orb web built by a spider.

BRISTLE A short spine or seta, usually macroseta, that is shorter than the diameter of the segment it is on.

CALAMISTRUM One or two rows of uniform curved spines on the metatarsi of the fourth legs of certain spiders used to comb out cribellate silk.

CARAPACE The part of the exoskeleton that covers the top of the cephalothorax.

CEPHALOTHORAX The front portion of the body of a spider, including the fused head (cephalic) region and the thoracic region. The mouthparts, palps, and legs are attached to the cephalothorax.

CHELICERA (CHELICERAE) The principal mouthparts of an arachnid, with two parts: the cheliceral base and the moveable fang.

CHELICERAL BASE The bulky basal portion of the chelicera to which the fangs are attached. Also known as the paturon.

CHEMOSENSORY Referring to a chemical sense such as taste or smell.

CLASS A higher taxonomic category including many similar organisms. A class may include many orders. For example, the class Arachnida includes eleven orders in North America. The next larger category, the phylum, may include many classes.

CLYPEUS The region of the face of a spider between the anterior median eyes and the edge of the carapace in front. This is roughly the space between the front eyes and the jaws. If this space is taller than the space between the anterior median eyes and the posterior median eyes, the clypeus is said to be "high."

COLULUS A fleshy lobe, sometimes reduced to a pair of setae that represents a nonfunctional cribellum. If present, it is located between the anterior spinnerets.

COXA (COXAE) The first short segment of the leg, closest to the body.

CRIBELLATE SILK WEB A web composed of silk from the cribellum, often called hackle-banded silk. Webs made of this silk sometimes have a bluish appearance.

CRIBELLUM A silk-producing plate near the end of the abdomen in some spiders. When present, it is located just in front of the spinnerets. It is a modified form of spinneret with dozens of tiny spigots, where very fine silk fibers are produced. The fine silk from this structure is combed out by the spider using a brush of specialized spines on the metatarsi of the fourth legs called the calamistrum. Thus spiders with a cribellum also have a calamistrum on each fourth leg.

CRYPTOZOIC An inconspicuous lifestyle. This may infer life in dark, hidden, or inaccessible spaces or activity only at night.

CUTICLE The outer layer of the body covering. This part of the exoskeleton is usually covered with a waxy substance that helps to reduce water loss as well as prevent entanglement with sticky silk.

CYMBIUM The spoon-shaped structure of the tarsus of a male spider's palp that covers the dorsal surface.

DISTAL An adjective referring to a structure farther from the center of the body.

ECRIBELLATE Spiders that do not possess a cribellum.

EMBOLUS The syringe-like structure on the palp of a male spider through which the seminal fluid is injected into the female's reproductive opening.

ENDITES The bases of the coxae of the palps.

EPIGASTRIC FURROW A side-to-side groove that marks the front portion of the underside of the abdomen. In most spiders it includes the openings to the book lungs and reproductive ducts. Typically the book lung covers and the epigynum lies in front of this groove.

EPIGYNAL SCAPE A projecting point, hook, lump, or other process that extends outward from the plate covering the reproductive openings in a female spider.

EPIGYNUM A hard plate that covers the reproductive openings in a female spider.

EYESHINES The pinpoint reflections from the eyes of a spider when illuminated with a flashlight.

FEMUR (FEMORA) The largest of the leg segments in most spiders. It is the first long leg segment, attached to the body by two relatively short segments (the coxa and trochanter), forming a very flexible joint.

FOLIUM A contrasting colored marking on the center of the abdomen, typically darker than the surrounding parts of the abdomen.

FOVEA (THORACIC FURROW) A distinct pit or groove on the upper surface in the thoracic part of the cephalothorax near the center, usually just behind the head region.

FRAME The part of an orb web that surrounds the main circular catching area; composed of dry or nonsticky silk.

FREE ZONE The open spaces of an orb web between the hub and the sticky spiral.

GENITALIA The reproductive parts of a spider. In the male these consist of the inconspicuous openings to the reproductive ducts in the abdomen as well as the enlarged tarsi of the palps. In the female these are the openings to the reproductive ducts as well as the elaborate plate (epigynum) that covers them in most spiders.

GENUS The taxonomic category that includes a number of closely related species. The generic name is the first part of the "binomial" Latin name given to each species.

GONOPORE A spider's genital opening. The gonopore is at the back edge of the epigastric furrow, in the center.

GRAVID An individual female spider that has a large egg mass developing in the swollen abdomen.

HACKLED, HACKLE-BANDED The unusual fluffy line composed of many fibers of very fine silk from the cribellum and combed out by the calamistrum.

HAIRS The smaller, less robust setae (technically microsetae).

HEAD The front region (cephalic region) of the cephalothorax or carapace. Sometimes demarcated from the thorax by a cervical groove.

HEART MARK A distinctive marking, either light or dark, on the upper surface of the abdomen. It is directly over the place within the abdomen where the heart is suspended by muscles.

HARVESTMEN A member of the order Opiliones, not spiders; aka daddylonglegs, harvesters.

HUB The central part of an orb web where the radial lines converge. There may be a meshlike platform, a series of circular lines, or even a hole here.

INSTAR One of the growth stages of a spider. At each molt the spider grows, the steps between each molt are referred to as instars.

LAMELLIFORM SETAE The unusual leaf-shaped hairs of the claw tufts in ghost spiders (Anyphaenidae).

LATERIGRADE Having legs, particularly the femora, that are twisted so that the front surface becomes the top. The curve created by the successive segments of the legs gives the spider a crablike appearance.

LEAF LITTER The accumulated fallen dead leaves that cover the soil, most evident in forested areas but present to some extent almost everywhere.

LEG The walking or weight-bearing appendages, as opposed to the palps.

LINYPHIID A spider in the family Linyphiidae.

MACROSETA (MACROSETAE) Often called spines, these are relatively robust structures extending from a socketlike structure on the surface of an arthropod.

MASTIDION (MASTIDIA) A process, projection, or bump on the chelicerae or jaws.

MEDIAL OR MEDIAN An adjective referring to a structure near the center of the body.

MEDIAN OCULAR AREA (MOA) Also called the median ocular quadrangle, this is the trapezoidal region demarcated by the four median eyes.

MEDIAN SPINNERETS (MS) The spinning organs in the middle, usually small.

METATARSUS (METATARSI) Also known as the basitarsus, this is the sixth of the seven segments of the walking legs, the next-to-last segment. The palps lack metatarsi.

MICROSETA (MICROSETAE) Often called hairs, these are the numerous small, flexible extensions of the exoskeleton of arthropods.

MITE A member of the order Acari, the most diverse arachnid group.

MOLT The process of replacing the exoskeleton with a new larger one during growth.

MYGALOMORPH A spider in the infraorder Mygalomorphae. These are mostly relatively large burrowing spiders with jaws that close parallel to each other, striking downward.

ORB WEB A flat circular web consisting of a frame, hub, radial spokes, and spiral circular strands.

ORDER The taxonomic level in a classification scheme that includes many families.

OVIDUCT The tubelike passageway for the eggs when laid by a female spider.

PALP (PEDIPALP) The small six-segmented appendage found at the front of the body arising on either side of the mouthparts in a spider. The palps are sensory organs and function somewhat like the antennae of insects. In adult male spiders the last segment of the palps, the tarsus, is modified to hold and transfer the sperm.

PATELLA (PATELLAE) The fourth part of a leg, between the femur and tibia.

PEDICEL A stalk connecting the cephalothorax and the abdomen, containing the digestive tube, nerves, breathing tubes, and blood vessels.

PHEROMONE A chemical communication substance; these chemical compounds are perceived as odors and are usually related to sexual attraction.

POSTERIOR EYE ROW (PER) The row of (usually) four eyes that are farthest back.

POSTERIOR LATERAL EYES (PLE) The outermost eye pair in the back row.

POSTERIOR MEDIAN EYES (PME) The innermost eye pair in the back row.

POSTERIOR SPINNERETS (PS) The pair of spinning organs farthest back.

PRINCIPAL EYES The anterior median eyes in spiders that are connected by a major bundle of nerves directly to the brain.

PROCURVED EYE ROW A row of eyes that is curved so that eyes at the outer ends of the row are farther forward than the ones near the center.

PSYCHODID FLIES Flies (insects) that are members of the family Psychodidae, a group including tiny gnats, sand flies, and the hairy drain flies.

RADIAL GROOVES (ALSO RADIAL FURROWS) The indentations in the carapace that reflect the separate legs.

RADIUS, RADIAL LINE, (RADII) The silk lines in an orb web that extend from the center toward the edges.

RASTELLUM A patch of heavy blunt setae in the chelicerae of burrowing spiders used in digging.

RECURVED EYE ROW A row of eyes that is curved so that eyes at the outer ends of the row are farther behind than the ones near the center.

RETINA Sensory layer in the eye.

RETREAT A spider's hiding place, often lined with silk.

SCALES Short flat setae.

SCAPE A fingerlike projection on the epigynum of a female spider.

SCLERITE A rigid plate of hard exoskeleton, usually surrounded by softer, flexible exoskeleton.

SCLEROTIZED A hardened area of the exoskeleton; usually referring to a spot or plate on an otherwise softer portion of the body covering.

SCOPULA (SCOPULAE) A dense brush of soft hairs usually on the tips of the legs. They can extend from the tarsi up to the tibiae. They are usually found in spiders with extraordinary climbing ability; these spiders may also have claw tufts.

SCORPION Member of the order Scorpionida, not spiders.

SCUTE OR SCUTUM (SCUTES OR SCUTA) Hard parts or plates of shiny exoskeleton. They are noted when covering all or part of the abdomen; the distinction is made because the abdomen has a soft exoskeleton in most spiders.

SETA (SETAE) The spines or hairs on a spider's body are actually extensions of the exoskeleton. See also microsetae (hairs), macrosetae (spines), or scales.

SIGILLA Small depressions in the surface of a spider's body, sometimes with a harder or different-colored patch.

SIGNAL LINE A line of silk attached to the web, usually at the hub, and held by a spider.

SPECIES The members of a group composed of one type of organism. Most biologists would define it as comprising all potentially interbreeding individuals.

SPERMATHECA (SPERMATHECAE) A sperm storage sac in the abdomen of a female spider.

SPIDERLING Tiny immature spiders in the first few instars.

SPINATION The pattern of arrangement of spines on the body of a spider.

SPINES Relatively thick and rigid hairlike extensions of the exoskeleton.

SPIRACLE (SPIRACULAR OPENINGS OR TRACHEAL SPIRACLE) An opening to the respiratory (breathing) tubes. These are usually found on the underside of the abdomen.

SPIRAL, STICKY SPIRAL The circular portion of an orb web built with adhesive silk. It is either wet gluey silk from the aggregate glands or hackled silk from the cribellum.

STABILIMENTUM A form of decoration found in some orb webs. It usually consists of reflective bands or a spiral of white silk. Sometimes it is built in a zigzag pattern above and below the hub, or four bands extending away from the hub. Some spiders include debris or egg cases in these web decorations.

STERNUM The hard plate of exoskeleton forming the bottom of the cephalothorax.

STRIDULATING ORGAN A hard structure used to make sounds or vibratory signals by rubbing against another hard part.

SUBADULT An immature spider that is one molt away from maturity. Sometimes called penultimate.

SUBMARGINAL BAND A band of contrasting color around the carapace but separated from the edge by another, marginal, band of a different color.

SYSTEMATICS The study of relationships among organisms. This work is related to taxonomy, the naming of organisms.

TAPETUM A reflective layer at the back of the retina in the eye. This layer is responsible for the appearance of eyeshines when illuminated at night.

TARSAL CLAWS The small hooklike structures at the tips of the legs and sometimes palps.

TARSUS (TARSI) The last (distal) segment of a leg or palp.

TEMPORARY SPIRAL, HOLDING SPIRAL A spiral silk line that is added during orb web construction, then replaced as the sticky spiral is completed.

TERGITE A rigid plate of hard exoskeleton, usually surrounded by softer, flexible exoskeleton. This term is used for such plates located on the dorsal surface of the organism.

THORAX The thoracic region of the carapace. The hind part of the cephalothorax characterized by the attachment of the legs, demarcated from the front or head portion by the cervical groove.

TIBIA (TIBIAE) The fifth segment of the legs.

TOTAL BODY LENGTH (TBL) This is the combined length of the cephalothorax and abdomen (without the legs).

TRACHEAE The breathing tubes with their opening (tracheal spiracle) are respiratory organs that supplement and sometimes replace the book lungs.

TRICHOBOTHRIUM (TRICHOBOTHRIA) Very thin hairs arising from a pit in the exoskeleton. They are exquisitely sensitive motion detectors, sensing the slightest vibrations in the air.

TROCHANTER (TROCHANTERS) The second segment of the legs.

TUBERCLE A bump or prominence.

TURNING PLACE A pattern of silk lines created when the spider reversed direction while spinning an orb web.

VISCID SILK The wet, sticky, or adhesive type of silk. Usually produced by the aggregate silk glands.

VOUCHER A specimen or photograph that serves as proof. Its presence establishes the authenticity of a record.

REFERENCES

Barth, F. G. 2002. *A Spider's World: Senses and Behavior.* Berlin: Springer Verlag. 394 pp.

Beatty, J. A., and J. M. Berry. 1988. "The Spider Genus *Paratheuma* (Araneae, Desidae)." *Journal of Arachnology* 16: 47–54.

Bennett, R. G. 2005. "Cybaeidae." In *Spiders of North America: An Identification Manual.* Edited by D. Ubick, P. Paquin, P. E. Cushing, and V. Roth, pp. 85–90. Stockton, Calif.: American Arachnological Society. 377 pp.

Binford, G. J. 2001. "An Analysis of Geographic and Intersexual Chemical Variation in Venoms of the Spider *Tegenaria agrestis* (Agelenidae)." *Toxicon* 39: 955–68.

Branch, J. H. 1942. "Notes on California Spiders: 2. A Spider Which Amputates One of Its Palpi." *Bulletin of the Southern California Academy of Sciences* 41: 139–40.

Bristowe, W. S. 1958. *The World of Spiders.* London: Collins. 304 pp.

Carico, J. E. 1973. "The Nearctic Species of the Genus *Dolomedes* (Araneae: Pisauridae)." *Bulletin of the Museum of Comparative Zoology* 114 (7): 435–88.

Castaneiras, A. 1995. "Natural Enemies of *Bemisia tabaci* (Homoptera: Aleyrodidae) in Cuba." *Florida Entomologist* 78: 538–40.

Clark, D. L. 1994. "Sequence Analysis of Courtship Behavior in the Dimorphic Jumping Spider *Maevia inclemens* (Araneae, Salticidae)." *Journal of Arachnology* 22: 94–107.

———, and C. L. Morjan. 2001. "Attracting Female Attention: The Evolution of Dimorphic Displays in the Jumping Spider *Maevia inclemens*

(Araneae: Salticidae)." *Proceedings of the Royal Society London* series B, biological sciences 268: 2461–65.

Coddington, J. A. 2005. "Anapidae." In *Spiders of North America: An Identification Manual.* Edited by D. Ubick, P. Paquin, P. E. Cushing, and V. Roth, pp. 64–65. Stockton, Calif.: American Arachnological Society. 377 pp.

———. 1987. "Notes on Spider Natural History: The Webs and Habits of *Araneus niveus* and *A. cingulatus* (Araneae: Araneidae)." *Journal of Arachnology* 15: 268–70.

Comstock, J. H. 1912. *The Spider Book.* Garden City, N.Y.: Doubleday, Page and Co. 721 pp.

Crawford, R. L., P. M. Sugg, and J. S. Edwards. 1995. "Spider Arrival and Primary Establishment on Terrain Depopulated by Volcanic Eruption at Mount St. Helens, Washington." *American Midland Naturalist* 133: 60–75.

Crews, S. C. 2005. "Homalonychidae." In *Spiders of North America: An Identification Manual.* Edited by D. Ubick, P. Paquin, P. E. Cushing, and V. Roth, pp. 118–19. Stockton, Calif.: American Arachnological Society. 377 pp.

Cushing, P. E., and R. G. Santangelo. 2002. "Notes on the Natural History and Hunting Behavior of an Ant Eating Zodariid Spider (Arachnida, Araneae) in Colorado." *Journal of Arachnology* 30: 618–21.

Cutler, B. 1992. "Experimental Microhabitat Choice in *Pseudicius piraticus* (Araneae: Salticidae)." *Entomolgical News* 103 (4): 145–47.

———. 1988. "A Revision of the American Species of the Antlike Jumping Spider Genus *Synageles*

(Araneae, Salticidae)." *Journal of Arachnology* 15 (3): 321–48.

Dondale, C. D., and J. H. Redner. 1990. *The Wolf Spiders, Nurseryweb Spiders, and Lynx Spiders of Canada and Alaska (Araneae: Lycosidae, Pisauridae, Oxyopidae)*. Part 17 of *The Insects and Arachnids of Canada*. Publication 1856. Ottawa: Agriculture Canada. 383 pp.

Dondale, C. D., J. H. Redner, P. Paquin, and H. W. Levi. 2003. *The Orb-weaving Spiders of Canada and Alaska: Uloboridae, Tetragnathidae, Araneidae, and Theridiosomatidae (Araneae)*. Part 23 of *The Insects and Arachnids of Canada*. National Research Council publications, NRC 44466. Ottawa: Agriculture Canada. 371 pp.

Draney, M. L., and D. J. Buckle. 2005. "Linyphiidae." In *Spiders of North America: An Identification Manual*. Edited by D. Ubick, P. Paquin, P. E. Cushing, and V. Roth, pp. 124–61. Stockton, Calif.: American Arachnological Society. 377 pp.

Edwards, G. B. 2004. "Revision of the Jumping Spiders of the Genus *Phidippus* (Araneae: Salticidae)." *Occasional Papers of the Florida State Collection of Arthropods* 11: 1–156. 156 pp.

———. 1984. "Mimicry of Velvet Ants (Hymenoptera: Mutillidae) by Jumping Spiders (Araneae: Salticidae)." *Peckhamia* 2 (4): 46–49.

Edwards, R. J., E. H. Edwards, and A. D. Edwards. 2003. "Observations of *Theotima minutissimus* (Araneae, Ochyroceratidae): A Parthenogenetic Spider." *Journal of Arachnology* 31: 274–77.

Emerton, J. H. 1888. "New England Spiders of the Family Ciniflonidae." *Transactions of the Connecticut Academy* 7: 443–61.

Eubanks, M. D., and G. L. Miller. 1993. "Sexual Differences in Behavioral Response to Conspecifics and Predators in the Wolf Spider *Gladicosa pulchra* (Aranea: Lycosidae)." *Journal of Insect Behavior* 6: 641–48.

Gertsch, W. J. 1979. *American Spiders*. Second edition. New York: Van Nostrand Reinhold Co. 274 pp.

Hayashi, C. Y., N. H. Shipley, and R. V. Lewis. 1999. "Hypotheses That Correlate the Sequence, Structure, and Mechanical Properties of Spider Silk Proteins." *International Journal of Biological Macromolecules* 24: 271–75.

Hentz, N. M. 1875. *The Spiders of the United States*. Boston: Boston Society of Natural History. 171 pp.

Hölldobler, B. 1970. "*Steatoda fulva* (Theridiidae): A Spider That Feeds on Harvester Ants." *Psyche* 77 (2): 202–8.

Hormiga, G. 1994. "A Revision and Cladistic Analysis of the Spider Family Pimoidae (Araneoidea:

Araneae)." *Smithsonian Contributions to Zoology*, no. 549: 1–104. 104 pp.

———, D. Dimitrov, J. A. Miller, and F. Alvarez-Padilla. 2008. *LinyGen: Linyphioid Genera of the World (Pimoidae and Linyphiidae): An Illustrated Catalog*. Version 2.0. George Washington University. Online at http://www.gwu.edu/~linygen/index.cfm.

Jakob, E. M. 2004. "Individual Decisions and Group Dynamics: Why Pholcid Spiders Join and Leave Groups." *Animal Behaviour* 68 (1): 9–20.

Jocqué, R., and A. S. Dippenaar-Schoeman. 2006. *Spider Families of the World*. Tervuren, Belgium: Royal Museum for Central Africa. 336 pp.

Kaston, B. J. 1978. *How to Know the Spiders*. Third edition. Boston: WCB/McGraw-Hill. 272 pp.

Leech, R. 1972. *A Revision of the Nearctic Amaurobiidae (Arachnida: Araneida)*. Memoirs of the Entomological Society of Canada No. 84. Ottawa, Canada: Entomological Society of Canada. 182 pp.

Levi, H. W. 1978. "The American Orb-weaver Genera *Colphepeira*, *Micrathena* and *Gasteracantha* North of Mexico (Araneae, Araneidae)." *Bulletin of the Museum of Comparative Zoology* 148 (9): 417–42.

———. 1973. "Small Orb-Weavers of the Genus *Araneus* North of Mexico (Araneae: Araneidae)." *Bulletin of the Museum of Comparative Zoology* 145 (9): 473–552.

———. 1971. "The Diadematus Group of the Orb-Weaver Genus *Araneus* North of Mexico (Araneae: Araneidae)." *Bulletin of the Museum of Comparative Zoology*, 141 (4): 131–78.

———, and L. Levi. 1968. *A Guide to Spiders and Their Kin*. New York: Golden Press. 160 pp.

Maddison, W., and M. Hedin. 2003. "Phylogeny of *Habronattus* Jumping Spiders (Araneae: Salticidae), with Consideration of Genital and Courtship Evolution." *Systematic Entomology* 28: 1–21.

Main, B. Y. 1976. *Spiders*. Sydney: Collins. 296 pp.

McCrone, J. 1963. "Taxonomic Status and Evolutionary History of the *Geolycosa pikei* Complex in the Southeastern United States (Araneae, Lycosidae)." *American Midland Naturalist* 70 (1): 47–73.

Miller, J. A., A. Carmichael, M. J. Ramírez, J. C. Spagna, C. R. Haddad, M. Rezác, J. Johannesen, J. Král, X. P. Wang, and C. E. Griswold. 2010. "Phylogeny of Entelegyne Spiders: Affinities of the Family Penestomidae (NEW RANK), Generic Phylogeny of Eresidae, and Asymmetric Rates of Change in Spinning Organ Evolution (Araneae, Araneoidea, Entelegynae)." *Molecular Phylogenetics and Evolution* 55: 786–804.

Morse, D. H. 2007. *Predator upon a Flower: Life History and Fitness in a Crab Spider.* Cambridge: Harvard University Press. 377 pp.

Opell, B. D., and J. A. Beatty. 1976. "The Nearctic Hahniidae (Arachnida: Araneae)." *Bulletin of the Museum of Comparative Zoology* 147: 392–433.

Peckham, G. W., and E. G. Peckham. 1909. "Revision of the Attidae of North America." *Transactions of the Wisconsin Academy of Sciences, Arts, and Letters* 16, Part 1 (5): 355–646

Penniman, A. J. 1985. "Revision of the *britcheri* and *pugnata* groups of *Scotinella* (Araneae, Corinnidae, Phrurolithinae) with a Reclassification of Phrurolithine Spiders." Ph.D. dissertation, Ohio State University, Columbus, Ohio. 247 pp.

Platnick, N. I. 2012. *The World Spider Catalog.* Version 12.5. American Museum of Natural History. Online at http://research.amnh.org/iz/spiders/catalog. DOI: 10.5531/db.iz.0001.

Platnick, N. 1974. "The spider family Anyphaenidae in America north of Mexico." *Bulletin of the Museum of Comparative Zoology* 146(4): 205–266.

Platnick, N., and C. Dondale. 1992. "The Ground Spiders of Canada and Alaska (Araneae: Gnaphosidae)." Part 19 of *The Insects and Arachnids of Canada.* Publication 1875. Ottawa: Agriculture Canada. 297 pp.

Platnick, N., and D. Ubick. 2005. "A Revision of the North American Spider Genus *Anachemmis* Chamberlin (Araneae, Tengellidae)." *American Museum Novitates* 3339: 1–25.

Ramirez, M. G. 1995. "Natural History of the Spider Genus *Lutica* (Araneae, Zodariidae)." *Journal of Arachnology* 23: 111–17.

Reed, D. H., and A. C. Nicholas. 2008. "Spatial and temporal variation in a suite of life-history traits in two species of wolf spider." *Ecological Entomology* 33: 488–96.

Reichert, S. E. 1988. "The Energetic Costs of Fighting." *American Zoologist* 28 (3): 877–84.

Reiskind, J. 1969. "The spider subfamily Castianeirinae of North and Central America (Araneae, Clubionidae)." *Bulletin of the Museum of Comparative Zoology* 138(5): 163–325.

Richman, D. 2008. "Revision of the Jumping Spider Genus *Sassacus* (Araneae, Salticidae, Dendryphantinae) in North America." *Journal of Arachnology* 36: 26–48.

Rovner, J. 1980. "Morphological and Ethological Adaptations for Prey Capture in Wolf Spiders (Araneae, Lycosidae)." *Journal of Arachnology* 8: 201–15.

———. 1975. "Sound Production by Nearctic Wolf Spiders: A Substratum-coupled Stridulatory Mechanism." *Science* 190: 1309–10.

Stowe, M. K., J. H. Tumlinson, and R. R. Heath. 1987. "Chemical Mimicry: Bolas Spiders Emit Components of Moth Prey Species Sex-Pheromones." *Science* 236: 964–67.

Tarsitano, M. S., and R. R. Jackson. 1997. "Araneophagic Jumping Spiders Discriminate between Detour Routes That Do and Do Not Lead To Prey." *Animal Behaviour* 53: 257–66.

Taylor, R. M., and W. A. Foster. 1996. "Spider Nectarivory." *American Entomologist* 42: 82–86.

Theuer, B. 1954. "Contributions to the life history of *Deinopis spinosus.*" M.S. thesis, University of Florida, Gainesville, Florida. 75 pp.

Ubick, D., P. Paquin, P. E. Cushing, and V. Roth, eds., 2005. *Spiders of North America: An Identification Manual.* Stockton, Calif.: American Arachnological Society. 377 pp.

Uetz, G. W., and G. Denterlein. 1979. "Courtship Behavior, Habitat, and Reproductive Isolation in *Schizocosa rovneri* Uetz and Dondale (Araneae: Lycosidae)." *Journal of Arachnology* 7: 121–28.

Uetz, G. W., and J. A. Roberts. 2002. "Multisensory Cues and Multimodal Communication in Spiders: Insights from Video/Audio Playback Studies." *Brain, Behavior, and Evolution* 59 (4): 222–30.

Vetter, R. S., and Bush S. P. 2002. "The Diagnosis of Brown Recluse Spider Bite Is Overused for Dermonecrotic Wounds of Uncertain Etiology." *Annals of Emergency Medicine* 39: 544–46.

INDEX

Bold page numbers indicate the main discussion of a family or species.